CIRCUIT ANALYSIS

CIRCUIT ANALYSIS

By

Dr. K. PADMANABHAN
B.E., M.Sc. (Engg.), Ph.D., F.I.E.E., F.I.E.T.E., S.M.I.E.E.E.

P. SWAMINATHAN
B.E., M.I.E.T.E.

UNIVERSITY SCIENCE PRESS
(An Imprint of Laxmi Publications Pvt. Ltd.)
An ISO 9001:2008 Company
BENGALURU • CHENNAI • COCHIN • GUWAHATI • HYDERABAD
JALANDHAR • KOLKATA • LUCKNOW • MUMBAI • RANCHI • NEW DELHI
BOSTON (USA) • ACCRA (GHANA) • NAIROBI (KENYA)

CIRCUIT ANALYSIS

Printed and bound in India
First Edition : 2006, *Second Edition* : 2008, *Reprint* : 2009, *Third Edition* : 2017
ISBN : 978-81-318-0490-2

Published in India by
UNIVERSITY SCIENCE PRESS
(An Imprint of Laxmi Publications Pvt. Ltd.)
An ISO 9001:2008 Company
113, GOLDEN HOUSE, DARYAGANJ, NEW DELHI-110002, INDIA
Telephone : 91-11-4353 2500, 4353 2501
Fax : 91-11-2325 2572, 4353 2528
www.laxmipublications.com info@laxmipublications.com

Branches

Bengaluru	080-26 75 69 30
Chennai	044-24 34 47 26
Cochin	0484-237 70 04, 405 13 03
Guwahati	0361-254 36 69, 251 38 81
Hyderabad	040-27 55 53 83, 27 55 53 93
Jalandhar	0181-222 12 72
Kolkata	033-22 27 43 84
Lucknow	0522-220 99 16
Mumbai	022-24 91 54 15, 24 92 78 69
Ranchi	0651-220 44 64

C—12120/016/06
Printed at : Repro India Ltd., Mumbai.

CONTENTS

PREFACE TO THE THIRD EDITION

The third edition of this book presents the students of engineering an updated and an extensively revised subject. It covers basic Circuit Analysis, which is required in the first course of the subject.

It is written in a lucid, easy and crystal-clear language. All the necessary information on Circuit Analysis has been provided. The book is written in such a way as if a teacher is teaching the subject to the students in front of him and the students are receiving the same with spontaneity and without any extra effort.

The book is divided into ten chapters – DC Circuit Analysis, Sinusoidal Steady State Analysis, Network Theorems, Transients and Resonance in Circuits, Coupled Circuits, Non-sinusoidal Waves and Fourier Analysis, Fourier Transforms, Laplace Transforms and Transient Analysis, Frequent Domain and Time Domain, and Networks Including Two-ports. At the end of the book, exercises (elements of synthesis, two-port networks and simple) and index have been provided.

Professors and students are welcome to send us their suggestions regarding the betterment of this book. Those will be duly considered in the subsequent edition.

Dr. K. Padmanabhan
P. Swaminathan

PREFACE TO THE FIRST EDITION

The subject of Electric Circuit Analysis requires a thorough understanding of the basic principles and regular study through working out problems in each topic. In this book, each topic is followed by typical worked out examples. Exercises are given for practice and have been provided with hints and answers. Short questions are also given at the end of each chapter.

The book covers basic Circuit Analysis as required in the first course of the subject. It covers transient analysis also, but for single time constant networks. Coupled circuits and resonance are covered for series and parallel circuits. Duality and topology are introduced with typical examples on tree, cut-sets, etc.

The material covered in this book will meet the requirements of a first course in circuits prior to study of network theory, digital signal processing, active network theory and control system theory.

Suggestions for improvement are welcome.

–AUTHORS

SYLLABUS

EC 2102	CIRCUIT ANALYSIS	L T P M 2 0 2 100

AIM

To provide knowledge to the students to understand R, L, C circuits and transformer components with the aid of standard theorems.

OBJECTIVES

On completion of this course,

- The students must be capable to analyze complex RL, RC, RLC and transformer coupled circuits using network theorems.
- To study the transient analysis of RC, RL and RLC circuits.
- To experimentally verify the network theorems.

UNIT I : BASICS OF CIRCUIT ANALYSIS—DC CIRCUITS **12**

Charge-Current–Voltage–Power–Resistor–Ohm's Law–Resistors in series and parallel Network–Circuit–Star Delta conversion–Voltage source–Current source–Source transformation–Series and parallel connected sources–independent and dependent sources–Kirchoff's current law–Kirchoff's voltage law–Voltage and current division–linearity and superposition–nodal analysis–mesh analysis–supermesh–Thevenin and Norton equivalent circuits–Maximum power transfer.

UNIT II : AC CIRCUITS—STEADY STATE AND TRANSIENT ANALYSIS **12**

Definition of Laplace transform–Transform of simple time functions–inverse transform techniques–basic theorems–initial and final value theorems–nodal and mesh analysis in the s-domain–poles, zeros and transfer functions–convolution–complex frequency plane–natural response and s-plane.

Capacitor–Inductor–energy storage–duality–RL, RC and RLC Circuits–their DC transient and steady state response–Laplace transform in solving differential equations for complete response in RL, RC and RLC Circuits–Damping–Over damped, Critically damped and Under damped response–their DC transient and steady state response.

UNIT III : SINUSOIDAL STEADY—STATE ANALYSIS **12**

Characteristics of sinusoids–forced response to sinusoidal functions–Complex forcing functions–Phasor–Phasor relationship for R, L and C–impedance–admittance–Phasor diagram–AC power circuit analysis–instantaneous power–average power–effective values of current and voltage–apparent power and power factor–Complex power.

Nodal and Mesh analysis of AC circuits–Use of Superposition–Thevenin and Norton theorems.

UNIT IV : COUPLED CIRCUITS, RESONANCE **12**

Self inductance–Mutual inductance–Coupling coefficient–linear transformer–T and π equivalent networks–ideal transformer–turns ratio–impedance matching-series resonance–parallel resonance–their frequency response, bandwidth and quality factor–other resonant forms–Mesh analysis for coupled circuits.

UNIT V : TWO PORT NETWORKS **12**

One port and two port networks–impedance parameters–admittance parameters–hybrid parameters–transmission parameters–g parameters–network parameter transformations–combination of two port networks–Thevenin and Norton equivalent circuits for two port networks.

Total : 60 Periods

1

DC CIRCUIT ANALYSIS

1.1 BASIC COMPONENTS RESISTANCES

Resistances are the simplest of circuit elements. The equation determining the current through and the p.d. (potential difference) across a resistor is given by Ohm's law

$$v = R\,i \tag{1.1}$$

Such an equation relating v and i for an element, here a resistor, is called the *terminal equation* of the element.

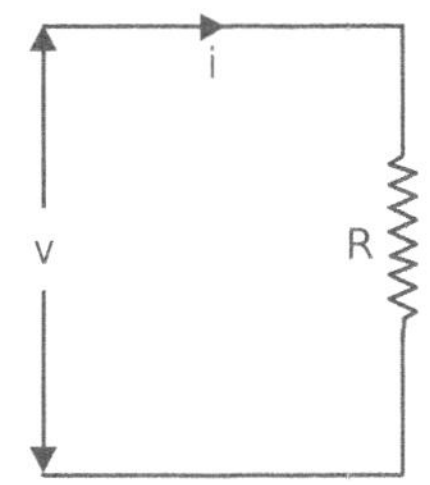

Fig. 1.1

Practical resistors are :

1. Wires or coils of wire, of copper, aluminium etc. High resistances are obtainable with alloys like Eureka, Nichrome etc.
2. Carbon resistors are used in different ohmic values and capacities (watts) in electron tube and semiconductor circuits.
3. Metal film resistors are films of metal evaporated on a base; they give good stability.
4. Liquid resistors, used in high voltage work.

The relation $v = Ri$ is true at any instant of time. So,

$$v(t) = R\,i\,(t) \tag{1.2}$$

The power consumed in a resistor is i^2R watts. Note that if i varies with time, the power also varies. If i takes positive and negative values, (*i.e.*, an alternating current), the quantity i^2 is positive and hence power is non-negative.

1.2 MAGNETIC FLUX AND FARADAY'S LAWS

A magnetic field can be produced either by a current carrying coil or by a permanent magnet. The influence of a magnetic field on circuits is determined by Faraday's laws. By Faraday's law an e.m.f. is induced in any conductor situated in a magnetic field when there is a relative motion between the flux lines and the conductor. If we take a single conductor and move it in a magnetic field, so as to cut the flux, then an e.m.f. e is induced in the conductor, given by

$$e = Blv \text{ volts} \tag{1.3}$$

Here B is the density of magnetic flux, l the conductor length and v is its velocity of movement. The e.m.f. is called "motionally induced e.m.f."

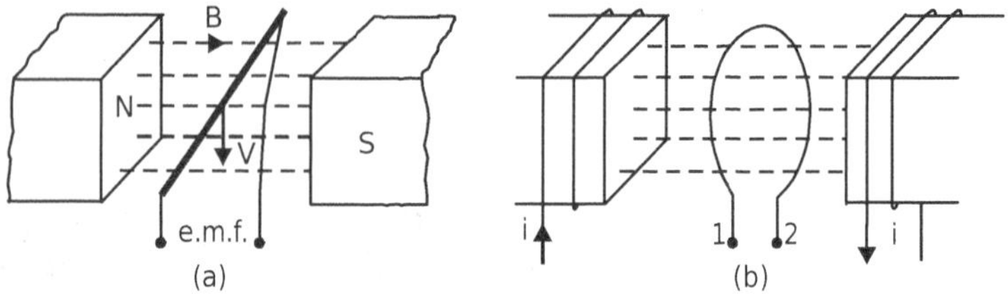

Fig. 1.2 (*a*) Motional induced e.m.f. (*b*) e.m.f. due to flux change.

Alternatively, the magnetic flux can be changed and then also an e.m.f. is induced. In this case, we look at Fig. 1.2(*b*) in which we have a circular coil facing the flux. If φ is the flux lines passing through the coil, (also called 'linking' the coil), and if the flux φ is changed, an e.m.f.' is induced at the ends 1 and 2.

$$e = -d\varphi/dt \text{ volts}$$

Here the coil has one turn only. If we take a coil of N turns,

$$e = -\text{N}\frac{d\varphi}{dt} \text{ volts} \tag{1.4}$$

The *minus* sign is introduced by experiment. It is also considered as a law, called Lenz's law. The e.m.f. induced tends to oppose the change of flux.

In equations (1.3) and (1.4), the quantities magnetic flux φ and flux density B were introduced. The unit of flux is the *weber*. The unit of flux density is then *weber / sq. metre*. From (1.4), *weber* can be defined as that flux, which linking a single turn coil and changing at the rate of 1 weber/second, induces 1 volt at the terminals of the coil. Equation (1.4) is important in circuit theory.

1.3 INDUCTANCE

A coil carrying current is known to produce a magnetic field. We consider a coil of N turns. The more the current, the more is the flux. For coils wound over non-magnetic substances, the flux and the current are found by experiment, to be proportional.

$$\varphi \propto i \tag{1.5}$$

$$\varphi = ki$$

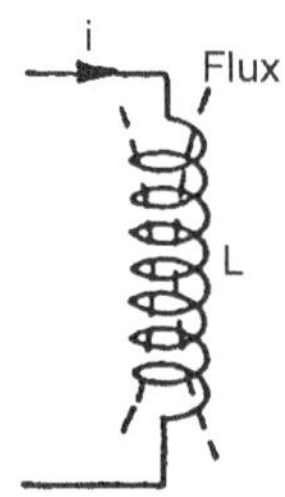

Fig. 1.3

The ratio Nφ/i is called the *inductance* L of the coil of N turns and is

$$\text{L} = \frac{\text{N}\varphi}{i} \tag{1.6}$$

A coil has one Henry inductance if φ = 1 weber, N = 1 and i = 1 ampere. If the coil has a magnetic core, a soft iron core for example, then the proportionality relation (1.5) does not hold good for large currents, because the iron gets saturated beyond a certain flux. In that case also, the inductance L is defined as in (1.6), but L then will vary with current.

Now, if a current i through the inductance changes (with time of course)

$$e = -N\frac{d\varphi}{dt}$$

$$= -N\frac{d(Li/N)}{dt} \qquad \text{from (1.6)}$$

$$e = -L\frac{di}{dt} \qquad (1.7)$$

Equation (1.7) is an important relation, the terminal equation of an inductance. In an inductance of 1 henry, a current changing at the rate of 1 ampere per second will induce an e.m.f. of 1 volt.

In Laplace transform variable s, (1.7) can be written

$$E = LsI \qquad (1.8)$$

where E and I denote the Laplace transforms of e and i.

Practical inductors are made of coils of wire wound over an air core or iron core. Air core inductances are of the order of a few hundred microhenries, while iron core inductances have values of a few millihenries. A henry inductor will be a very heavy iron cored coil. Inductances are used for starting discharge lamps, in smoothing circuits and network filters.

Because every conductor carrying current has a magnetic field produced by it (though small), every conductor (whether in the form of a coil or not), possesses an inductance. Since inductances are made of wire, they possess resistance also. So, it is not possible to have only inductance without resistance. Thus a pure resistance is only theoretical.

Power in an inductance. If a current i is to be sent through an inductance, there is in an induced e.m.f. $e(t)$ which opposes it at every instant t. So, the applied voltage $v(t)$ must be equal and opposite to $e(t)$. The power expended at any instant is

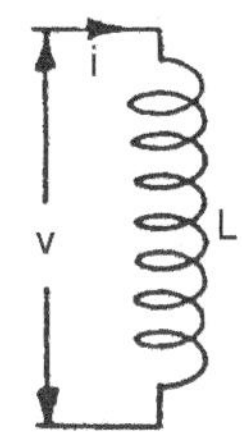

Fig. 1.4

$$p(t) = v(t)\, i(t)$$

$$= -e(t)\, i(t)$$

$$= L\frac{di(t)}{dt} i(t)$$

So, the energy spent from time $t = 0$ to time t is

$$w(t) = \int_0^t [Li(t)\, di(t)/dt]\, dt$$

$$= \frac{1}{2}L\,[i^2(t) - i^2(0)]$$

If we take time $t = 0$ when there was no current through L, $i(0) = 0$. So the energy spent upto time t is

$$W(t) = \frac{1}{2}Li^2(t)$$

This energy is said to be stored in the magnetic field. The relation indicates a loss of energy as far as the source is concerned. So, energy in L is

$$W_L = \frac{1}{2}Li^2 \qquad (1.9)$$

1.4 CAPACITANCE

The influence of the magnetic field in a circuit has been taken care of as inductance, a new element. Similarly the effect of the electric field is considered by the third element, 'Capacitance' In Fig. 1.5, two plates *p*, *q* are shown schematically, which are separated and which are charged with free electric charges. If we apply charge, there is an electric field between the plates. Electric flux lines can be imagined to start from the *p* plate to go to the *q* plate. The more the charge we put, the greater is the field. The effect of the field is to raise the potential difference between the plates. The charge and p.d. between the plates are generally linearly related.

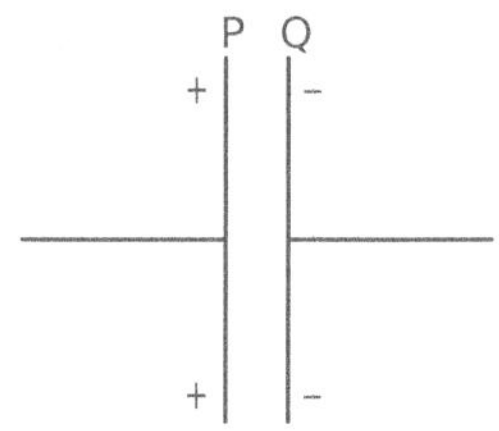

Fig. 1.5

$$q \propto v$$

$$q = Cv$$

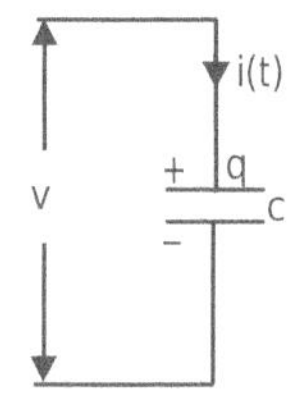

Fig. 1.6

The constant of proportionality C is called 'Capacitance'. The unit is the *Farad* being the value of a capacitor which stores a charge of $q = 1$ Coulomb for a p.d. of 1 volt between its electrodes.

The terminal equation of a capacitance is got by differentiating the above equation

$$i(t) = C\frac{d(v)}{dt} \tag{1.10}$$

The Laplace transform of the above equation is (if v (0) at $t = 0$),

$$I = CsV \tag{1.11}$$

Practical capacitors are generally of the parallel plate type with a dielectric medium in between the plates. Capacitors, have values in microfarads, since Farad is a big unit. Commercial capacitors are :

1. Paper capacitors—with paper dielectric, have low values of capacitance but capable of withstanding high voltage.
2. Mica capacitors, value in a few hundred pF (pF = 10^{-12} farads).
3. Ceramic capacitors, value in a few pF.
4. Electrolytic capacitors, in which a very thin oxide layer is the medium between the capacitor plates, the plates immersed in a solution of an electrolyte, give values in microfarads.

Energy stored in a capacitor :

$$i(t) = C\frac{dv(t)}{dt}$$

$$\therefore \quad p(t) = v(t)\, i(t) = Cv(t)\, dv(t)/dt$$

Integrating, the energy

$$w(t) = \int_0^t Cv(dv/dt) = \tfrac{1}{2}C\,[v^2(t) - v^2(0)]$$

Taking time $t = 0$ to start at the instant when v was zero,

$$w(t) = \tfrac{1}{2}Cv^2(t) \tag{1.12}$$

This energy is stored in the capacitor at time t.

Example 1. *A voltage v(t) varies with time in the form of a trapezium wave, increasing, remaining steady and then decreasing to zero. Find the current flowing through a capacitor C when this voltage is applied to it.*

$$i(t) = C\,dv(t)/dt.$$

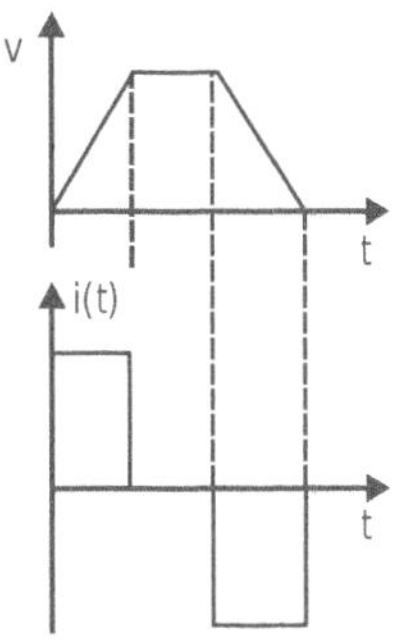

Fig. 1.7

Solution. In the straight-line positions, $v(t) \propto \pm t$

$$\therefore \qquad i(t) \propto \pm C$$

Therefore the current wave changes with time, as shown. The current is + when v increases and is – when v decreases.

The dielectric medium in a capacitor consumes some power when the capacitor passes changing currents and this power loss which is dissipated as heat in the medium is called *dielectric loss*. The effect of dielectric loss is represented by a resistor in parallel with C (Fig. 1.8).

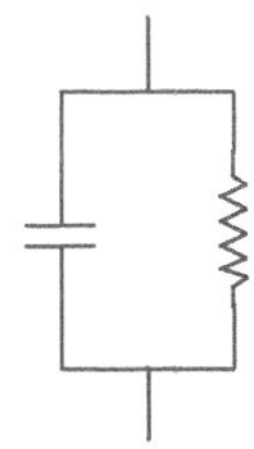

Fig. 1.8

1.5 ELEMENTS IN SERIES AND PARALLEL

Circuit elements—resistance, inductance, capacitance—can be connected in series or parallel. When they are in series, the applied voltage is the total of the voltages V_1, V_2, V_3 across them. The current i is the same through all. When they are connected in parallel, the current i is the sum of the currents i_1, i_2, i_3 drawn by them. The voltage across each element is the same.

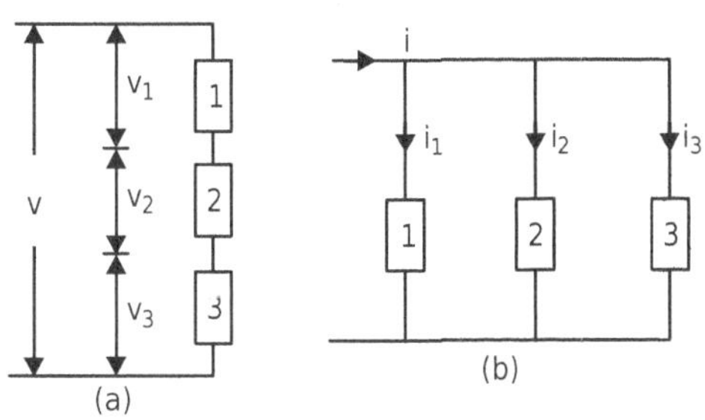

Fig. 1.9 (*a*) Elements in series. (*b*) In parallel.

For inductors : L_1, L_2, L_3 *in series :*

$$v = v_1 + v_2 + v_3$$

$$v = L_1\frac{di_1}{dt} + L_2\frac{di_2}{dt} + L_3\frac{di_3}{dt}$$

$$= (L_1 + L_2 + L_3)\frac{di}{dt} = L\frac{di}{dt}$$

Hence the series connection is equivalent to the sum of $L_1 + L_2 + L_3$.

For inductances in parallel. $i = i_1 + i_2 + i_3$

$$\frac{di}{dt} = \frac{di_1}{dt} + \frac{di_2}{dt} + \frac{di_3}{dt}$$

The voltage v is the same across the three inductors.

$$v = L_1 \frac{di_1}{dt} = L_2 \frac{di_2}{dt} = L_3 \frac{di_3}{dt}$$

$$\therefore \qquad \frac{di}{dt} = \frac{v}{L_1} + \frac{v}{L_2} + \frac{v}{L_3} \tag{1.13}$$

The equivalent inductance is given by $v = L\, di/dt$. Putting v/L for di/dt,

$$\frac{1}{L} = \frac{1}{L_1} + \frac{1}{L_2} + \frac{1}{L_3} \tag{1.14}$$

a relation similar to parallel resistors.

Capacitors in series. When capacitors are in series,

$$v = v_1 + v_2 + v_3$$

Since $i = C\, dv/dt$ for a capacitor, and since i is same for all,

$$\int \frac{i}{C} dt = \int \frac{i}{C_1} dt + \int \frac{i}{C_2} dt + \int \frac{i}{C_3} dt$$

$$\frac{1}{C} = \frac{1}{C_1} + \frac{1}{C_2} + \frac{1}{C_3} \tag{1.15}$$

a relation different than that for resistors and inductors.

Capacitors in parallel. Here $i = i_1 + i_2 + i_3$

$$C \frac{dv}{dt} = C_1 \frac{dv}{dt} + C_2 \frac{dv}{dt} + C_3 \frac{dv}{dt}$$

giving

$$C = C_1 + C_2 + C_3 \tag{1.16}$$

Example 2. *The following circuit has the voltage e(t) and current i(t) given by*

$$e(t) = 2 \sin \omega t$$

$$i(t) = 4 \cos \omega t$$

and for $t < 0$, both e and i are zero. Find the current delivered by the voltage e(t) at time t and the voltage across i(t) at time t. Also find the energy expended in R, and the energies stored in L and C.

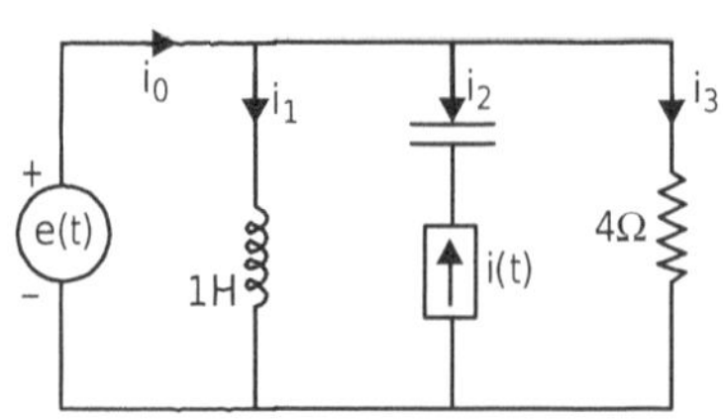

Fig. 1.10

Solution. Let i_1, i_2, i_3 be the currents. Then $i_0 = i_1 + i_2 + i_3$.

$$L\frac{di_1(t)}{dt} = e(t)$$

$$\therefore \quad i_1(t) = \int_0^t \frac{e(t)}{L}\,dt = \frac{2}{\omega}[1 - \cos \omega t].$$

$$i_2(t) = -i(t) = -4 \cos \omega t$$

$$i_3(t) = e(t)/4 = \frac{1}{2}\sin \omega t$$

$$\therefore \quad i_0 = \frac{1}{2}\sin \omega t - 4\cos \omega t + (2/\omega)[1 - \cos \omega t]$$

Note that $i_2(t) = -i(t)$ irrespective of the capacitor in series with it.

To find the voltage across the current source, *viz.* e_i,

$$e(t) = e_{2F} + e_i$$

$$2 \sin \omega t = \frac{1}{C}\int i_2(t)dt + e_i$$

$$2 \sin \omega t = \frac{1}{2}\int_0^t -4\cos \omega t\, dt + c_t$$

$$\therefore \quad e_t = 2\sin \omega t + (2/\omega)\sin \omega t$$

The power in the 4 Ω resistor is $P = Ri_3^2 = 4\left(\frac{1}{2}\sin \omega t\right)^2$

The energy spent upto time t is

$$\int p\,dt = \int \sin^2 \omega t\, dt$$

$$= \frac{1}{2}[t - \sin 2\omega t / 2\omega]$$

The energy in $L = \frac{1}{2}Li^2 = \frac{1}{2} \times 1 \times \left(\frac{2}{\omega}\right)^2 [1 - \cos \omega t]^2$

The energy in $C = \frac{1}{2}Cv^2(t) = \frac{1}{2} \times 2 \times [-(2/\omega)\sin \omega t]^2$

$$= (4/\omega^2)\sin^2 \omega t.$$

1.6 CIRCUITS AND ANALYSIS

A circuit is one which carries an electric current and has a pathway via resistors, inductances, capacitance and even active devices like the transistor. A power source is needed to send a current through a circuit. There may be more than one source and there are also types of sources like voltage source, current source and induced power source.

The circuits involving a single voltage source (say, a battery) and two resistors in parallel form a simple circuit given in Example 1 below. What one can analyse by an example will be indicative of general principles. We study the currents in the different paths of the circuits, the power consumed and so on. Here we consider a D.C. circuit because the voltage source is D.C.

Example 1 also considers how resistances wound by wires (of copper or 'aluminium' or nichrome, as the case may be) will change in value with heating, since the value of a resistor at t°C is related to its 0°C resistance by the approximate formula

$$R_t = R_0 (1 + \alpha t)$$

This fact is used in the example.

Energy is the product of power and time. The unit of energy in electricity distribution is the Kilo-Watt-hour or KWH. This is also illustrated in Example 1.

Ohm's law is used for each resistor, as $V_a = I_a R_a$ and $V_b = I_b R_b$. In the example 1, the voltage across R_a and R_b are both 200 V (D.C). Knowing R_a, R_b, we find I_a, I_b.

The voltage, current and resistance value are related by Ohm's law.

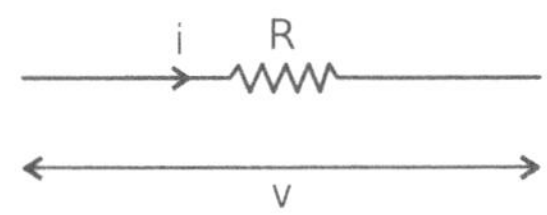

$$v = R\, i$$

If v is in volts, i is in Amperes, R is in ohms. Since this is true for any instant of time t,

$$v(t) = R\, i(t)$$

Example 3. *Two resistors A and B, made of different material and connected in parallel, consume 5 units of energy per hour from a 200 V supply source at 0°C with the resistance A passing a current of 10 amperes. Then, the temperature was raised to 30°C. Find the currents now drawn by A and B and also the energy spent per hour. α_0 for A = 1/240; α_0 for B = 1/270.*

Solution. Energy spent per hour $= VIt =$ KWH

$$200\, I = 5000 \text{ or } I = 25 \text{ A}$$

$$I_a = 10 \text{ A}$$

$$I_b = 15 \text{ A}$$

$$R_a = 200/I_a = 20\ \Omega$$

$$R_b = 200/15 = 13.33\ \Omega$$

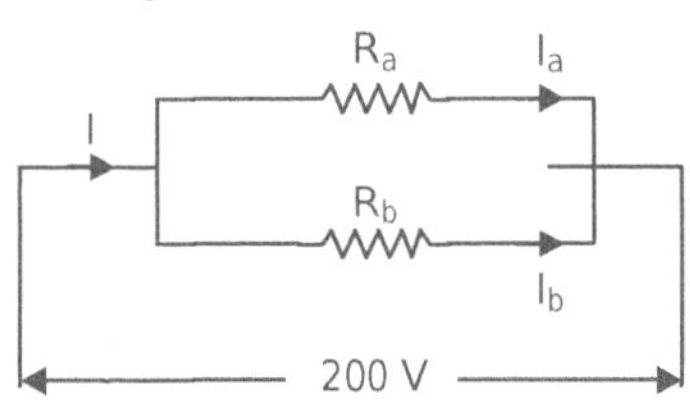

Fig. 1.11

At 30°C,

$$R_{30} = R_0 (1 + \alpha t) = R (1 + 30\alpha)$$

$$R_a = 20 [1 + (1/240)\, 30] = 180/8 = 22.5\ \Omega$$

$$R_b = 13.3 [1 + (1/270)\, 30] = 133.3/9 = 14.81\ \Omega$$

Currents at 30°C :

$$I_a = 200/22.5 \text{ A}$$

$$I_b = 200/14.81 \text{ A}$$

$$\therefore \quad I_L = (200/22.5) + (200/14.81) = 22.41 \text{ A}.$$

Energy spent in one hour

$$= 200 \times 22.41 \times 1 \times 10^{-3} \text{ KWH}$$
$$= 4.482 \text{ KWH.}$$

Example 4. *In the parallel circuit arrangement shown, the currents through the resistor x is 1 ampere and the total current drawn from the supply is 15 A. Find the current in each of the other resistors, the value of x and the potential across PQ.* (B. Tech. Nov. 1968)

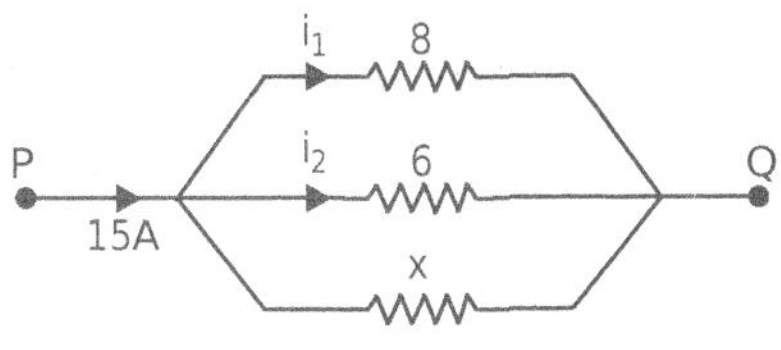

Fig. 1.12

Solution. The current through x is 1 amp. Hence the other currents $i_1 + i_2 = 15 - 1 = 14$ A. The 8 Ω and 6 Ω resistors in parallel can be combined into an equivalent resistor of value (8 × 6)/[8 + 6] or 24/7 ohms. Hence, the 14 A flowing through 24/7 ohms will produce a drop. Which is also the potential across PQ, as

$$V = (24/7) \times 14 = 48 \text{ volts}$$

Hence this 48 V also appears across x and so

$$R_\alpha = V/I_\alpha = 48/1 = 48 \text{ ohms.}$$

1.7 KIRCHHOFF'S LAWS

Kirchhoff's Current Law (KCL). The sum of the currents entering a junction must be equal to the currents leaving the junction*. Or, the algebraic sum of the currents entering (or leaving) a junction is zero.

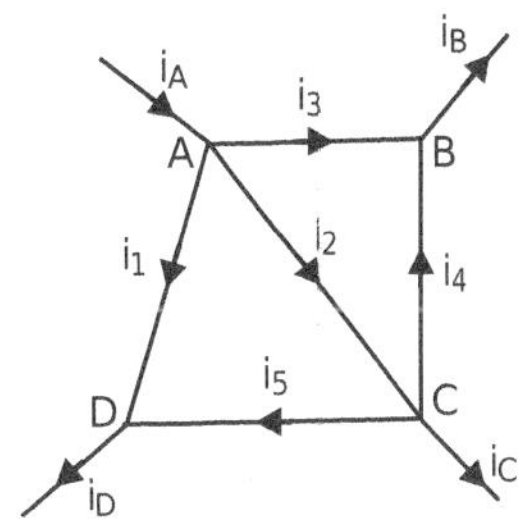

Fig. 1.13

In Fig. 1.13, at junction A,

$$i_A = i_1 + i_2 + i_3$$

At junction B, $i_B = i_3 + i_4$

At junction C, $i_C = i_2 - i_4 - i_5$

Kirchhoff's Voltage Law (KVL). The sum of the voltage drops around any mesh (*i.e.*, loop, circuit) is zero.

Thus, in loop ABCA,

$$V_{AB} + V_{BC} + V_{CA} = 0$$

If the resistances between A and B, B and C, C and A are R_{ab}, R_{bc} and R_{ca}, respectively, then by this law,

$$R_{ab}\, i_{ab} + R_{bc}\, i_{bc} + R_{ca}\, i_{ca} = 0$$

Since $i_{ab} = i_3$, $i_{bc} = -i_4$, $i_{ca} = -i_2$,

$$R_{ab}\, i_3 - R_{bc}\, i_4 - R_{ca}\, i_2 = 0.$$

---

* Junction, node, vertex are synonymous terms in circuit theory.

For the mesh ABDA also, we can write a similar equation. In finding the voltage drop *via.* a source of e.m.f. (a battery, for example) note that from the +ve to the –ve terminal, there is a drop; from negative to positive terminal, there is a rise. Using these three laws, namely.

1. Kirchhoff's Current Law (KCL)
2. Kirchhoff's Voltage Law (KVL)
3. Ohm's Law.

We can find the current in all the branches of a circuit, *i.e.*, solve a circuit problem.

Example 5. *In an unbalanced Wheatstone bridge network, the resistances of the arms and the detector are : AB = 1 Ω, BC = 2 Ω, CD = 3 Ω, DA = 2 Ω, AC = 4 Ω. The battery is 2 V with an internal resistance of 2 ohms. Also determine the P.D. across BD and the resistance from B to D.* (AMIE 1954)

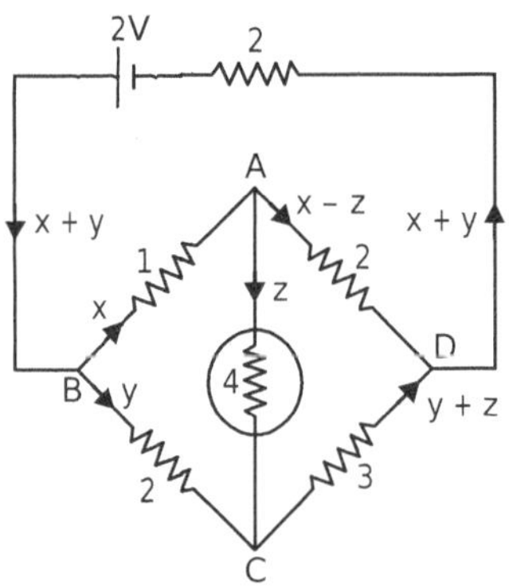

Fig. 1.14 Wheatstone bridge network.

Solution. Assume current x, y, z along BA, BC and AC. Using Kirchhoff's law 1, we may mark the currents in terms of these, as shown. For example, at A the current along AD is given by $x = z + i_{AD}$, so that current in AD is $x - z$. Then we write the voltage drop equations involving x, y, z.

Loop ACBA :

$$4z - 2y + 1x = 0 \tag{1}$$

Note the negative sign in $-2y$ since current y flows against the loop direction ACBA.

Loop ACDA :

$$4z + 3(y + z) - 2(x - z) = 0$$

or

$$9z + 3y - 2x = 0 \tag{2}$$

Loop EBADE :

$$-2 + 2(x + y) + x + 2(x - z) = 0$$

$$5x + 2y - 2z = 2 \tag{3}$$

Solve equations (1), (2) and (3) for x, y, z.

$$x = 30/91 \text{ A};\ y = 17/91 \text{ A};\ z = 1/91 \text{ A}.$$

The p.d. (potential difference) across BD is given by

$$1 \text{ ohm} \times x + 2 \text{ ohm} \times (x - z) = (30/91) + 2\ (29/91) = 88/91 \text{ volts}$$

The resistance between B and D can be found.

$$R_{BD} = \frac{\text{Volt drop B to D}}{\text{current thro' B and D}}$$

$$= \frac{88/91}{x+y} = \frac{88}{47} \text{ ohms.}$$

1.8 DEFINITIONS

In network problems, a *voltage source* is a source maintaining the voltage e across its ends as a function (independent of the current supplied) given by

either $e = f(t)$

or $e = \text{constant}$ (for all i)

where i and t denote current, time. For a constant voltage source, (a battery with no internal resistance for *e.g.,*)

$$e = \text{constant} \tag{1.17}$$

An alternating voltage source is given by

$$e = f(t) \quad (e = E \sin \omega t \text{ for example}) \tag{1.18}$$

In electrical network problems, we take voltage sources as *ideal, i.e.*, not dependent on current i. But practical sources of voltage such as the battery, for example, have internal resistances and so they will drop in voltage as they supply current. In such cases, $e = f(i)$. In Fig. 1.15 (a), the arrow shows the direction of the voltage rise.

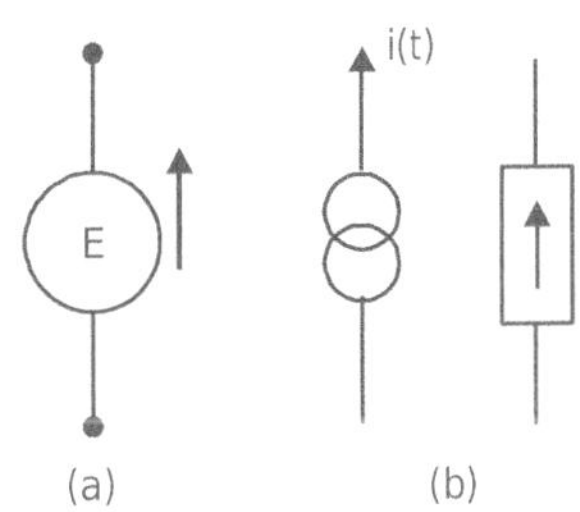

Fig. 1.15 Voltage and current sources—representation of.

A *current source* for network problems is defined on similar lines as a function independent of the voltage across it, given by

either $i - \text{constant}$ (1.19)

or $i = f(t)$ (1.20)

The current arrow shows the direction of the current at all times in case (a) and at $t = 0$ for case (b). Practical current sources (such as the simple battery for a voltage source) are not available, however. The nearest practical current source is the solar cell.

We shall deal with only constant sources in this chapter.

Active network is one containing sources.

Passive network is one which does not contain a source. Currents flow in passive networks if they are connected to sources, external to the networks.

Example 6. *In the network of Fig. 1.16, A is a constant voltage source with a rise of E volts from p to q and B is a constant current source with a current of 1 amps. Find the current through A and the voltage across B in terms of E, I, R_1, R_2 and R. Hence find the current through R if B is replaced by a short circuit.* (B.E. Nov. 1968)

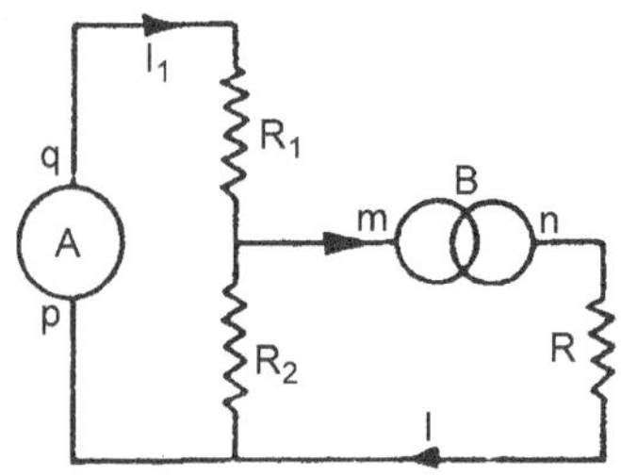

Fig. 1.16

Solution. Assume I_1 to be the current from *q*. Mark the current through R_2 using KCL (Kirchhoff's current law), as $I - I_1$. Apply KVL (Kirchhoff's voltage law) to the loop *pq* R_1R_2 :

$$-E + R_1 I_1 - R_2 (I - I_1) = 0$$

$$I_1 = (E + R_2 I)/(R_1 + R_2)$$

For loop $mnRR_2$:

$$V_{mn} + RI + R_2 (I - I_1) = 0$$

$$V_{mn} = R_2 I_1 - (R + R_2) I$$

$$= [R_2/(R_1 + R_2)] [E + R_2 I] - (R + R_2)I \qquad (a)$$

If B is replaced by a short circuit, $V_{mn} = 0$. Hence, for this case, we get

$$0 = [R_2/(R_1 + R_2)] [E + R_2 I] - (R + R_2)I$$

$$\therefore \qquad I = \frac{ER_2}{R_1(R + R_2) + RR_2}$$

Example 7. *In the circuit below, prove that*

1. $I_L = 50\ I_b$
2. *Voltage* V_{12} *between points 1 and 2 is 1475* I_b *volts.*

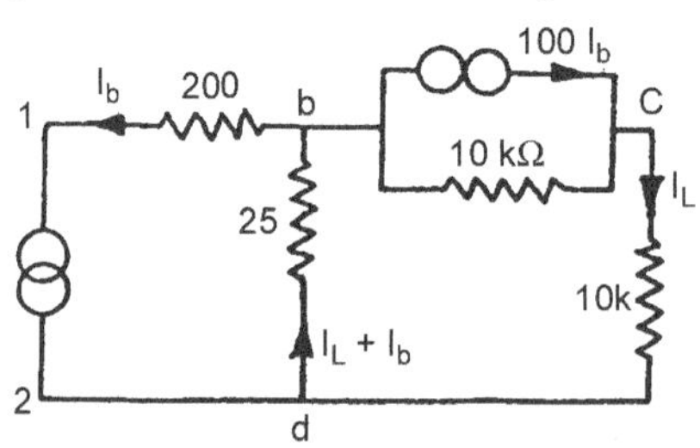

Fig. 1.17

Solution. The current in *db* is $I_L + I_b$.

$$V_{cd} = 10{,}000\ I_L \text{ in branch } cd.$$

$$V_{bc} = 10{,}000\ (I_L - 100\ I_b)$$

$$V_{bd} = 25\ (I_L + I_b) = V_{bc} + V_{cd}$$

$$V_{12} + V_{b1} = V_{bd}$$

$$V_{12} = V_{bd} - V_{b1} = -25 (I_L + I_b) - 200 I_b$$

$$= -25 (51 \ I_b) - 200 I_b$$

$$= -1475 \ I_b$$

This is an example containing a so-called 'dependent source', for 100 I_b is a current source whose current depends upon I_t, the current in another branch of the circuit. Such dependent sources are characteristic of electronic devices. The circuit given is the equivalent circuit of a transistor.

1.9 INPUT RESISTANCE

The resistance between the two terminals of a passive network where the input source is connected to it is called 'input resistance'.

Example 8. *Find the input resistance between terminals a and b of the following network.*

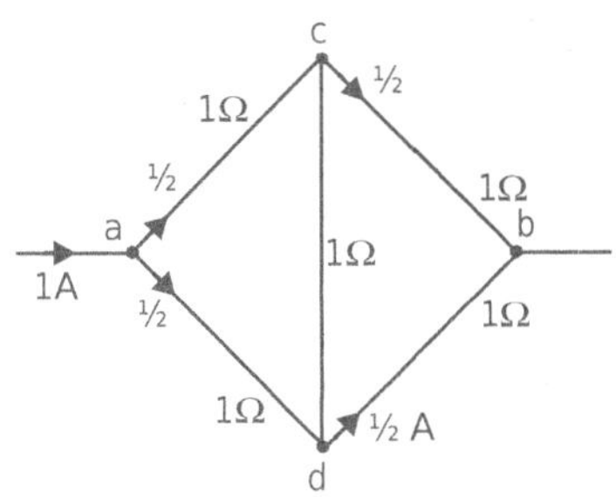

Fig. 1.18

Solution. Assume a current of 1 A being sent *via. a* and exit at *b*. By symmetry, the currents divide into $\frac{1}{2}$ amps. each as shown along *ac, ad* and *cb, bd*. So, at junction *c*, the current i_{ed} = 0 by KCL. Hence, branch *cd* carries no current and may be omitted. So, the circuit between *a* and *b* is just two parallel branches of 2 ohms each so that they are equivalent to 2 × 2 / (2 + 2), or 1 ohm only. So, the input resistance between *a* and *b* is 1 ohm.

Example 9. *The cube of Fig. 1.19 represents a network each of the sides ab, bc etc., of the cube having one ohm resistance. Find using the symmetry of the network (or using network transformations directly), the equivalent resistance between the terminals a and g of the network.*

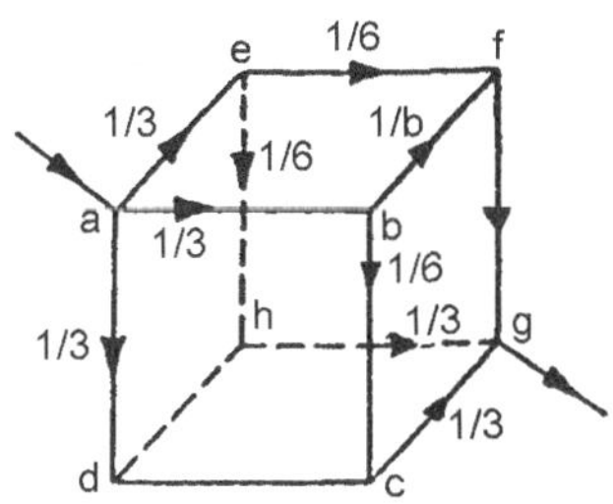

Fig. 1.19

Solution. As before, assume a current of 1 A entering A. By symmetry and since all sides have equal resistances this 1 Ampere at a and g divide into $\frac{1}{3}$ A in the branches ae, ab, ad, fg, cg, hg. The $\frac{1}{3}$ ampere passing along ae has to divide into the two branches ef, eh and by symmetry, the division will be equal so that ef and eh carry 1/6 amperes each.

So, the drop in voltage from a to g

$$= \text{Drop } ae + \text{Drop } ef + \text{Drop } fg$$

$$= 1 \times 1/3 + 1 \times 1/6 + 1 \times 1/3 = 5/6 \text{ volts.}$$

$$\therefore R_{eq.} = 5/6\ \Omega$$

1.10 THE CIRCULATING OR LOOP CURRENTS METHOD

In the wheatstone bridge circuit, we saw that it has three loops. Likewise, any circuit contains one or more loops. Though currents flow only in the resistances, the loop itself is imagined to be circulating a current. So, if a loop has three resistance, then, instead of three separate currents, we can visualise a single loop currents.

Maxwell introduced this method. The method depends on choosing a set of *independent* loops for a network. There are 3 loops down in Fig. 1.21. Currents i_1, i_2, i_3 are assumed to circulate through the loops. The branch currents depend on the loop currents. For example, the current in branch ac is $i_1 - i_2$; the minus sign is because the direction of i_2 is against i_1. In branch ad, the current is $i_3 – i_1$. Since a circulating current entering a junction (node) leaves it immediately we do not need to use KCL in this method explicitly.

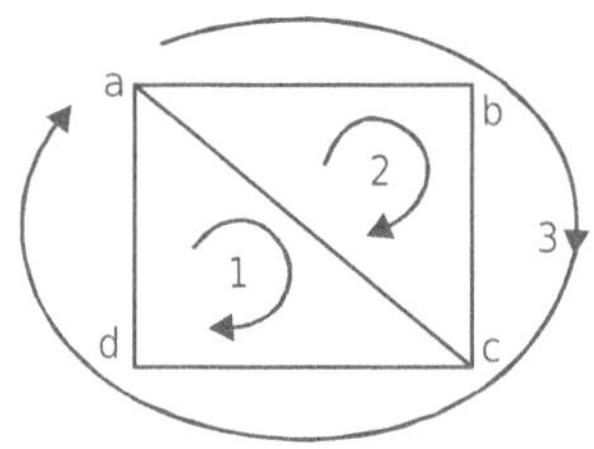

Fig. 1.20

Example 10. *In the network shown, determine the voltage across the 1 ohm resistor. (Loops were not marked in the question).*

(II B. Tech. April 1969)

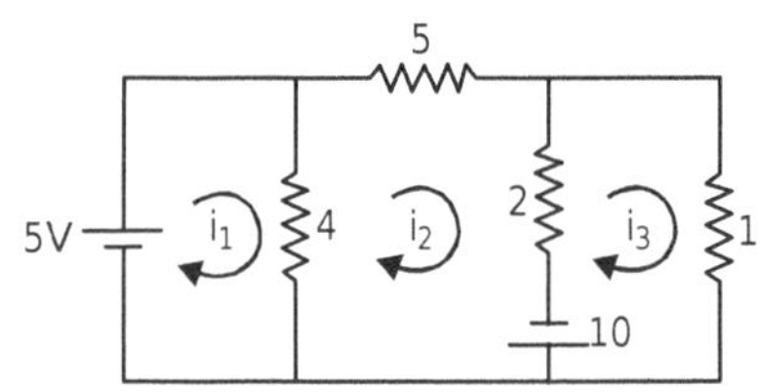

Fig. 1.21 Circulating current method.

Solution. We can choose arbitrarily the three loops marked as i_1, i_2, i_3 . Then we have to write KVL equations for each of the loops.

Loop 1. The current in the 4 ohms resistor is $i_1 - i_2$.

$$4\ (i_1 - i_2) - 5 = 0 \qquad (1)$$

Loop 2. The current in the 2 ohms resistor is $i_2 - i_3$.

$$4\ (i_2 - i_1) + 5i_2 + 2\ (i_2 - i_3) - 10 = 0 \qquad (2)$$

Loop 3. $2\ (i_3 - i_2) + 1i_3 + 10 = 0 \qquad (3)$

We can solve the three equations and get the currents.

$$i_1 = 185/168,\ i_2 = 25/17,\ i_3 = -\ 40/17\ i_3.$$

Matrix Form

However, let us rewrite each of these *loop equations* in the following form :

$$4i_1 - 4i_2 = 5$$

$$-\ 4i_1 + i_2(4 + 5 + 2) - 2i_3 = 10$$

$$-\ 2i_2 + i_3(2 + 1) = -10$$

Note that the equations can be written in matrix form as

$$\begin{vmatrix} 4 & -4 & 0 \\ -4 & (4+5+2) & -2 \\ 0 & -2 & (2+1) \end{vmatrix} \begin{vmatrix} i_1 \\ i_2 \\ i_3 \end{vmatrix} = \begin{vmatrix} 5 \\ 10 \\ -10 \end{vmatrix}$$

The right hand side column matrix [5, 10, – 10] consists of source voltages and the [3 × 3] matrix consists of resistance values and so the above is written*

$$RI = E \tag{1.21}$$

where R is the resistance matrix and E, I are column vectors.

Writing the Matrix Directly

The matrix could be written down from inspection of the loops. For that, note the following :

(*a*) The diagonal entries of the R matrix consist of the sum of all the resistances in that loop. For example, loop 2 contains 4, 5 and 2 ohms resistors and hence the entry (4 + 5 + 2).

(*b*) The off-diagonal entries in the R matrix are those resistance values which are common to the loops. Thus, the [*i, j*] entry (*i*th row and *j*th column entry) is given by the resistances common to the loops *i* and *j*, with a *negative sign* prefixed. For example, the [2, 1] entry is – 4, since 4 ohms is common to the loops, i_2 and i_1.

(*c*) The column matrix E has its entry equal to the e.m.f. sources in the respective loops 1, 2, 3... with a *positive* sign if the loop current leaves the positive terminal of the e.m.f. source; with a *negative* sign if the loop current leaves the negative terminals. For example, in loop 3, the current i_3 leaves the negative terminal of the 10 volt battery and hence the entry –10.

From the matrix, the simultaneous equations can be written and the currents i_1, i_2, i_3 can be found as their solution. Alternatively, from

$$RI = E$$

we can find the inverse R^{-1} of the matrix R and get

$$I = R^{-1} E \tag{1.22}$$

by multiplication of the matrices R^{-1} and E.

Example 11. *The network shown in Fig. 1.22 below has the resistances of values marked in the figure. Find the current in the ab, ad and aA branches when (i) 20 volts act in Aa from A to a and 20 volts in Cc from C to c, (ii) as in 1, but the voltage in Cc is reversed.*

Solution. (*i*) Let us choose the five loop currents as marked, *x, y, z, w, u*. Then the loop equations can be directly written in matrix form.

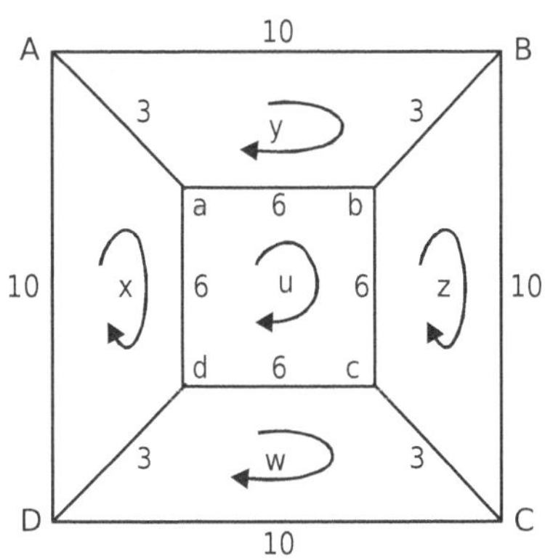

Fig. 1.22

* In general, as we shall see later, we shall have the impedance matrix Z in place of the resistance matrix R.

$$\begin{vmatrix} 22 & -3 & 0 & -3 & -6 \\ -3 & 22 & -3 & 0 & -6 \\ 0 & -3 & 22 & -3 & -6 \\ -3 & 0 & -3 & 22 & -6 \\ -6 & -6 & -6 & -6 & 24 \end{vmatrix} \begin{vmatrix} x \\ y \\ z \\ w \\ u \end{vmatrix} = \begin{vmatrix} 20 \\ -20 \\ 20 \\ -20 \\ 0 \end{vmatrix}$$

Note for example in the first row (3 + 6 + 3 + 10) = 22, the sum of all resistors in the x loop. (–3, 0, –3, –6) refer to the resistors common to the x loop and the remaining loops y, z, w, u. The 20 volts is written in the right hand side E matrix as + 20, since it aids to send the loop current x.

Inverting the matrix and finding the currents for the 5 × 5 matrix will be tedious. Let us eliminate the variable u. So, take the last equation :

$$-6x - 6y - 6z - 6w + 24u = 0$$

$$-6u = (x + y + z + w)$$

The first four rows contain $-6u$ and so the u can be eliminated from them by adding $(x + y + z + w)$. Then the equations become.

$$\begin{vmatrix} 23 & -2 & 1 & -2 \\ -2 & 23 & -2 & 1 \\ 1 & -2 & 23 & -2 \\ -2 & 1 & -2 & 23 \end{vmatrix} \begin{vmatrix} x \\ y \\ z \\ w \end{vmatrix} = \begin{vmatrix} 20 \\ -20 \\ 20 \\ -20 \end{vmatrix}$$

We observe that the network is symmetrical left to right and top to bottom. So we infer that $x = z$, $y = w$. Substituting these, from the first two rows,

$$24x - 4y = 20$$

$$-4x + 24y = -20$$

giving, $x = -y = 5/7.$

So the current in ab is 5/7; in aA, it is 10/7.

(*ii*) In the second part, the e.m.f. in Cc is reversed. So, we infer that the symmetry requires that $x = -z$ and $y = -w$. So, the first two loop equations now become.

$$(23 - 1)\,x + (-2 + 2)\,y = 20 \qquad 22x = 20$$

$$(-2 + 2)\,x + (23 - 1)\,y = -20 \qquad 22y = -20$$

giving $x = -y = 20/22.$

Hence current in ab 10/11; in ad 10/11; in aA it is 20/11 :

For networks containing current sources, the loop current method requires a modification. Current sources have a constant current passing through them and this restricts us in the choice of loop currents arbitrarily. The current source can be converted into a voltage source by the following principle.

Source equivalence principle. A positive voltage e in series with a resistor R, is equivalent to (*i.e.*, can be replaced by) a current source of current e/R in parallel with a resistor R.

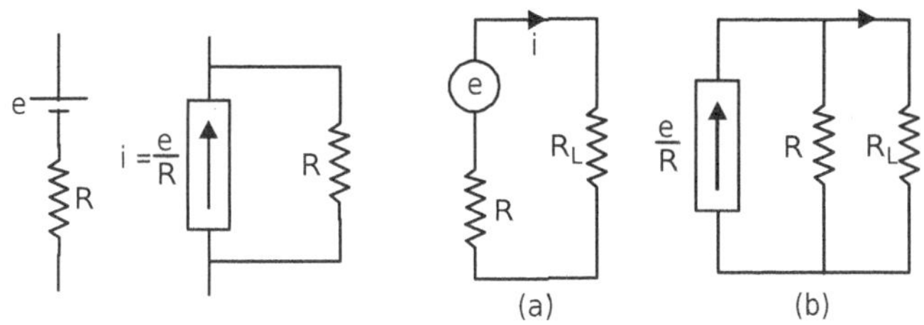

Fig. 1.23 Equivalence between voltage and current sources.

We would have proved this if we show that any load connected across both draw the same current i. Thus in the case of Fig. 1.23(a),

$$i = \frac{e}{R + R_L}$$

and in Fig. 1.23 (b), the current flowing through load R_L is

$$i = \frac{iR}{R + R_L} = \frac{e}{R + R_L}$$

so that the result that the two sources in Fig. 1.23 are equivalent is verified.

1.11 MESH CURRENT METHOD WITH CURRENT SOURCES

So, using the above principle, a current source i in parallel with a resistor R can be replaced by a voltage source $e = i$R in series with the same R. Most often networks with current sources will have parallel resistors across them.

Example 12. *In the following figure, find the currents in the branches. Also find the power supplied by the 6 V source.*

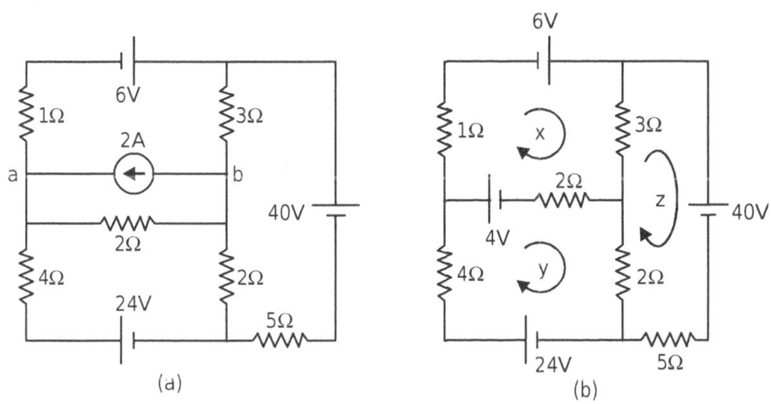

Fig. 1.24 Example with a current source.

Solution. The 2 ohms in parallel with the 2 A current source can be replaced by an e.m.f. source of 2 × 2 = 4 V in series with 2 ohms as in Fig. 1.24 (b). Then choose loop currents x, y, z as shown.

The matrix equation is :

$$\begin{vmatrix} 6 & -2 & -3 \\ -2 & 8 & -2 \\ -3 & -2 & 10 \end{vmatrix} \begin{vmatrix} x \\ y \\ z \end{vmatrix} = \begin{vmatrix} 6+4 \\ 24-4 \\ -40 \end{vmatrix} = \begin{vmatrix} 10 \\ 20 \\ -40 \end{vmatrix}$$

The equations have been written down by inspection. The e.m.f. in loop x is + 6 + 4 = 10 V; that in the y loop is 24 – 4 V and in the z loop is –40.

Thus the right hand side matrix gives [10, 20, –40].

The resistance matrix R is written following the rules enumerated before. Solving, we get.

$$x = \tfrac{1}{2},\ y = 1.33,\ z = -3.50.$$

The currents in the branches are :

1 ohm : $\tfrac{1}{2}$. 3 ohm : $(x - z) = 4.0$.

5 ohm : –3.5 4 ohm : 1.33.

The current in the 2 Ω between ab is to be found by calculating the p.d. between ab and dividing by 2.

$$\text{P.D. between } ab = 4\text{ V} - 2(x - y) = 4 - 2(-0.83) = 5.66$$

$$\text{Current in 2 ohm} = 5.66/2 = 2.83\text{ A}$$

The power supplied by the 6 V battery is

$$6x = 6\left(\frac{1}{2}\right) = 3.0\text{ W}.$$

1.12 NODE VOLTAGE METHOD

In this method, the potentials of all nodes are first labelled with respect to some fixed node. Thus in Fig. 1.25 below, there are 1, 2, 3, 4, 5, n nodes. We can fix node n as the ground or reference node and measure the potentials of all nodes above that of n. Let V_1, V_2, V_3, V_4, V_5 be the potentials of nodes 1 to 5.

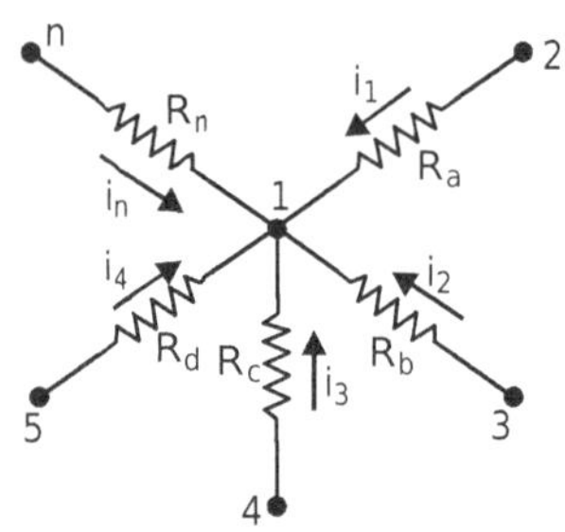

Fig. 1.25

The figure shows part of a network in which all the nodes from 1 to n are shown but the branches connecting node 1 only are shown. The current i_1 flowing into node 1 can be written as

$$i_1 = -(V_1 - V_2)/R_a$$

Similarly

$$i_2 = -(V_1 - V_3)/R_b$$

$$i_3 = -(V_1 - V_4)/R_c \text{ etc.}$$

$$i_n = -\left(\frac{V_1}{R_n}\right) \qquad (\because V_n \text{ is taken as zero})$$

Since KCL demands the sum $i_1 + i_2 + \ldots + i_n = 0$

$$\frac{V_1 - V_2}{R_a} + \frac{V_1 - V_3}{R_b} + \frac{V_1 - V_4}{R_c} + \ldots + \frac{V_1}{R_n} = 0$$

Rewriting this as below :

$$V_1\left(\frac{1}{R_a} + \frac{1}{R_b} + \ldots + \frac{1}{R_n}\right) - \frac{V_2}{R_a} - \frac{V_3}{R_b} \ldots - \frac{V_5}{R_d} = 0$$

Like this, equations can be written for the other nodes 2 to 5 also. The simultaneous solution of these node voltage equations give us V_1, V_2, V_3 etc. from which the currents in the branches may be calculated. This method is easily applied to current source networks. In this method, the KVL equations are implicitly obeyed and hence equations need not be written.

Example 13. *In the network shown, there are 3 current sources. Let us write the KCL equations for each node. Let us take node 4 as the reference, i.e., measure the voltages of nodes 1, 2, 3, with respect to KVL node. Then, we get.*

Solution. Node 1.

$$\frac{V_1 - V_2}{1} + \frac{V_1}{2\frac{1}{2}} = 4 - 3$$

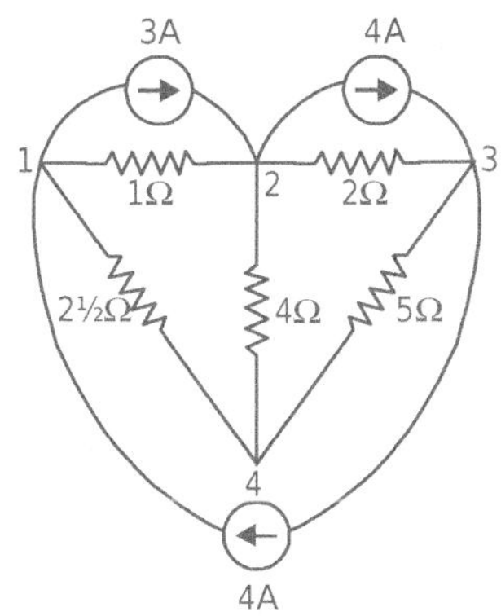

Fig. 1.26

Hence the current flowing from node 1 to node 2 is $(V_1 - V_2)/1$; that from 1 to 4 is $V_1/2\frac{1}{2}$. Their sum must be equal to the net current entering node 1 which is 4A – 3A = 1A.

So, $$V_1\left(\frac{1}{1} + \frac{1}{2\frac{1}{2}}\right) - \frac{V_2}{1} = 1$$

Note that the term multiplying V_1 consists of the sum of the reciprocals of the resistances in the branches joining node 1. The term multiplying V_2 is the negative of the reciprocal of the resistance connecting nodes 1 and 2.

For node 2.

$$\frac{V_2 - V_1}{1} + \frac{V_2 - V_3}{2} + \frac{V_2}{4} = 3 - 4$$

So, $$-\frac{V_1}{1} + V_2\left(\frac{1}{1}+\frac{1}{2}+\frac{1}{4}\right) - \frac{V_3}{2} = -1$$

Note that the coefficient of V_2 is the sum of the reciprocals of the resistors 1, 2 and 4 ohms. The coefficient of V_1 is the negative reciprocal of the 1 ohm resistor joining 2 and 1. The co-efficient of V_3 is the negative reciprocal of the 2 Ω resistor joining node 2 and 3.

For node 3. $$-\frac{V_2}{2} + V_3\left(\frac{1}{2}+\frac{1}{5}\right) = 0$$

The above three nodal voltage equations can be written in matrix form.

$$\begin{vmatrix} 1+0.4 & -1 & 0 \\ -1 & 1+0.5+0.25 & -0.50 \\ 0 & -0.5 & 0.5+0.2 \end{vmatrix} \begin{vmatrix} V_1 \\ V_2 \\ V_3 \end{vmatrix} = \begin{vmatrix} 1 \\ -1 \\ 0 \end{vmatrix}$$

Solving for v_1, v_2 and v_3, we get

$$v_1 = -0.3,\ v_2 = 0.41,\ v_3 = -0.42.$$

Example 14. *Find the voltages across the current sources in figure 1.27.*

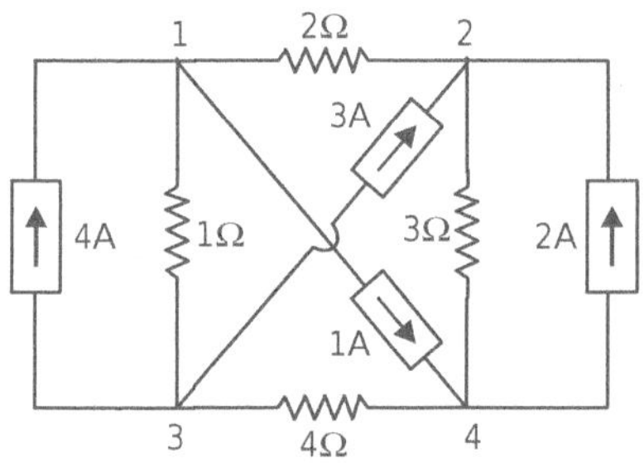

Fig. 1.27

Solution. The node voltage equations can be written for nodes 1, 2, 3, 4. The current entering node 1 is 4 – 1 = 3 A, node 2 is 2 + 3 = 5 A, node 4 = 1 – 2 = –1A, node 3 is –3 – 4 = –7 amperes. The node 1 has 1 Ω and 2 Ω resistors joined to it and has a resistor of 2 ohm connecting to node 2. The node voltage equations can be written directly in matrix form, choosing node 3 as reference node, *i.e.*, measuring all voltages above node 3.

$$\begin{vmatrix} 1+\frac{1}{2} & -\frac{1}{2} & 0 \\ -\frac{1}{2} & \frac{1}{2}+\frac{1}{3} & -\frac{1}{3} \\ 0 & -\frac{1}{3} & \frac{1}{3}+\frac{1}{4} \end{vmatrix} \begin{vmatrix} v_1 \\ v_2 \\ v_4 \end{vmatrix} = \begin{vmatrix} 3 \\ 5 \\ -1 \end{vmatrix}$$

Solving, $$v_1 = 5.8,\ v_2 = 11.4,\ v_4 = 4.8.$$

The voltages across the current sources are thus v_1, v_2, $(v_1 - v_4)$ and $v_2 - v_4$; giving 5.8 11.4, 1.0 and 6.6 volts, respectively.

EXERCISES

1. In the circuit shown, find the currents via R, L and C if

(*a*) $v(t) = \sin t$

(*b*) $v(t) = e^{-t}$ for $t > 0$

(*c*) $v(t) = 2$, volts d.c.

[**Ans.** (*a*) $\sin t$, $\sin (t - \frac{1}{2}\pi)$, $\sin (t + \frac{1}{2}\pi)$, (*b*) e^{-t}, $-e^{-t}$, $-e^{-t}$, (*c*) 2, ∞, 0]

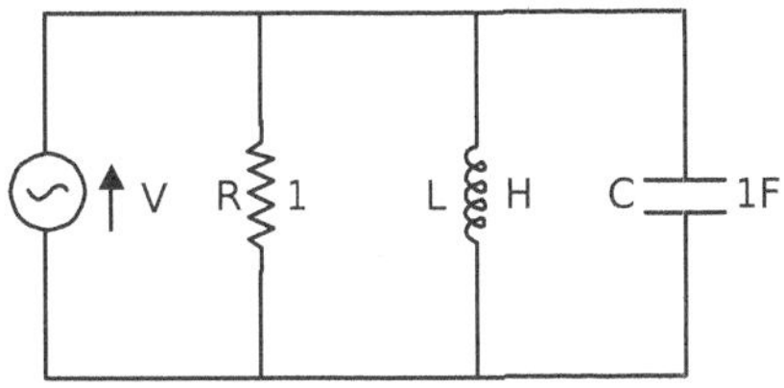

Fig. 1

2. The current through an inductance of 1 H is as in Fig. 2.

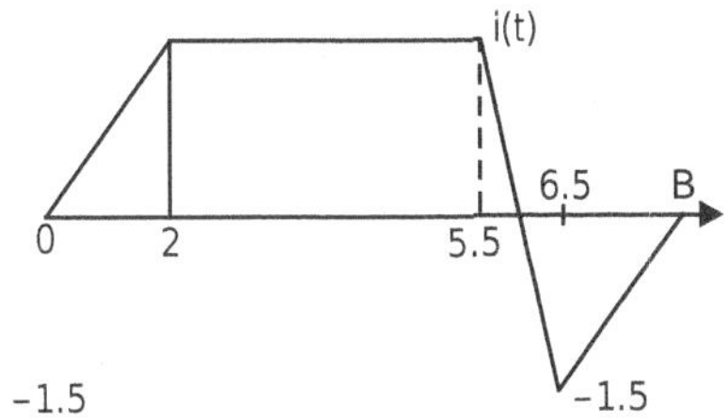

Fig. 2

Find the voltage across it at $t = 0$, $t = 2^-$, $t = 5.5^-$, $t = 5.5^+$, $t = 6.5^-$, $t = 6.5^+$ and $t = 8^-$.

[**Ans.** $\frac{3}{4}$, $\frac{3}{4}$, 0, –3, –3, –3, +1, +1 volts]

3. In the TV sweep coil, the sawtooth wave of current shown must be sent. If so, draw the wave form of the voltage source $e(t)$ (Fig. 3).

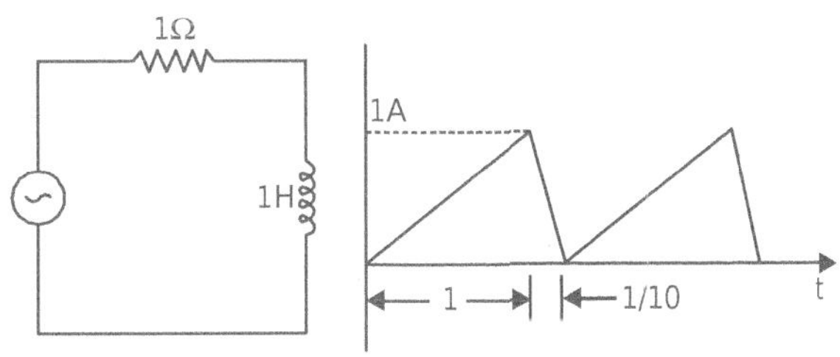

Fig. 3

[**Ans.**

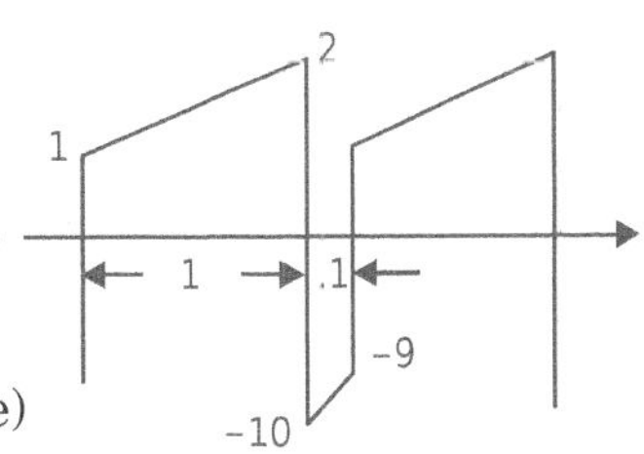

]

(Not to scale)

4. Three resistors are connected in the form of a triangle ABC. Across AB, the resistance is measured to be 2 ohms, across BC to be 20/9 ohms and across CA 14/9 ohms. The sum of the three resistors is 9 ohms. Find the resistors.

[**Ans.** 2, 3 and 4 ohms]

5. Find the value of the resistor R in Fig. 4 if the source supplies a total power of 6 W to the network. [**Ans.** 2 ohm]

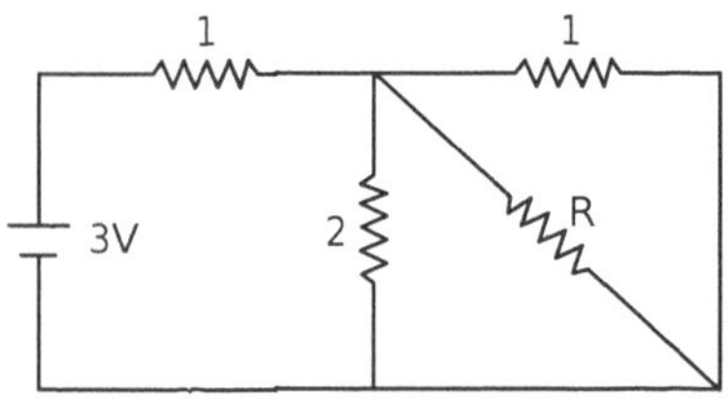

Fig. 4

6. Find the input resistances between AB, BC and CA of the square network shown if the sides and diagonals have one ohm each. [**Ans.** $\frac{1}{2}\ \Omega$, $\frac{1}{2}\ \Omega$, $\frac{1}{2}\ \Omega$].

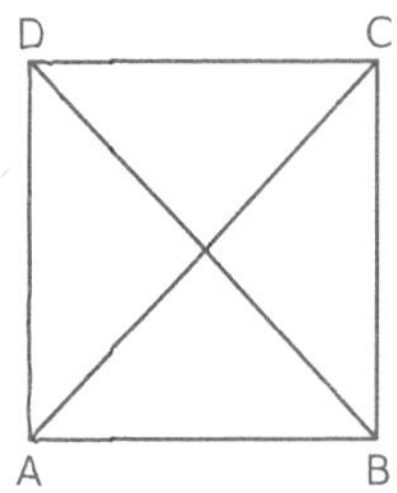

Fig. 5

7. Find the value of R in Fig. 6 such that the power supplied to it by the voltage source is double that supplied by current source. [**Ans.** $E_0 / 3I_0$]

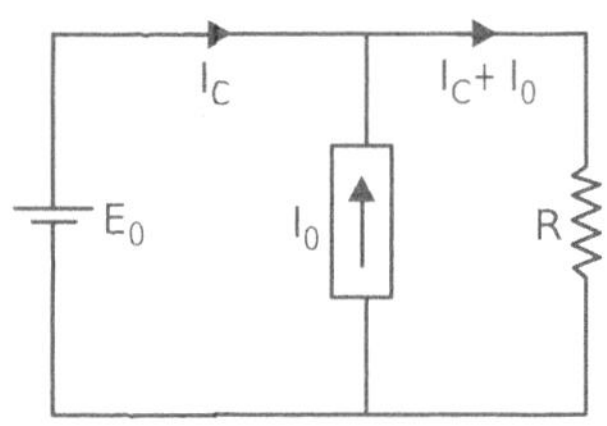

Fig. 6

8. In the hexagon (Fig. 7), the diagonals across at their junction. Each side and each semi-diagonal has one ohm. The battery is of 1 V and has internal resistance 1 Ω. Find the currents in the sides and diagonals.

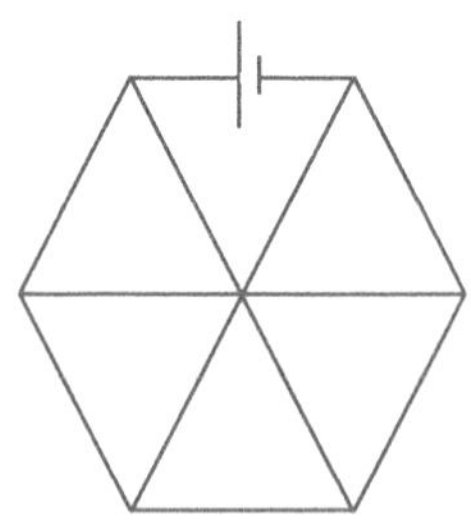

Fig. 7

[**Ans.** In the battery 0.45 A ; sides : 0.175, 0.075, 0.05; Diagonals : 0.275, 0.1, 0.025]

9. Find the resistance between CD of the cube shown if each side is one ohm.

[**Ans.** $\frac{1}{4}$ Ω]

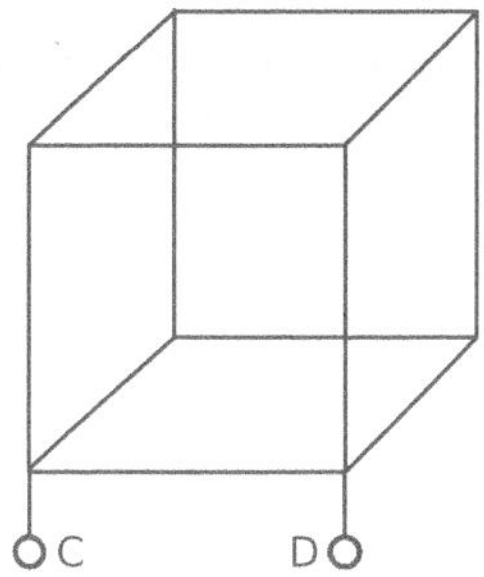

Fig. 8

10. In Fig. 9, S_1 is a 10 volt source and S_2 is a 5 A source. Find the voltage across S_2 and current through S_1.

[**Ans.** 26.67 A, – 1.667 V]

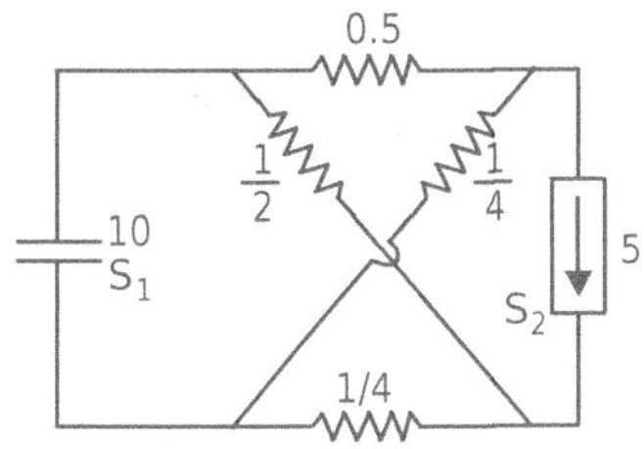

Fig. 9

11. In Fig 2, (page 21) one ampere is entering A and leaving B. Assume current x and y along BA and BC; write an expression for the total power in the network in terms of x and y and hence the input resistance.

[**Ans.** $8x^2 + 4y^2 + 8xy - 8x - 6y + 3$; $x = \frac{1}{4}$, $y = \frac{1}{2}$, $R_{ab} = \frac{1}{2}$]

12. Determine the current in the 4Ω resistor using (*a*) loop method (*b*) a node method (Fig. 10). [**Ans.** 0.1A]

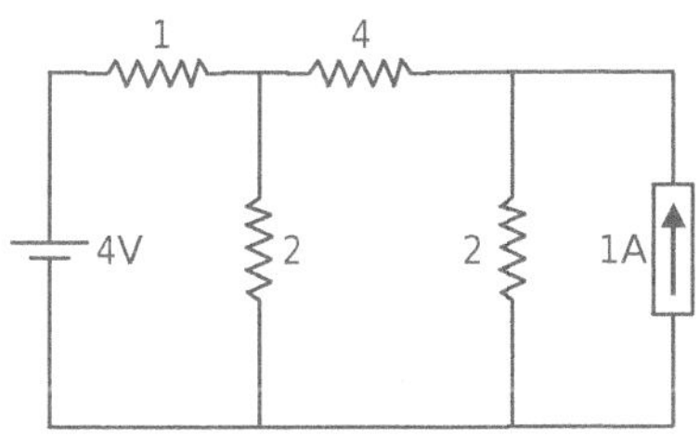

Fig. 10

13. Write the loop equations and find the matrix R for the currents i_1, i_2,...shown in Fig. 11. Each resistor is 1 ohm.

$$\left[\textbf{Ans.} \begin{vmatrix} 4 & -1 & 0 & -1 & -1 \\ -1 & 4 & -1 & -1 & -1 \\ 0 & -1 & 4 & -1 & -1 \\ -1 & -1 & -1 & 4 & 0 \end{vmatrix}\right]$$

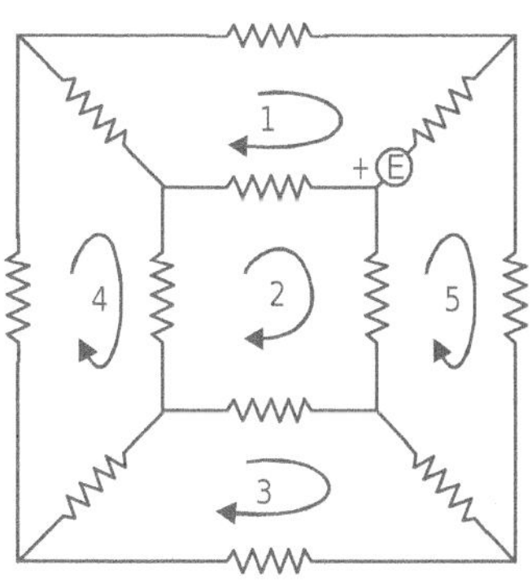

Fig. 11

14. Write the loop equations and by elimination of all other variables except i_1, i_2, find the current in branch AB. [**Ans.** 12/21, B to A]

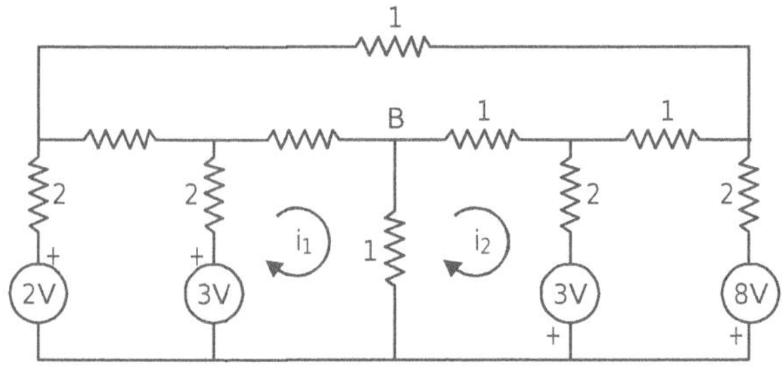

Fig. 12

15. Solve the voltage across AB and CD after writing the node admittance matrix choosing node B or D as reference. (Fig. 13).

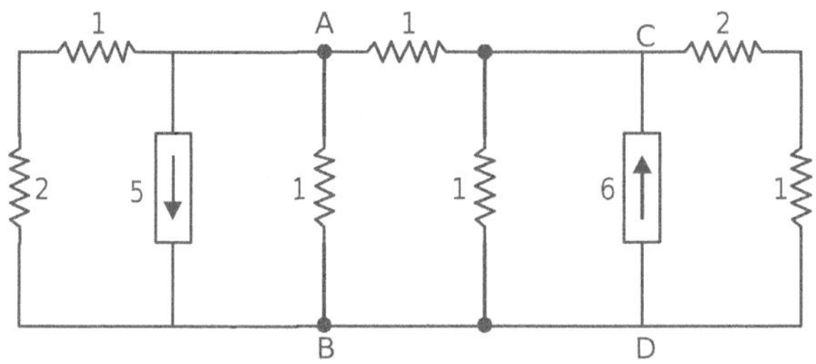

Fig. 13

$$\left[\textbf{Ans. } Y = \begin{vmatrix} 2\frac{1}{3} & -1 \\ -1 & 2\frac{1}{3} \end{vmatrix} \begin{matrix} V_{AB} = -3.977 \text{ V} \\ V_{CD} = -4.275 \text{ V} \end{matrix} \right]$$

16. Choose node D as reference and then write the node equations in matrix form. Find voltages V_{AB}, V_{CD} (Fig. 14). [**Ans.** 3/8 V drop A to B; $1\frac{1}{8}$ V drop C to D]

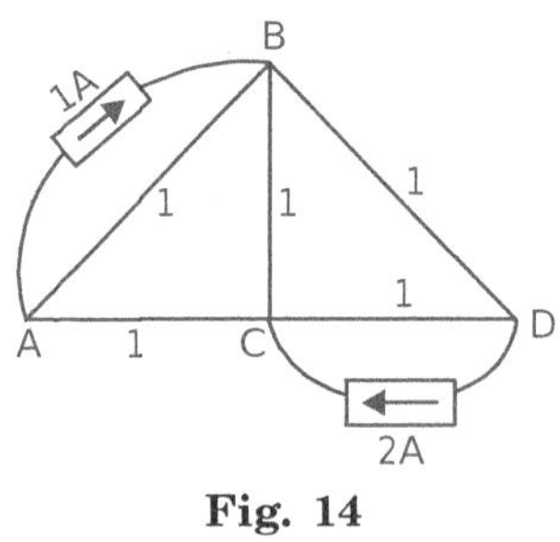

Fig. 14

17. Write node equations and find the voltage across the current source (Fig. 15).

$$\left[\textbf{Ans. } \frac{14}{15} \text{ V} \right]$$

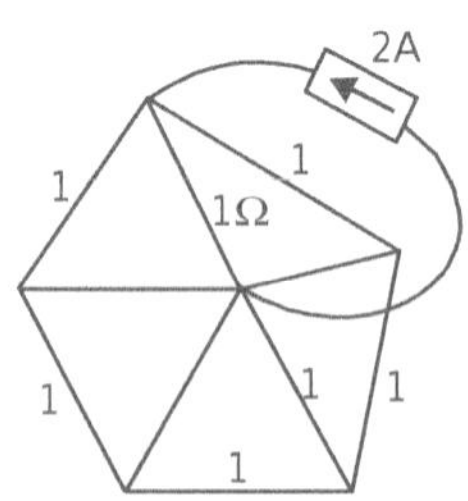

Fig. 15

18. For the constant-resistance 'bridge-T' network in Fig. 16, show that the input resistance is R_0 if $R_a R_b = R_0^2$.

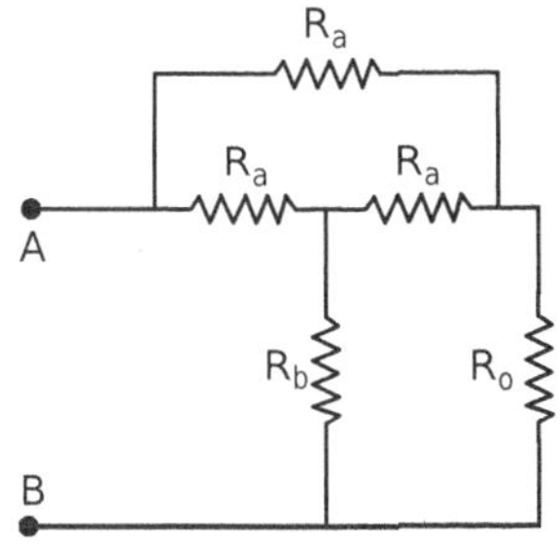

Fig. 16

19. Find the equivalent impedance of the bridge circuit of Fig. 17. [Anna University, May 2007]
[**Hint:** Combine resistors two at a time from the right end.]

[**Ans.** R]

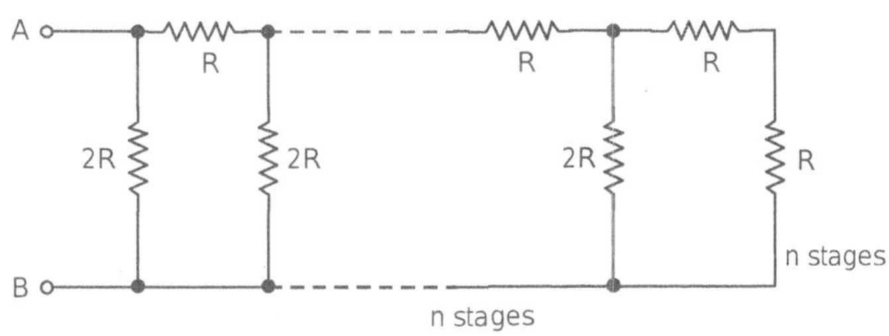

Fig. 17

20. Write down the Mesh equation and Nodal equation for the circuit below:

$$\begin{vmatrix} 15 & -10 & 0 \\ -10 & 45 & 20 \\ 0 & -20 & -45 \end{vmatrix} \begin{vmatrix} i_1 \\ i_2 \\ i_3 \end{vmatrix} = \begin{vmatrix} 25 \\ 0 \\ 50 \end{vmatrix}$$

$$\begin{vmatrix} 11/15 & -1/15 \\ -1/15 & 17/300 \end{vmatrix} \begin{vmatrix} V_1 \\ V_2 \end{vmatrix} = \begin{vmatrix} 5 \\ -2 \end{vmatrix}$$ [Anna University, May 2007]

(Currents are marked in question, note i_3 is counter clockwise).

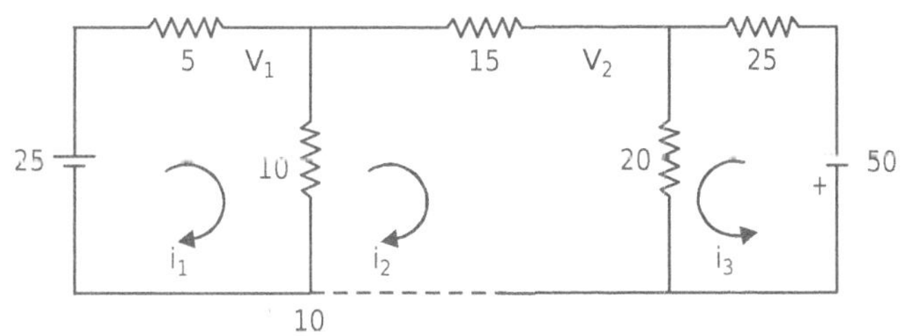

Fig. 18

SIMPLE EXERCISES

1. What is the equivalent resistance for N resistors in (*a*) series (*b*) parallel.
[Anna University, May 2007]
[**Ans.** RN, R/N]

2. What are the Differences between loop and node analysis ?

1. Loop analysis uses circuitous paths, one path per equation.

   Node analysis chooses a node junction for one equation.

2. The number of loop equations for a circuit with e elements, n nodes is $e - n + 1$.

   The number of node equations for the circuits is just $(n - 1)$ where n is the number of nodes in it.

3. Loop analysis forms the equation for each loop using Kirchhoff's voltage law for loop.

   Node analysis forms the equation by summing up the currents entering/leaving the node using Kirchhoff's current law.

4. Loop analysis is ideal for circuits having only voltage sources.

   Node analysis is ideal for circuits having only current sources.

3. State Kirchhoff's Laws.

The sum of the currents entering a junction must be equal to the currents leaving the junction.

—Current Law–KCL

The sum of the voltage drops around any mesh (or loop) in a circuit is zero.

—Voltage Law–KVL

2

SINUSOIDAL STEADY STATE ANALYSIS

2.1 SINUSOIDAL CURRENTS

Alternating currents are more widely used in the field of power and electronics than direct current. By the term alternating, it is meant that the current is not unidirectional but assumes alternately positive and negative values. Unlike the flow of direct current, there is no electron drift in a conductor carrying alternating current, but they merely oscillate about their average positions inside the conductor. If we plot an alternating current changing with time as a graph, with time in the X-axis and current in the Y-axis, we get the *waveform* of the alternating current. Sinusoidal wave alternating current is what we use in A.C. power circuits. Other waveforms like a triangular wave, square wave are used only in electronic circuits. Fig. 2.1(*c*) shows a random waveform. Such waveforms are not classified as alternating currents. Only those waves which are periodic *i.e.*, repeat themselves in a definite pattern are amenable for study. Even among these periodic waves, the sinusoidal (sine and cosine) waveform is of fundamental importance. *Fourier*, a mathematician, showed that all *periodic* waveforms (sinusoidal or not) can be split up into a series of sinusoidal waves. So, a study of the behaviour (*response*) of circuit elements to steady sinusoidal voltages (or currents) will be made in this chapter. The response to any other a.c. wave of voltage (or current) will be the sum of the responses of the Fourier components of the waveform.

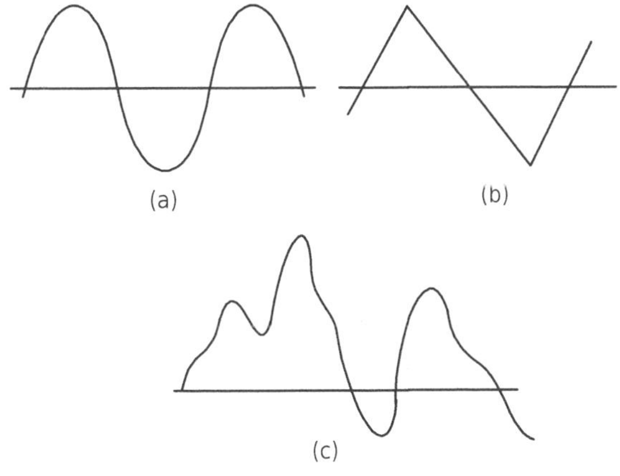

Fig. 2.1 Waveforms of alternating quantities (*a*) Sinusoidal waveform (*b*) Triangular waveform (*c*) Random wave.

2.2 PRODUCTION OF SINUSOIDAL VOLTAGES

To get an insight into sinusoidal voltages and currents, let us consider how such an a.c. waveform can be produced. Consider a uniform magnetic field in which a coil is rotated (Fig. 2.2). The coil is shown as rectangular and rotates at a velocity of ω radians per second. The ends of the coil are marked 1 and 2. The flux linking the coil changes due to rotation and so, by Faraday's law of electro-magnetic induction, an e.m.f. is induced in the coil, which appears as a p.d. between the ends 1 and 2.

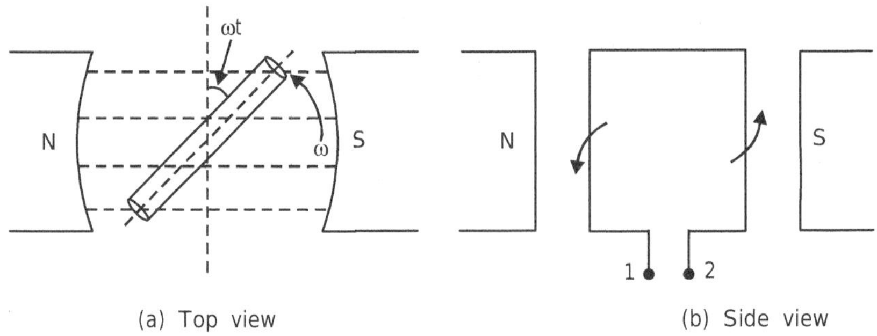

Fig. 2.2 Rotating coil in a magnetic field produces A.C. voltage.

Flux φ linking the coil = BA cos θ where A is the cross-sectional area of the coil.

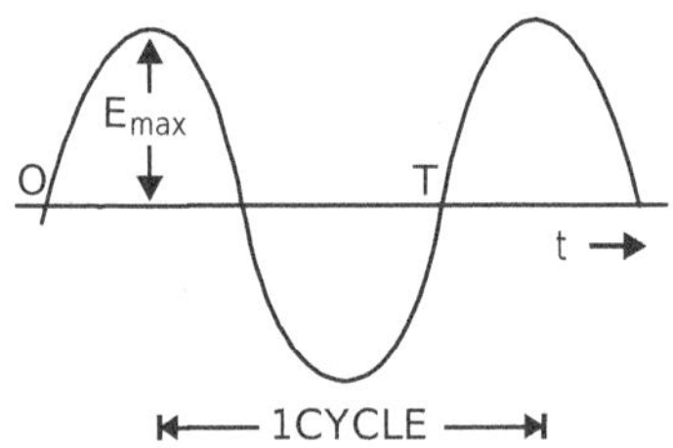

Fig. 2.3

$$\text{E.M.F. Induced} = -\,d\varphi/dt$$

$$= -\text{BA}\, d/dt\,[\cos\theta]$$

$$= -\text{ BA} \sin\theta\,(d\theta/dt)$$

Since $\theta = \omega t$ and $d\theta/dt = \omega$;

$$e = \text{BA}\,\omega \sin \omega t$$

$$= \text{E}_{max} \sin \omega t \qquad (2.1)$$

Here BAω is the maximum value of e, denoted E_{max}. The waveform is sinusoidal. We get one wave for each rotation of the coil. One complete *cycle* of the waveform is shown in Fig. 2.3. The time taken for one cycle is denoted T and called the *period* of the waveform.

$$\text{Period} = \text{T secs.}$$

$$\text{Number of cycles/sec} = 1/\text{T}$$

The number of cycles in a second is called the *frequency* of the wave and denoted as f.

$$f = 1/\text{T cycles/sec. or Hertz.} \qquad (2.2)$$

The name Hertz (abbr. Hz) is given in honour of Scientist, Engineer Heinrich Hertz.

Normal power frequency a.c. is of a sinusoidal waveform and frequency 50 cycles/second. The power supply is obtained only by the rotation of coils in a magnetic field as dealt with above. The mechanical energy is got from steam turbines or waterwheels.

2.3 CHARACTERISTICS OF SINUSOIDAL CURRENTS

We know that when a direct current of i amperes flows in a resistance R, the power is i^2R watts. An alternating current, on the other hand, produces an alternating power, i^2R. The power varies with time. For an alternating current $i = I_{max} \sin \omega t$ passing through R, the instantaneous power P is

$$P = i^2R = RI_{max}^2 \sin^2 \omega t$$

The total energy spent during one cycle, or T secs, will be

$$\text{Energy/Cycle} = RI_m^2 \int_0^T \sin^2 \omega t \, dt$$

Suppose a direct current of $I_{d.c.}$ flowing through the resistance R for the same T seconds produces the same energy. Then the d.c. current expends $I_{d.c.}^2$ RT *joules* and hence for equal power in d.c. and a.c.,

$$I_{d.c.}^2 RT = RI_m^2 \int \sin^2 \omega t \, dt$$

$$I_{d.c.} = I_m \sqrt{\left[\int \sin^2 \omega t \, dt\right] \div T} \qquad (a)$$

The integral $\int \sin^2 \omega t \, dt$ can be found to be

$$\int_0^T \sin^2 \omega t \, dt = \frac{1}{2\omega} \int_0^T (1 - \cos 2\omega t) \, d\omega t$$

$$= \frac{1}{2} T \text{ on integration.}$$

Note that $\omega T = 2\pi$, so we get

$$I_{d.c.} = I_m \div \sqrt{2} \qquad (2.3)$$

The above expression tells us that a *sinusoidal* alternating current is as *effective* in producing energy as a steady direct current of value $1/\sqrt{2}$ times its maximum value. Hence it is called the effective value of the A.C.

$$I_{eff.} = I_{max}/\sqrt{2} \qquad (2.3(a))$$

If we recall the steps used in deriving 2.3(a), we can apply the same steps to find the effective value of any other (non-sinusoidal) a.c. waveform also. Following the same steps for a general a.c. of $i(t)$,

$$I_{d.c.}^2 RT = R \int_0^T i^2(t) \, dt$$

$$\therefore \qquad I_{eff.} = I_{d.c.} = \sqrt{\frac{\int i^2(t) \, dt}{T}} \qquad (2.4)$$

Here $\int i^2(t) \, dt$ between the limits 0 and T represents the summation of the *squared function* of $f(t)$ over a cycle. Division by T means taking the average or *mean value*. Then $I_{eff.}$ is obtained by taking the *root*. Thus, we can rename $I_{effective}$ as the *root mean square value*, abbreviated as **R.M.S. value.**

2.4 FREQUENCY AND PHASE OF A SINE WAVE

Frequency f of a sine wave is defined as reciprocal of the period T.

$$f = 1/T$$

Since 1 cycle corresponds to a time of T seconds, the number of cycles in one second is 1/T and hence frequency f is expressed also as cycles/second (Unit is *Hertz*). The waveform $e = E_{max} \sin \omega t$ can be written :

$$e = E_{max} \sin \omega t$$

Put $\theta = \omega t$. Then, when $t = T$, $\theta = 2\pi$, since, in one period, one complete cycle of sine wave, *i.e.*, 2π radians of θ, have passed.

$$\therefore \quad 2\pi = \omega T$$

$$T = 2\pi/\omega \text{ and so } f = 1/T = \omega/2\pi$$

$$\therefore \quad \omega = 2\pi f \tag{2.5}$$

When two waveforms

$$e_1 = E_m \sin \omega t \text{ and } e_2 = E_m \sin (\omega t - \varphi)$$

are considered, the second wave is said to have a *phase difference* of φ radians. The plot of the two waves of e_1 and e_2 (Fig. 2.4) shows that the second wave reaches its maximum value φ radians later than the first. The second wave is said to *lag* the first wave by φ radians *i.e.*, φ/ω seconds). Similarly, the waveforms

$e_1 = E_m \sin \omega t$ and $e_3 = E_m \sin (\omega t + \varphi)$ have also the same phase difference, but the e_3 wave is said to *lead* the e_1 wave.

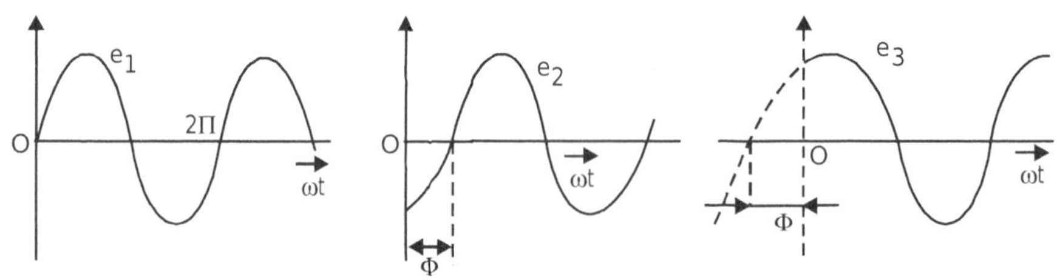

Fig. 2.4 The waveforms e_2 and e_3 have a phase difference of ϕ with respect to e_1.

2.5 AVERAGE VALUE, FORM FACTOR AND PEAK FACTOR OF A WAVE

The sinusoidal wave has its negative half-cycle exactly the same as the positive half-cycle and so its average value over the cycle is zero. Hence the average value is not measured as the total average over the whole cycle, but only for one half cycle. Thus, for a sine wave, the average value is

$$I_{av} = \frac{1}{\pi}\int_0^{\pi} E.\sin\theta \, d\theta$$

since the average value is found by finding the area under the sine curve (Fig. 2.3) for one half-cycle, 0 to π.

$$I_{av} = \frac{1}{\pi} E_m (-\cos\theta)_0^{\pi} = (2/\pi)\, E_m \tag{2.6}$$

Hence the average value of a sine wave = $(2/\pi) \times$ Max. value.

In general, for any A.C. wave, the *average value* is

$$\frac{1}{\pi}\int i(\theta)d\theta \tag{2.7}$$

The ratio of the R.M.S. value to the average value of an A.C. wave is defined as the form factor.

$$\text{Form factor} = \frac{\text{R.M.S. value}}{\text{Average value}} \tag{2.8}$$

For a *sine wave*, the ratio is $\left[E_{max}/\sqrt{2}\right] \div [2/\pi \, E_{max}]$ which is $\pi \div 2\sqrt{2}$ or 1.11.

The peak factor is defined as the ratio

$$\text{Peak factor} = \frac{\text{Max. value}}{\text{R.M.S. value}} \tag{2.9}$$

For a sine wave, it is $E_m \div E_m / \sqrt{2}$ or $\sqrt{2}$.

Example 1. *Explain what is meant by the root mean square value of an alternating current. Explain why this value is used.*

Calculate the r.m.s. value of a periodic voltage having the following values for equal intervals, changing suddenly from one value to the next : 0, 10, 30, 60, 30, 10, 0, – 10, – 30, – 60, – 30, – 10 etc.

Solution. If an alternating current and a direct current when passed through a resistance for a certain time expend equal amounts of energy in it, then the magnitude of that direct current is called the *effective value* of the alternating current. It is calculated by dividing the a.c. waveform into n intervals, finding the sum of the squares of the instantaneous current values for these n intervals, striking the average and then taking its root. Hence it is also called as the *root mean square* value.

$$I_{r.m.s.} = \sqrt{\frac{i_1^2 + i_2^2 + i_3^3 + \ldots + i_n^2}{n}}$$

When the current is expressible as a known mathematical function, it can be calculated by the formula,

$$I_{r.m.s.} = \sqrt{\frac{1}{2\pi}\int_0^{2\pi} i^2(\theta)d\theta}$$

Two alternating currents can be compared only by their effective (power producing) values and not by their maximum or average values.

The given values are tabulated.

n	1	2	3	4	5	6	7	8	9	10	11	12	13
i	0	10	30	60	30	10	0	– 10	– 30	– 60	– 30	– 10	0

The wave is symmetrical because the negative half cycle has values same as the positive ones.

Then we write the squares of all entries corresponding to one full cycle (up to 12 entries, the 13th entry belongs to the next cycle).

i^2	0	100	900	3600	900	100	0	100	900	3600	900	100

$$\text{Total sum of all } i^2 \text{ row entries} = 2 \times 5600$$

$$\text{Their average} = 2 \times 5600/12 = 933.33$$

$$\text{The r.m.s. values} = \sqrt{933.33} = 30.5$$

Example 2. *Find the effective value of the current obtained by applying a 100 volts A.C. voltage via a rectifier to a 10 ohms load, the rectifier offering zero resistance in the forward and infinite resistance in the reverse direction.*

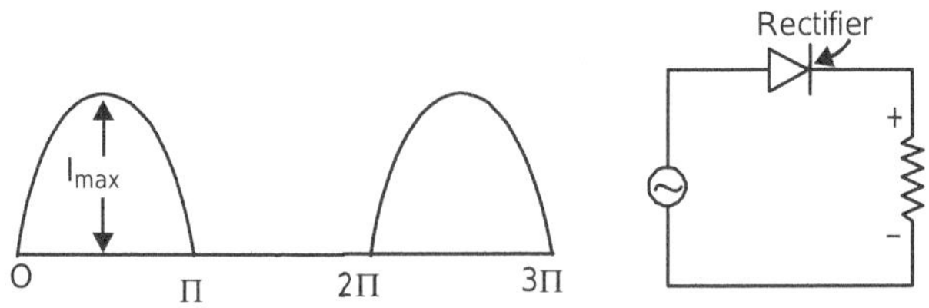

Fig. 2.5

Solution. The current in the forward direction will be having a maximum value of $100\sqrt{2}/10$; because the max. value of the voltage $100\sqrt{2}$, and the current is E/R. The rectifier blocks current flow in the negative half-cycles of the a.c. wave. The current waveform is as in Fig. 2.5.

The r.m.s. value of this current is obtained by finding the area of the square function curve, taking the average and finding its square root. The square function is

$$\left(10\sqrt{2}\right)^2 \sin^2\theta \text{ in the range of } 0 < \theta < \pi$$

$$\text{Zero in the range } \pi < \theta < 2\pi$$

$$\text{Area under square function graph} = \int_0^{\pi} \left(10\sqrt{2}\right)^2 \sin^2\theta \, d\theta$$

$$= 200 \int_0^{\pi} \frac{1}{2}(1 - \cos 2\theta) d\theta = 100 \left[\pi - 0\right] = 100\pi$$

$$\text{The average of the above} = 100\pi/2\pi = 50.$$

$$\text{The r.m.s. value} = \sqrt{50} = \sqrt{100/2} = 5\sqrt{2} = 7.07 \text{ amperes.}$$

The r.m.s. voltage across R = 10 × 7.07 = 70.7 volts.

Example 3. *A certain A.C. wave increases uniformly from 0 to 10 amperes during the first 0.005 seconds and then decreases uniformly to zero in the next 0.005 secs, the variations being similar in the negative half cycle. Calculate the frequency, r.m.s. value, the peak factor and form factor.*

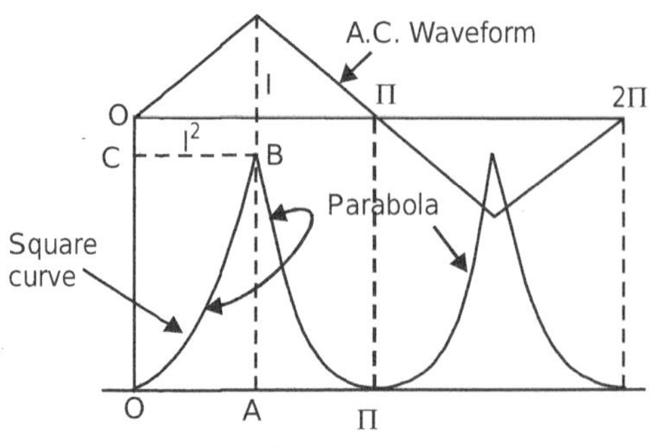

Fig. 2.6

Solution. One half cycle extends for 0.005 + 0.005 or 0.01 secs., and hence the period is 0.02 secs. Frequency is 1/0.02, or 50 cycles/second.

The waveform increases and decreases along straight lines. The square function is hence parabolic, and the plot of the square function graph consists of parabolic cusps as shown. The area under a parabola is 1/3rd of the area of the rectangle containing the parabola. This area OAB is 1/3rd of the rectangle OABC and since there are 4 such sections in one cycle.

$$\begin{aligned}\text{Area under square function} &= 4 \times (1/3) \times (100 \times 0.005)\\ &= 20/3 \text{ Amp}^2\text{-secs.}\\ \text{Average of this} &= (20/3) \times (1/0.02) = 100/3.\\ \text{The r.m.s. value} &= \sqrt{100/3} = 10/\sqrt{3} \text{ Amps.}\\ \text{The peak value} &= 10 \text{ Amps.}\end{aligned}$$

Average value of the waveform (*i.e.*, for a 1/2 cycle) can be easily found as the function is a triangle.

$$\begin{aligned}\text{Area under the wave for a } \frac{1}{2} \text{ cycle} &= \frac{1}{2} \times \text{base} \times \text{altitude}\\ &= \frac{1}{2} \times 0.01 \times 10 = 0.05\\ \text{Average value} &= 0.05/0.01 = 5 \text{ Amps.}\\ \text{Form factor} &= \text{R.M.S. value/Average value}\\ &= (10/\sqrt{3}) \times (1/5) = 1.152\\ \text{Peak factor} &= (\text{Max. value})/(\text{r.m.s. value}) = 10/(10/\sqrt{3}) = 1.732.\end{aligned}$$

2.6 RECAPITULATION OF COMPLEX ALGEBRA

A complex number $a + jb$ contains a real part a and an imaginary part b, where j is the imaginary quantity $\sqrt{-1}$. In mathematics the symbol i is used for $\sqrt{-1}$, but since i denotes current in Electrical Engineering, the symbol j will be used to denote $\sqrt{-1}$. The complex number $a + jb$ can be represented on the complex plane known as the Argand Diagram in mathematics.

The point P shown in Fig. 2.7 represents the numbers $a + jb$, having a component a on the real axis and the component b on the imaginary or j-axis.

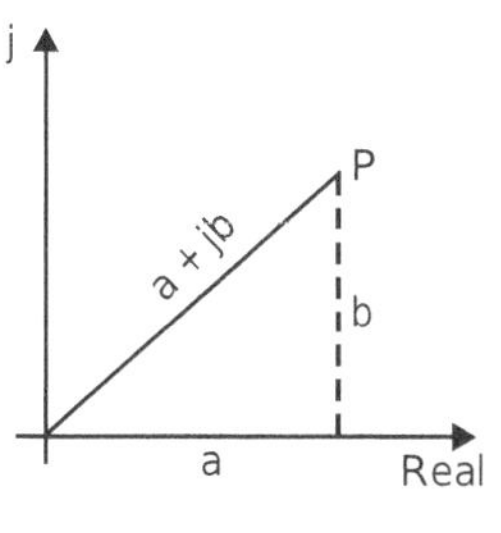

Fig. 2.7

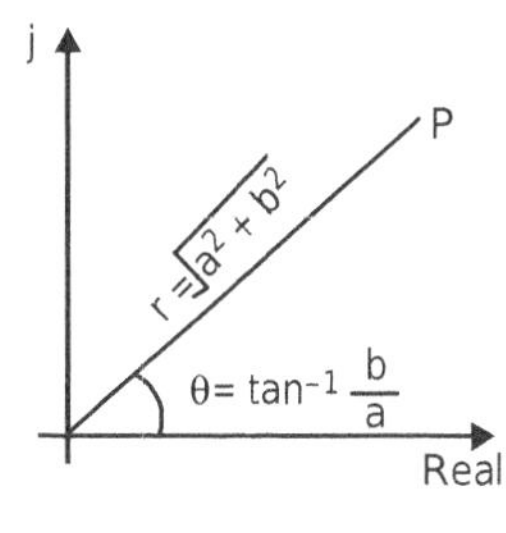

Fig. 2.8

The same point P can also be represented on a polar co-ordinate fashion, as P(r, θ) where r is seen to be $\sqrt{a^2+b^2}$ and $\theta = \tan^{-1}(b/a)$. The interlink between the *polar co-ordinates* and the $a + jb$ or (*real imaginary*) *co-ordinates* is by Euler's theorem,

$$re^{j\theta} = r\cos\theta + jr\sin\theta$$

$$= a + jb \qquad (\because r\cos\theta = a,\ r\sin\theta = b)$$

As an application of the above, the number $3 + j4$ can be written :

$$r = \sqrt{3^2+4^2} = 5,\ \theta = \tan^{-1}(4/3) = 0.93 \text{ radian.}$$

$$\therefore \qquad 3 + j4 = 5e^{j\theta}.$$

2.7 ADDITION AND SUBTRACTION OF COMPLEX NUMBERS

Two complex numbers $a + jb$ and $c + jd$ can be added by adding the real parts separately and the imaginary parts separately.

$$(a + jb) + (c + jd) = (a + c) + j(b + d).$$

The subtraction process is carried out thus :

$$(a + jb) - (c + jd) = (a - c) + j(b - d).$$

The addition process can be done on the complex plane also. Thus Fig. 2.9 shows addition made by treating the radius vectors OP and OQ as if they were *vectors* and adding them by parallelogram law of vectors. The subtraction process is done after taking the negative of $(c + jd)$ and adding it (*vectorially*) with $(a + jb)$ as in Fig. 2.10.

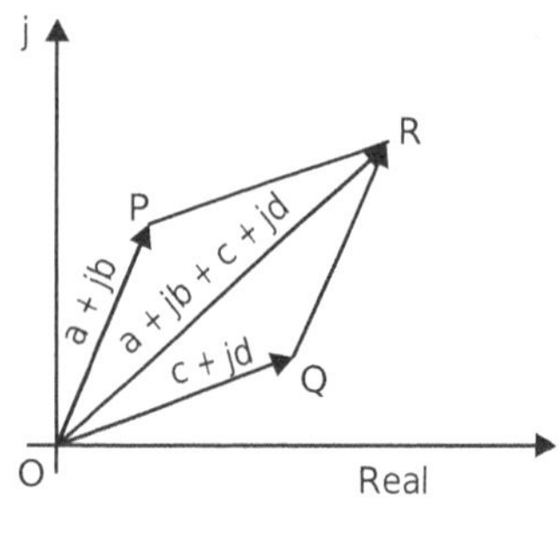

Fig. 2.9

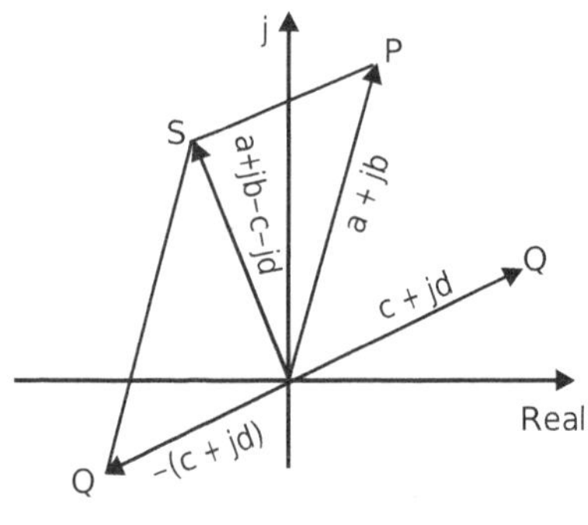

Fig. 2.10

Multiplication. Two complex numbers $(a + jb)$ and $(c + jd)$ can be multiplied as such giving :

$$(a + jb)(c + jd) = ac + jad + jbc + j^2bd$$

$$= (ac - bd) + j(ad + bc).$$

In *polar form*, if $r_1e^{j\theta} = a + jb$ and $r_2e^{j\phi} = c + jd$.

$$(a + jb)(c + jd) = r_1e^{j\theta}\, r_2e^{j\phi}$$

$$= r_1r_2e^{j(\theta + \phi)}.$$

Division. Division can be carried out more easily in polar form.

$$\frac{a + jb}{c + jd} = \frac{r_1e^{j\theta}}{r_2e^{j\phi}} = \frac{r_1}{r_2}e^{j(\theta - \phi)}$$

Alternatively, in real-imaginary form,

$$\frac{a+jb}{c+jd} = \frac{(a+jb)(c-jd)}{(c+jd)(c-jd)} = \frac{(ac+bd)+j(bc-ad)}{c^2+d^2}$$

$$= \left[(ac+bd)/c^2+d^2\right]+j\left[(bc-ad)/c^2+d^2\right]$$

Multiplication and division by $(c - jd)$ is called "Rationalisation".

Example 4. *Divide the number 3 + j4 by* $1+j\sqrt{5}$.

Solution.

$$3 + j4 = \sqrt{3^2+4^2}\ \text{exp.}\left(j\tan^{-1} 4/3\right) \text{ in polar form}$$

$$= 5 \text{ exp. } j(53° \ 8')$$

$$1+j\sqrt{3} = \sqrt{1+3}\ \text{exp.}\left(j\tan^{-1}\sqrt{3}\right) \text{ in polar form}$$

$$= 2\ \text{exp.}\ j(\pi/3) = 2\ \text{exp.}\ j\,60°$$

$$\therefore \quad (3+j4)/\left(1+j\sqrt{3}\right) = (5/2)\ \text{exp.}\ j(53°\,8' - 60°)$$

$$= 5/2 \text{ exp. } j(-6° \ 52')$$

$$= 5/2\ (\cos 6° \ 52' - j \sin 6° \ 52')$$

$$= 5/2\ (0.9926 - j0.1200)$$

$$= 2.48 - j0.3$$

Alternatively,

$$\frac{3+j4}{1+j\sqrt{3}} = \frac{(3+j4)\left(1-j\sqrt{3}\right)}{1^2+\left(\sqrt{3}\right)^2}$$

$$= \frac{\left(3+4\sqrt{3}\right)+j\left(4-3\sqrt{3}\right)}{4}$$

$$= 2.48 - j0.3 \text{ again.}$$

2.8 SINUSOIDAL QUANTITIES REPRESENTED AS PHASORS—COMPLEX FORCING FUNCTION

If an alternating quantity is of a sinusoidal waveform, it may be expressed as the real part (or, imaginary part) of a complex number. Thus the waveforms

$$E \cos \omega t = \text{Real part of } (Ee^{j\omega t}) = Re\ Ee^{j\omega t}$$

$$E \sin \omega t - \text{Imaginary part of } (Ee^{j\omega t}) - Im\ Ee^{j\omega t}$$

where 'Re' and 'Im' denote the real and imaginary parts. The above are valid by the relation

$$e^{j\omega t} = \cos \omega t + j \sin \omega t.$$

Thus a voltage $E_1 \sin \omega t$ added to another voltage $E_2 \sin \omega t$, treated as complex numbers, gives

$$E_1\, e^{j\omega t} + E_2\, e^{j\omega t} = (E_1 + E_2)\, e^{j\omega t}$$

So, the addition of voltages (and currents also) which are sinusoidal (of frequency ω) can be handled after converting the voltages (or currents) in the complex number form and adding the complex numbers. The result is another complex number, whose imaginary part gives the result.

So, Sinusoidal currents and voltages can be represented in the complex plane. The waveforms $E_1 \sin \omega t$ and $E_2 \sin (\omega t + \varphi)$ will be represented by the complex numbers :

$$E_1 \sin \omega t = \text{Im}\,(E_1 e^{j\omega t})$$

$$E_2 \sin (\omega t + \varphi) = \text{Im}\,(E_2 e^{j(\omega t + \varphi)})$$

The representation of $E_1 e^{j\omega t}$ and $E_2 e^{j\omega t + \varphi}$ on a complex plane is shown in Fig. 2.11(*a*) below. Such a representation of alternating voltages and currents in the complex plane is called 'phasor representation', and the diagram as a 'phasor diagram'.

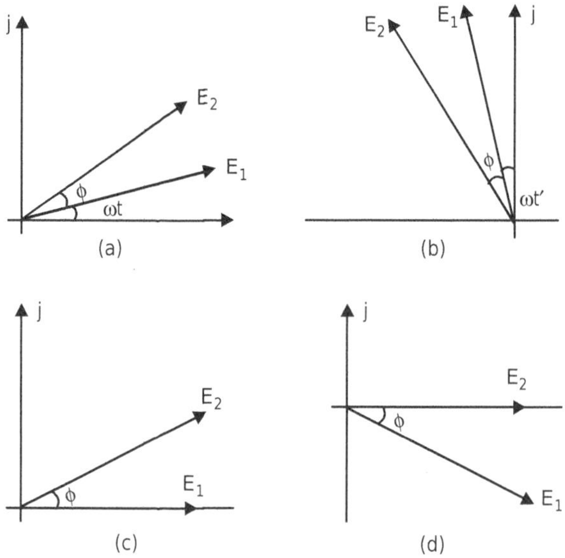

Fig. 2.11

The function $Ee^{j\omega t}$ is called the Complex Forcing Function.

As ωt changes with time, the angle which E_1 and E_2 make with the real axis changes with time, *i.e.,* the radius vectors E_1 and E_2 rotate with an angular velocity ω. Hence 'phasors' are also called 'rotating vectors'. Note that in Fig. 2.11 (*b*), they are shown after they have turned by a quarter of a revolution, but still the angular separation φ, (*i.e.,* the phase difference between E_1 and E_2) remains the same. Since we cannot be drawing the phasors for all instants of time, we usually draw for one instant (usually $t = 0$). Fig. 2.11 (c). In that case, the phasor E_1 lies along the real axis, and E_2 is an angle φ ahead of E_1. The figure shows that E_2 *leads* E_1 by φ. The phasor (here E_2) which is drawn along the real axis, is called as the *reference phasor.* The phasor E_2 leads the reference phasor E_1 by the angle φ. Alternatively if E_2 is taken as reference Fig. 2.11 (*d*), E_1 *lags* from it by the angle φ.

The advantages of the phasor diagram representing alternating quantities (voltages or currents) are :

1. They show clearly the relative phase difference between the quantities, whether leading or lagging.
2. Since the phasors (complex quantities) can be added or subtracted like vectors, a lot of trigonometrical manipulation is avoided (for example, addition of two voltages $E_1 \sin \omega t$ and $E_2 \sin (\omega t + \varphi)$ trigonometrically is cumbersome).

The addition and subtraction of phasors can be done either graphically or by resolving them along the real and imaginary axes as in the following example. Sometimes, the real and imaginary components are referred to as the *in-phase* and *quadrature components*. If a voltage E sin ωt is taken as reference, (*i.e.,* along the real or x-axis), the currents that may flow due to this voltage could be $I_1 \sin(\omega t - \varphi_1)$, $I_2 \sin(\omega t - \varphi_2)$ etc. The currents are lagging by φ_1 and φ_2 with respect to that voltage.

The resolved parts of I_1 and I_2 along the real axis are called 'in-phase' components and the resolved parts along the imaginary axis (along y-axis are called the 'quadrature' components).

Example 4 (*a*). *Four branches A, B, C, D in an A.C. circuit meet at a point P. The currents in A, B, C flow towards P while the current in D flows away from P. The currents in A, B, C are 10 sin 628t, 20 sin (628t + π/3) and 12 sin (628t – π/6). Find the current in branch D and the heat it produces when it flows in a 5 ohms resistance.*

Solution. By Kirchhoff's law,

Current in D = Current in A + Current in B + Current in C.

We can add the given currents trigonometrically and get the result. But we use a different method, namely, using phasor representation.

I_1, the current in A is taken to be the reference, *i.e.,* its phasor (10 Amps) is along the x-axis.

I_2 is the current in B and it leads I_1 by $\pi/3$ or $60°$, while I_3, the current in C lags I_1 by $\pi/6$ or $30°$ as shown. To get the total of I_1, I_2 and I_3, we add them like vectors resolving along the x and y directions.

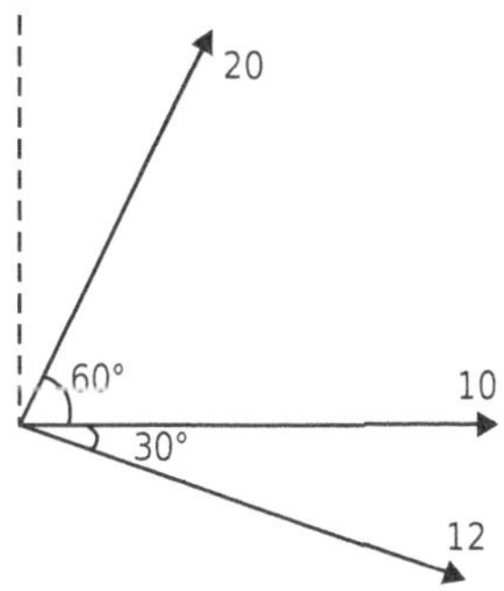

Fig. 2.12

Strictly speaking, the x and y directions in this case are referred to as 'real' and 'imaginary' axes, or 'in-phase' and 'quadrature' axes. Resolving

$$\text{In-phase components} = 10 + 20\cos 60° + 12\cos 30°$$
$$= 10 + 10 + 6\sqrt{3} = 9.608$$
$$\text{Quadrature components} = 0 + 20\sin 60° - 12\sin 30°$$
$$= 10\sqrt{3} - 6 = 11.32.$$

So, the current in D has 9.608 in phase and 11.32 in quadrature.

$$\text{Its value is got by } I_D^2 = 9.608^2 + 11.32^2$$
$$I_D = 14.85$$

Its phase angle is $\tan^{-1}$ (11.32/9.608) = 49° 43', or in radians it is (0.8677 from tables).

So $$I_D = 14.85 \sin (628t + 0.8677).$$

Evidently 14.85 is its maximum value and hence its r.m.s. value, the wave being sinusoidal, is $14.85/\sqrt{2}$. So the power in 5 Ω

$$= (14.85/2)^2 \times 5 = 550 \text{ Watts.}$$

2.9 A.C. THROUGH A RESISTANCE—PHASOR RELATIONSHIP FOR R

If an alternating voltage $v = V \sin \omega t$ (or in complex form $Ve^{j\omega t}$) is applied to a resistance, the current also is alternating and is got by Ohm's law as

$$i = v/R$$

$$i = V \sin \omega t/R = (V/R) \sin \omega t$$

Hence the current wave is also sin ωt; the current is in phase with the applied voltage. The phasor's of voltage and current are shown in the figure below. The power consumed is $I_{r.m.s.}{}^2R$ from the results of a previous article, 2.3.

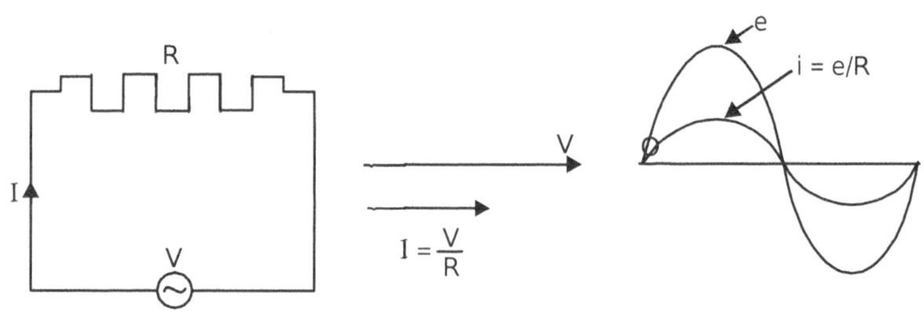

Fig. 2.13 Current and voltage drop are in-phase in a resistance.

2.10 A.C. THROUGH AN INDUCTANCE—PHASOR RELATION FOR L

Let an A.C. Voltage $V = Ve^{j\omega t}$ be applied to an inductance of L. Henries. Then we know that

$$v = L\frac{di}{dt}$$

$$\therefore \quad Ve^{j\omega t} = L\frac{di}{dt}$$

$$di/dt = (V/L)e^{j\omega t}$$

$$i = [V/j\omega L]e^{j\omega t} \text{ on integration.} \qquad (a)$$

Hence the ratio of voltage to current is

$$v/i = [Ve^{j\omega t}] \div [(V/j\omega L)e^{j\omega t}]$$

$$= j\omega L$$

which is called the *Inductive reactance.*

$$\textbf{Inductive Reactance} = j\omega L. \qquad (2.10)$$

For a sine voltage excitation, of $V \sin \omega t$, the current i is got as the imaginary part of (a), so, if

$$v = V \sin \omega t$$

$$
\begin{aligned}
i &= \text{Im}\left[(V/j\omega L)(\cos \omega t + j \sin \omega t)\right] \\
&= \text{Im}\left[(V/\omega L)(-j \cos \omega t + \sin \omega t)\right] \qquad (\because j^2 = -1) \\
&= -V/\omega L \cos \omega t \\
&= V/\omega L \sin(\omega t - \pi/2) \qquad (2.11)
\end{aligned}
$$

Hence there is a phase lag of 90° of the current behind the voltage. If the voltage V is taken as the reference phasor, the current phasor lags 90° behind it. The magnitude of the current is V/ωL.. The quantity ωL is the magnitude of the inductive reactance *j*ωL. The magnitude is denoted by X_L, the suffix L referring to inductance L.

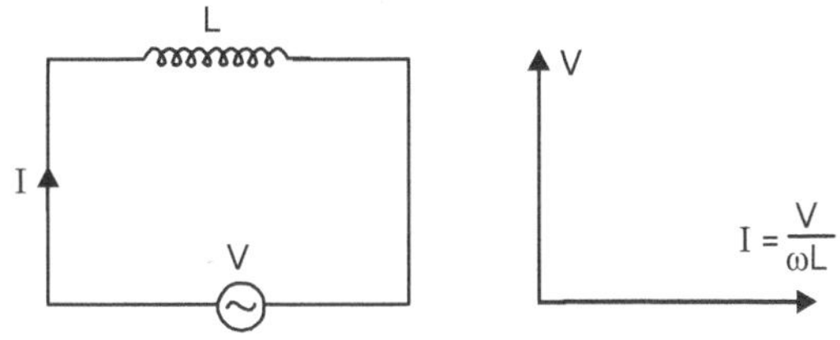

Fig. 2.14 A.C. through an inductance.

2.11 THE POWER CONSUMED BY AN INDUCTANCE

Power = Voltage × Current by Joule's law. The quantities voltage and current are changing with time, so does power.

$$
\begin{aligned}
p &= vi \\
&= V \sin \omega t\,(-I \cos \omega t) \\
&= -VI \sin \omega t \cos \omega t \\
&= -\frac{1}{2} VI \sin 2\omega t \qquad (2.12)
\end{aligned}
$$

The power has a double frequency variation.

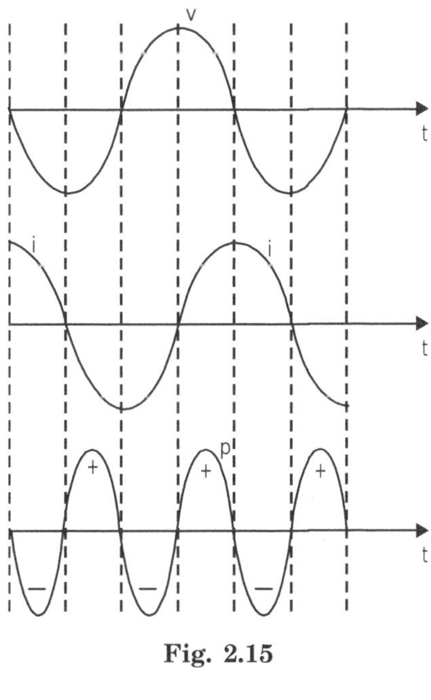

Fig. 2.15

The total power consumed in a cycle of the wave is zero. This is because the positive and negative waves cancel. The energy stored in the positive (quarter cycle) areas is given back to the voltage source in the next quarter cycle.

2.12 A.C. THROUGH A CAPACITANCE—PHASOR RELATION FOR C

For an excitation $v = Ve^{j\omega t}$, the capacitor response is found from the relation,

$$q = C \times \text{voltage}$$

$$q = CVe^{j\omega t}$$

$$\therefore \quad i = \frac{dq}{dt} = CV\, j\omega\, e^{j\omega t} = j\omega C v \tag{2.13}$$

Hence, for a sinusoidal excitation $v = V \sin \omega t$, the current

$$i = \text{Imag. part of } (CV\, j\omega\, e^{j\omega t})$$

$$= \text{Im}\left[CV\omega\,(j \cos \omega t - \sin \omega t)\right]$$

$$= CV\, \omega \cos \omega t = \omega CV \sin\left(\omega t + \pi/2\right)$$

The current is $I \sin\left(\omega t + \pi/2\right)$, where $I = VC\omega$.

The ratio of voltage v to current i is got from (2.13) as

$$\frac{v}{i} = \frac{1}{j\omega C} \tag{2.14}$$

The ratio v/i is called *capacitive* reactance. Its magnitude is $1/\omega C$ and is denoted by X_C.

$$\text{Capacitive reactance} = \frac{1}{j\omega C} = -jX_C \tag{2.15}$$

The current wave is sin ($\omega t + \pi/2$) and hence leads the voltage by 90°. The phasors of V and I are hence as shown.

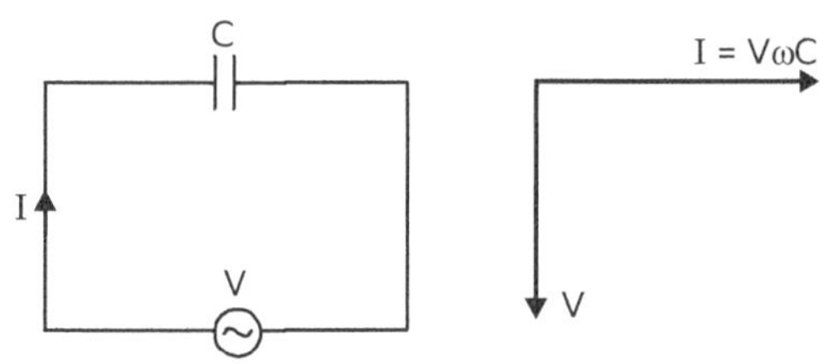

Fig. 2.16 Current through a capacitor and phasor diagram.

$$\therefore \quad V/I = -jX_C$$

$$V = -jIX_C \tag{2.16}$$

Hence the V phasor lies on the $-j$ axis.

Power consumed by a capacitance for an A.C.

$$p = vi = V \sin \omega t . I \sin (\omega t + \pi/2)$$

$$= VI \sin \omega t \cos \omega t = \frac{1}{2} VI \sin 2\omega t$$

Hence, as in the case of the inductance, there is a double frequency variation $2\omega t$ and similarly the

$$\text{Power in a capacitance} = \text{Zero.} \tag{2.17}$$

Example 5. *An alternating current of 3 sin (4t + π/4) passes through*

(a) an inductance of 0.1 H

(b) a capacitance of 100 μF.

Find the voltage applied in each case.

Solution. (*a*) $i = 3 \sin(4t + \pi/4)$

The voltage magnitude is given by

$$V/I = X_L$$

and it leads the current by π/2. Hence $V = IX_L$.

$$X_L = \omega L = 4 \times 0.1 = 0.4$$

$$\therefore \quad V = 0.4 \times 3 \text{ in magnitude.}$$

$$v = 1.2\sin(4t + \pi/4 + \pi/2) = 1.2\sin\left(4t + \frac{3}{4}\pi\right)$$

The addition of π/2 is since voltage leads current by $\frac{1}{2}\pi$ in an inductance.

(*b*)
$$V/I = X_C \text{ and } X_C = 1/\omega C.$$

$$X_C = 1/\omega C = 1/(4 \times 100 \times 10^{-6}) = 2500$$

$$V \text{ in magnitude} = IX_C = 2500 \times 3$$

$$v = 7500\sin\left(4t + \frac{\pi}{4} - \frac{1}{2}\pi\right)$$

Since voltage lags current by π/2 in capacitance.

2.13 RESISTANCE AND INDUCTANCE—IMPEDANCE

For a resistance and inductance in series, if the current flowing is i, the voltage applied will be the sum of the drops in R and L.

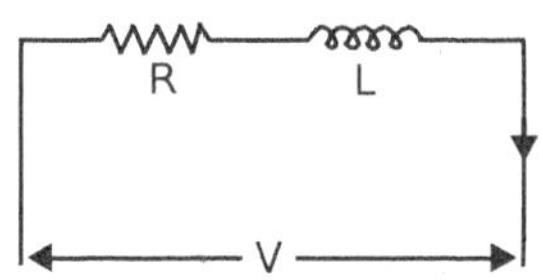

Fig. 2.17

$$\therefore \quad v = Ri + L\frac{di}{dt}$$

$$v = RIe^{j\omega t} + Ij\omega Le^{j\omega t}$$

$$= I(R + j\omega L)e^{j\omega t} = (R + j\omega L)i$$

$$v/i = R + j\omega L \tag{2.18}$$

R + *j*ωL is a complex number. It is denoted by the letter Z, and called ***'Impedance'***. The impedance is the ratio between the voltage and current and

$$\frac{V}{I} = Z = R + j\omega L \tag{2.19}$$

is known as the modified 'Ohm's law' for alternating excitations to the circuit elements R and L.

Z has the real part R and imaginary part $X_L = \omega L$. It can be written $Ze^{j\phi}$, where

$$Ze^{j\phi} = R + j\omega L$$

$$Z\cos\varphi + jZ\sin\varphi = R + j\omega L$$

$$R = Z\cos\varphi, \quad Z\sin\varphi = \omega L$$

$$\therefore \quad Z = \sqrt{Z^2 \cos^2 \varphi + Z^2 \sin^2 \varphi}$$

$$= \sqrt{R^2 + (\omega L)^2} \text{ and so}$$

$$Z = \sqrt{R^2 + (\omega L)^2}\, e^{j\phi}$$

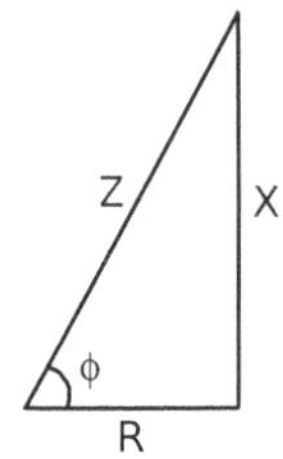

Fig. 2.18 Impedance triangle.

where $\tan \varphi = \sin \varphi / \cos \varphi = \omega L/R$

$$\varphi = \tan^{-1} \omega L/R$$

$$\therefore \quad Z = \sqrt{R^2 + (\omega L)^2} \angle \tan^{-1}(\omega L/R) \qquad (2.20)$$

For a sine wave of current, I sin ωt, we get the voltage as the imaginary part

$$v = (R + j\omega L)\, Ie^{j\omega t}$$

$$= \sqrt{R^2 + (\omega L)^2}\, e^{j\phi}\, I\, e^{j\omega t}$$

$$= ZIe^{j\omega t + j\phi}$$

The imaginary part gives $\quad v = ZI \sin(\omega t + \varphi)$

$$\therefore \quad v = V \sin(\omega t + \varphi) = ZI \sin(\omega t + \varphi)$$

The voltage leads the current by the angle $\varphi = \tan^{-1} X_L/R$ and its magnitude is $V = ZI$.

Example 6. *A choke coil has a resistance of 9 ohms and an inductance of 0.04 H. An alternating voltage of 150 sin 300t is applied. Find the current.*

Solution. $\quad v = 150 \sin 300t$

$\omega = 300$. So $\quad X_L = \omega L = 300 \times 0.04 = 12.$

$$Z = \sqrt{R^2 + X_L^2} \angle \tan^{-1}(X_L/R)$$

$$= \left[9^2 + 12^2\right]^{1/2} \angle \tan^{-1} 12/9 = 15\angle 0.93 \text{ radian}$$

$$V = ZI \text{ and so } I = V/Z$$

$$= \frac{150}{15} \sin(\omega t - \varphi)$$

$$= 10 \sin(300t - 0.93)$$

The $-\varphi$ is because the current lags behind the voltage by the angle φ.

The current can also be found from :

$$I = \frac{V}{Z} = \frac{V}{R + j\omega L} = \frac{150}{9 + j12} = \frac{150(9 - j12)}{9^2 + 12^2}$$

$$= \frac{150}{15^2}(9 - j12)$$

The current magnitude is $\frac{10}{15}(9^2 + 12^2)^{1/2} = 10 \times 15/15 = 10.$

The phase angle $\quad \varphi = \tan^{-1}(-12/9) = -0.93$ radians.

$$\therefore \quad i = 10 \sin(300t - 0.93) \text{ as before.}$$

2.14 RESISTANCE, INDUCTANCE AND CAPACITANCE IN SERIES—NET IMPEDANCE

Let a current $i = \mathrm{I}e^{j\omega t}$ flow in the circuit. The applied voltage will be the sum of the drops in R, L and C (Fig. 2.19).

$$v_\mathrm{R} = \mathrm{R}i,\ v_\mathrm{L} = \frac{di}{dt},\ v_\mathrm{C} = \frac{\mathrm{Q}}{\mathrm{C}}$$

$$\therefore \qquad v = \mathrm{R}i + \mathrm{L}\frac{di}{dt} + \frac{\mathrm{Q}}{\mathrm{C}}$$

$$= \mathrm{R}i + \mathrm{L}\frac{di}{dt} + \left(\frac{1}{\mathrm{C}}\right)\int i\,dt$$

$$= \mathrm{R\,I}e^{j\omega t} + \mathrm{LI}j\omega\, e^{j\omega t} + (1/j\omega\mathrm{C})\,\mathrm{I}e^{j\omega t}$$

$$= \mathrm{I}e^{j\omega t}\left[\mathrm{R} + j\omega\mathrm{L} + \frac{1}{j\omega\mathrm{C}}\right]$$

The ratio
$$v/i = \mathrm{R} + \mathrm{j}\omega\mathrm{L} + \frac{1}{j\omega\mathrm{C}} = \mathrm{Z} \tag{2.21}$$

Hence the impedance Z consists of three terms.

$$\mathrm{Z} = \mathrm{R} + j\mathrm{X_L} - j\mathrm{X_C} \qquad \text{(since } 1/j = -j\text{)}$$

$$= \mathrm{R} + j(\mathrm{X_L} - \mathrm{X_C})$$

As before, Z can be expressed

$$\mathrm{Z} = \sqrt{\mathrm{R}^2 + (\mathrm{X_L} - \mathrm{X_C})^2}\ \angle\tan^{-1}(\mathrm{X_L} - \mathrm{X_C})/\mathrm{R}.$$

If we write $\mathrm{X} = \mathrm{X_L} - \mathrm{X_C}$

$$\mathrm{Z} = \sqrt{\mathrm{R}^2 + \mathrm{X}^2}\ \angle\tan^{-1}\mathrm{X/R}, \tag{2.22}$$

is a general result. X will be $\mathrm{X_L}$ if the circuit has inductance only and no capacitance. X will be $\mathrm{X_C}$ if the circuit has capacitance only and no inductance. If both L and C are in series, $\mathrm{X} = \mathrm{X_L} - \mathrm{X_C}$.

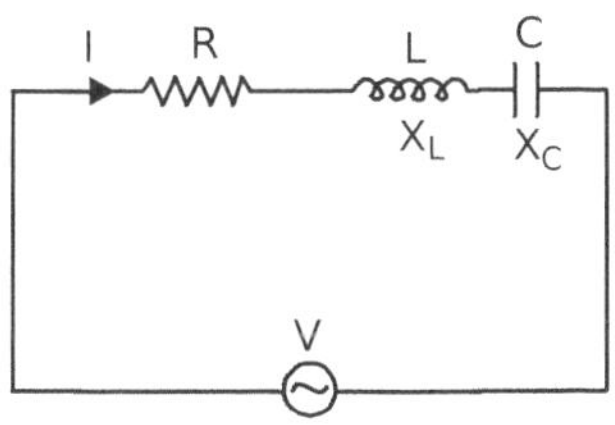

Fig. 2.19 Current in R-L-C series circuits.

If the current is sinusoidal, $i = \mathrm{I}\sin\omega t$, the voltage is got as the imaginary part of :

$$v = (\mathrm{R} + j\mathrm{X})\,\mathrm{I}e^{j\omega t}$$

$$= \mathrm{Z}\,e^{j\phi}\,\mathrm{I}e^{j\omega t} \qquad \text{(where } \varphi = \tan^{-1}\mathrm{X/R}\text{)}$$

$$= \mathrm{Z\,I}e^{j\omega t + j\phi} \tag{2.23}$$

the imaginary part giving

$$V = V \sin(\omega t + \varphi) = ZI \sin(\omega t + \varphi)$$

$$V = ZI \text{ and } V/I = Z \text{ as before.} \tag{2.24}$$

Note that φ is the phase difference between voltage V and current I. Since $X = X_L - X_C$, it may be positive or negative depending upon $X_L > X_C$ or $X_L < X_C$ [from 2.21].

If $X_L > X_C$, X/R is positive, φ positive, V leads I by φ.

If $X_L < X_C$, X/R is negative, φ s –ve, V lags I by φ. (2.25)

2.15 THE PHASOR DIAGRAM—IMPEDANCE

If the current is I sin ωt, its complex form is $Ie^{j\omega t}$. Let us draw it along the real axis as OI.

Voltage across resistance V_R = OA = RI (in phase with I)

Voltage across inductance V_L = OB = jX_L I (Leads I by 90°)

Voltage across capacitance V_C = OC = – jX_C I (lags I by 90°)

$$Z = \frac{V}{I}$$

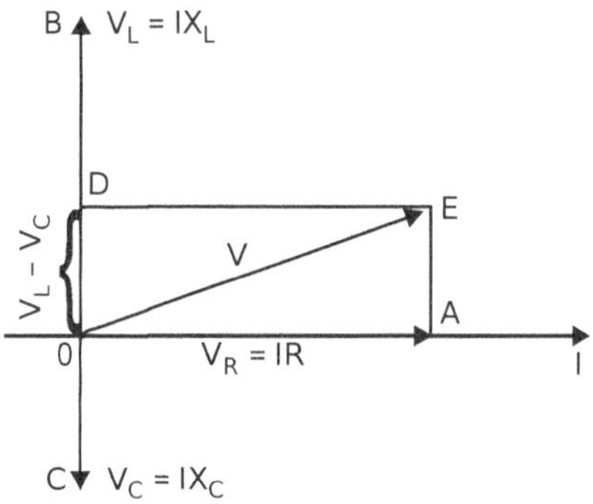

Fig. 2.20 Phasor diagram of R-L-C series circuit.

The sum of V_R, V_L and V_C will be the applied voltage V. Note that

$$OB - OC = OD = X_L I - X_C I$$

and OD combines with OA to give OE, the applied volts V.

$$\tan \varphi = \frac{AE}{OA} = \frac{X_L I - X_C I}{RI} = \frac{X_L - X_C}{R} = \frac{X}{R} \tag{2.26}$$

$$V^2 = OE^2 = (RI)^2 + (X_L I - X_C I)^2$$

Dividing by I^2,

$$(V/I)^2 = R^2 + X^2 = Z^2$$

$$V/I = Z$$

and in complex form,

$$V = ZI \angle\varphi. \tag{2.27}$$

2.16 POWER IN THE A.C. CIRCUIT

An alternating current flowing through a resistance produces a power of $I^2_{r.m.s.}R$ or simply I^2R, the r.m.s. value being meant by I. The inductance and capacitance consume no power. Hence in a circuit of R, L and C, the total power is only

$$P = I^2R$$

Since

$$I = V/Z,$$

$$P = (V/Z)\, IR = VI\,(R/Z)$$

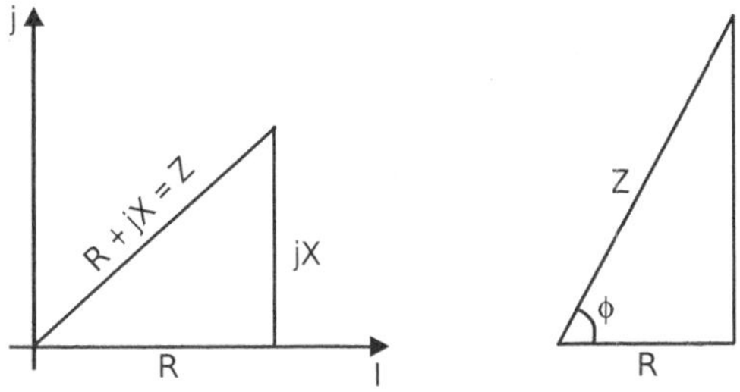

Fig. 2.21 Impedance as a complexor.

Considering Z, it is a complex number.

$$Z = R + jX$$

and hence cos φ is got from the *impedance triangle* as R/Z.

$$\therefore \quad P = VI \cos \varphi \tag{2.28}$$

The factor cos φ is the ratio.

cos φ = Power/[RMS Volts × RMS Amperes] and is called Power Factor.

$$\text{Power Factor} = \cos \varphi = R/Z \tag{2.29}$$

The same relation can be proved by considering the instantaneous power and averaging it over one A.C. cycle. Thus, since

$$v = V_m \sin \omega t$$

$$i = I_m \sin (\omega t - \varphi)$$

$$p = vi = V_m \sin \omega t \, I_m \sin (\omega t - \varphi)$$

$$= \frac{1}{2} V_m I_m [\cos \varphi - \cos (2\omega t - \varphi)],$$ thus comprising of

a steady power $\frac{1}{2} V_m I_m \cos\phi$ and time varying power $\frac{1}{2} V_m I_m \cos(2\omega t - \phi)$.

The average power P is thus the average of $\int p \, d\theta$.

$$P = \frac{1}{2\pi} \int_0^{2\pi} vi \, d\theta = \frac{1}{2} V_m I_m \cos[2\pi - 0] \div 2\pi$$

$$= \frac{1}{2} V_m I_m \cos \varphi \tag{2.30}$$

Since $\quad V_{r.m.s.} = V_m/\sqrt{2}$ and $I_{r.m.s.} = I_m/\sqrt{2}$,

$$P = VI \cos \varphi \tag{2.31}$$

when V and I denote R.M.S. values. Hence power in an A.C. circuit is not given by the product VI as in D.C., but is given by V × I × cos φ. If we consider the phasor diagram of V and I, we see that the current I has a component I cos φ in phase with the phasor of V and a component I sin φ at 90° to V (*i.e.*, in quadrature with V). The power is the product of the voltage and the in-phase component of the current. The product of V and the quadrature component I sin φ is called "Reactive Power". I cos φ and I sin φ are also called *active component* and *reactive component*, respectively.

$$Q = \text{Reactive power} = VI \sin \varphi \tag{2.32}$$

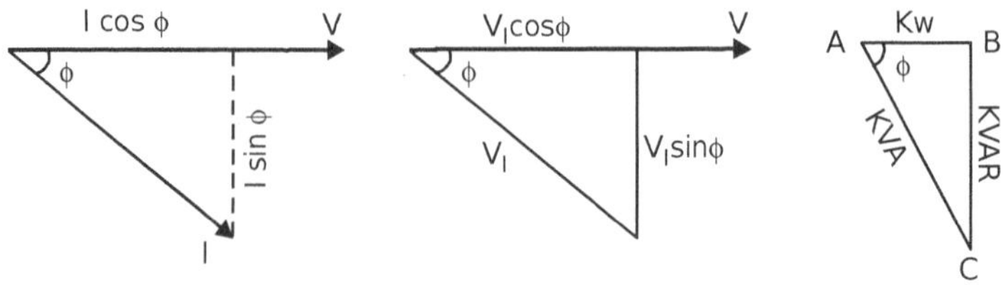

Fig. 2.22 Power and reactive power illustration.

The reactive power is expressed as so many volt-amps-reactive, or VARs. The reactive power is positive or negative according as φ (and hence sin φ), is positive or negative. When φ is positive, I leads V and so it is referred to as *leading reactive power*. If φ is negative, or I lags V, it is called *lagging reactive power*. (Fig. 2.22 shows lagging p.f.). The power factor is also similarly defined:

Leading power factor → Current leads voltage. (2.33)

Lagging power factor → Current lags voltage.

$$\text{The reactive power } Q = VI \sin \varphi$$

$$= VI.\ X/Z \qquad (\sin \varphi = X/Z)$$

$$\approx (V/Z)\ IX = I\,(IX)$$

$$= \text{Current} \times (\text{Volts across reactance})$$

So the product of the voltage across the reactance (inductance and capacitance) and the current is reactive power. The product of the voltage across the resistance and the current through it, as we know, gives the power (in the resistance). So, resistances consume *active* power; reactances consume *reactive* power. Reactive power is lagging or leading if current is lagging or leading.

Example 7. *Determine the inductance, reactance impedance and power factor of a circuit which takes 10 amps and dissipates 1390 watts when connected to a 250 V 50 c/s mains.*

Solution. Power $= V\,I \cos \varphi = 1390$

Power factor $= \cos \varphi = 1390/VI = 1390/V\ (250 \times 10) = 0.555$

Impedance Z $= V/I = 250/10 = 25$ ohms.

R $= Z \cos \varphi,\ X = Z \sin \varphi \qquad \varphi = \cos^{-1} 0.555 = 56°\ 18'$

X $= Z \sin \varphi = 25 \sin \varphi = 25 \sin 56°\ 18'$

$= 25 \times 0.832 = 20.85$ ohms

L $= X/\omega = 20.85/(2\pi f) = (20.85/2\pi \times 50) = 0.0663$ Henries

Inductive reactance = 20.85 ohms, Impedance = 25 ohms.

Example 8. *A capacitance of 20 μF and a resistance of 100 ohms are connected in series across a 120 volt 60 c/s mains. Determine the average power in the circuit.*

Solution. Reactance of a capacitor $X_C = 1/\omega C = 1/(2\pi f C)$

$$= \frac{1}{2\pi\,(60)\,(20 \times 10^{-6})} = \frac{10^4}{24\pi} = 132.8$$

Impedance $= (R^2 + X^2)^{1/2} = (100^2 + 132.8^2)^{1/2} = 166.5$ ohms

Current $= V/Z = 120/166.5 = 0.721$ Amperes

Circuit power = Power consumed by resistance only (as capacitance consumes no net power)

$$= (0.721)^2 \times 100 = 52 \text{ watts.} \qquad \textbf{Ans.}$$

Example 9. *A 100 volt 80 watt lamp is to be operated on 300 V 60 Hz A.C. supply. Calculate the inductance of the choke required to be connected in series with the lamp for this operation. The lamp can be taken as equivalent to a non-inductive resistance.*

Solution. Resistance of lamp is got from $V^2/R = W$ as

$$R = V^2/W = 100^2/80 = 125 \text{ ohms}$$

The lamp and an inductance are to be connected in series. From a 300 V supply, the lamp must get 100 V only and the inductance drops some voltage. Taking the inductance of choke as L, the lamp resistance as R, the circuit of R and L in series has the impedance

$$Z = \sqrt{(R^2 + L^2\omega^2)}$$

and $$I = V/Z = 300/Z.$$

$$\text{Power of lamp} = 100 \text{ W} = \text{volt across lamp} \times \text{current I}$$
$$= 100 \times I.$$

So $$I = 1 \text{ and } Z = V/I = 300 \text{ ohms.}$$

Hence $$300^2 = 125^2 + (\omega L)^2$$

$$\omega L = \sqrt{300^2 - 125^2} = 273$$

$$L = 273/\omega = 273/2\pi\ 60 = 0.724 \text{ Henries.}$$

Example 10. *A coil when connected in series with a non-reactive resistor across 250 V 50 Hz supply takes 5 amperes. The voltages across the coil and resistor are 200 V and 125 V, respectively. Find the inductance, power factor and power dissipated by the coil.*

Solution. Total impedance Z of coil and resistor in series = 250 volt/5 Amp = 50 ohms.

Voltage across non-inductive resistor

$$= RI = 125 \text{ V};$$

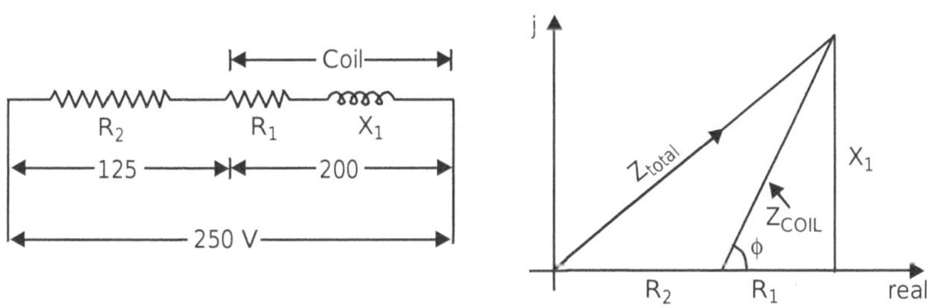

Fig. 2.23

So, $$R = 125/5 = 25 \text{ ohms.}$$

Let R and X be the coil's resistance and reactance. Then, total impedance Z is given by

$$Z^2 = (R + 25)^2 + X^2 = 50^2 \qquad (i)$$

The voltage across the coil is the product of the current and the impedance of the coil only. It is 200 V.

$$\text{The impedance of coil} = (R^2 + X^2)^{1/2}$$

So, $200^2 = 5^2 (R^2 + X^2)$; $40^2 = R^2 + X^2$ (*ii*)

Hence the coil impedance is 40 ohms. From equations (*i*) and (*ii*),

$$50^2 - 40^2 = (R + 25)^2 - R^2 = 25^2 + 50R$$

This gives $R = 5.5$ ohms.

$$X^2 = 40^2 - 5.5^2, \text{ or } X = 39.6$$

So $L = 39.6/\omega = 39.6/(2\pi\ 50) = 0.136$ Henries

Power factor of the coil is the ratio (Resistance/Impedance) = 5.5/40 = 0.138.

The power dissipated by the coil is due to the power in its resistance of 5.5 ohms only and hence is I^2R.

$$\text{Power} = 5^2 \times 5.5 = 138 \text{ Watts.}$$

Example 11. *A coil of resistance 20 ohms and an inductance of 60 mH is connected in series with a 130 μF capacitor across a 230 V 50 Hz supply. Calculate the impedance and the p.f. of the circuit, the current flowing in it, the voltage across the coil, and the capacitor.*

(M.U., B.E. 1967)

Solution. Reactance of the inductance,

$$X_L = 2\pi f L$$
$$= 2\pi \times 50 \times 60 \times 10^{-3}$$
$$= 18.84\ \Omega$$

Reactance of the capacitor, $X_C = 1/(2\pi f C)$

$$= 1/(2\pi \times 50 \times 130 \times 10^{-6})$$
$$= 10^{-6}/(314 \times 130)$$
$$= 24.5\ \Omega$$

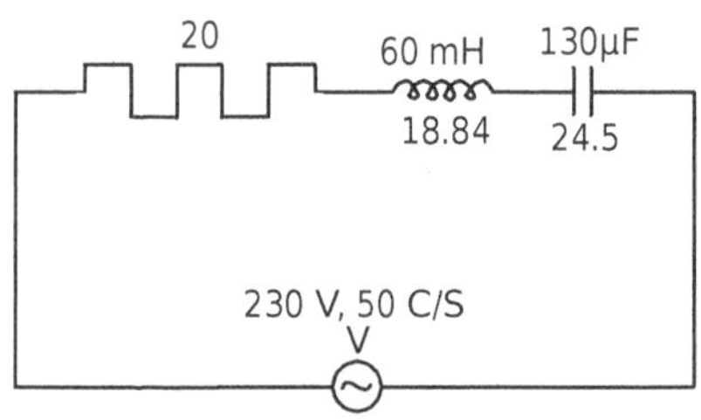

Fig. 2.24

Total impedance of the circuit,

$$Z = \sqrt{R^2 + (X_L - X_C)^2}$$
$$= \sqrt{20^2 + (24.5 - 18.84)^2}$$
$$= 20.8 \text{ ohms}$$

Power factor of the circuit, $\cos\varphi = R/Z = 20/20.8 = 0.96$; the p.f. is leading because $X_C > X_L$.

$$\text{Current flowing} = V/Z = 230/20.8 = 11.05 \text{ amps.}$$

$$\text{Voltage across coil} = IX_L = 11.05 \times 18.84 = 208.5 \text{ volts}$$

$$\text{Voltage across capacitor} = IX_C = 11.05 \times 24.5 = 271 \text{ volts}$$

Power consumed by the circuit :

(1) $\quad VI \cos \varphi = 230 \times 11.05 \times 0.96$

$= 2440$ Watts

(2) $\quad I^2R = 11.05^2 \times 20 = 2440$ W.

2.17 POWER IN COMPLEX FORM

Power and reactive power can be combinedly represented as the real and imaginary parts of a complex number S.

$$S = \text{Real Power} \pm j\,(\text{Reactive power})$$
$$= p \pm jQ$$
$$= VI \cos \varphi \pm j\, VI \sin \varphi \qquad (2.34)$$

The + sign is for leading power factor, or leading reactive power;

The – sign is for lagging p.f. or lagging reactive power.

We can represent the voltage V sin ωt and the current in a lagging p.f. circuit, of I sin $(\omega t - \varphi)$ [V and I denote R.M.S. values here], as complex numbers, *i.e.,* phasors, as done previously.

$$V = V\, e^{j\omega t};\ I = I\, e^{j\omega t - j\phi}$$

We now define the conjugate of a complex number V $e^{j\omega t}$. The conjugate of any complex number is obtained by replacing j by $-j$. Thus, the conjugate of $3 + j4$ is $3 - j4$. The conjugate of $10\, e^{j\theta}$ is $10\, e^{-j\theta}$. The conjugate is denoted by using a symbol *. Thus

$$(3 + j4)^* = 3 - j4$$

Similarly, $\quad V^* = V\, e^{-j\omega t}$

Now consider the product of the complex numbers V* and I. We get

$$\boxed{\text{Complex Power} = V^*I}$$

$$V^*I = V\, e^{-j\omega t}\, I^{j(\omega t - \phi)} = VI\, e^{-j\phi}$$
$$= VI \cos \varphi - j\, VI \sin \varphi$$
$$= P - jQ$$

The $-jQ$ is because the current is lagging by φ behind the voltage.

So we see that $\quad S = V^*I \qquad (2.35)$

Example 12. *An alternating voltage 80 + j60 is applied to a circuit and the current following is – 4 + j10 amperes. Find (i) the circuit impedance (ii) power consumed (iii) the phase angle.* (M.U., III B.Tech. April 1969)

Solution.

$$Z = \frac{V}{I} = \frac{80 + j60}{-4 + j10} = \frac{(80 + j60)(-4 - j10)}{(-4)^2 + 10^2}$$
$$= \frac{[(-320 + 600) - 240j - 800j]}{116} = 2.41 - j8.95$$

i.e., a resistance of 2.41 and a capacitance reactance of 8.95.

$$Z \text{ in magnitude} = \left[(2.41)^2 + (8.95)^2\right]^{1/2} = 9.27 \text{ ohms}$$
$$\cos \varphi = R/Z = 2.41/9.27 = 0.26 \text{ leading.}$$
$$\text{Power} = V^*I = (80 - j60)(-4 + j10)$$
$$= 280 + 1040j$$

giving power as 280 W and reactive power as 1040 VARs leading.

2.18 IMPEDANCES IN SERIES

When two (or more) impedances $Z_1 = R_1 + jX_1$ and $Z_2 = R_2 + jX_2$ are in series, the total impedance is simply got by adding the resistances and reactances.

$$Z_1 = R_1 + jX_1 \qquad Z_2 = R_2 + jX_2$$

$$Z \text{ total} = (R_1 + R_2) + j(X_1 + X_2)$$

$$= Z_1 + Z_2 \tag{2.36}$$

Example 13. *An impedance of resistance 4 ohms and inductive reactance 3 ohms is connected in series with another having 5 ohms resistance and a capactive reactance of 2 ohms. Find the total impedance in series.*

Solution. $R_1 + R_2 = 4 + 5 = 9$

$X_1 + X_2 = 3 - 2 = 1$ ohm, inductive.

Note that inductive reactance is positive, capacitive reactance – ve.

2.19 IMPEDANCES IN PARALLEL—ADMITTANCE

When two impedances Z_1 and Z_2 are connected in parallel, the total current I is the sum of I_1 and I_2.

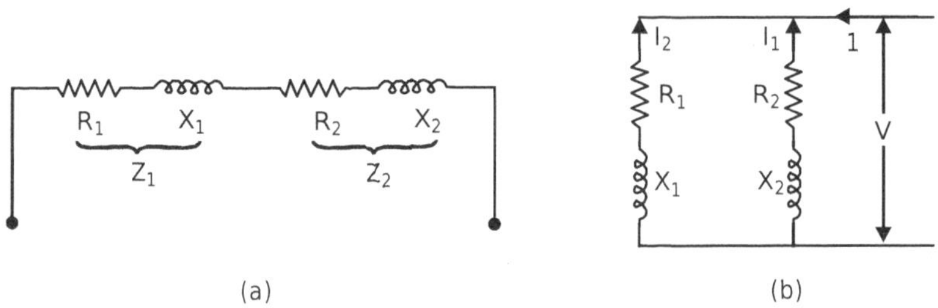

Fig. 2.25 Series and parallel circuits.

Referring Fig. 2.25 (*b*),

$$I_1 = V/(R + jX_1) = V/Z_1$$

$$I_2 = V/(R_2 + jX_2) = V/Z_2$$

$$I = I_1 + I_2 = \frac{V}{Z_1} + \frac{V}{Z_2}$$

Hence

$$\frac{I}{Z} = \frac{I}{Z_1} + \frac{I}{Z_2} \tag{2.37}$$

The total impedance is the ratio of applied volts V to the current I. Hence V/I = Z

$$\frac{1}{Z} = \frac{1}{Z_1} + \frac{1}{Z_2}$$

or,

$$Y = Y_1 + Y_2$$

Note that Z_1, Z_2, Z are complexors. The reciprocals of Z are called admittances and denoted by the symbol Y. Hence, the admittance of a parallel circuit is the sum of the admittances of its branches. In general, for *n* parallel branches, the total admittance Y is

$$Y = Y_1 + Y_2 + ... + Y_n \tag{2.38}$$

The *admittance* Y is a complex number. Thus

$$Y = \frac{1}{Z} = \frac{1}{R + jX} = \frac{R - jX}{R^2 + X^2}$$

$$= \frac{R}{R^2+X^2} - j\frac{X}{R^2+X^2}$$

$$= G - jB \qquad (2.39)$$

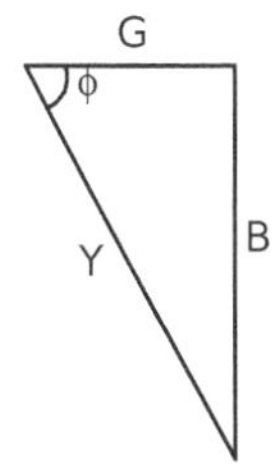

Fig. 2.26

The admittance has G for its real part and B for its imaginary part.

$$G = R/(R^2 + X^2) = R/Z^2$$

$$B = -X/(R^2 + X^2) = -X/Z^2 \qquad (2.40)$$

Note that B has the opposite sign to that of X. Thus inductive susceptance is negative. (The susceptance of a branch having two branches for example),

$$Y = Y_1 + Y_2 = G_1 + jB_1 + G_2 + jB_2$$

$$= (G_1 + G_2) + j(B_1 + B_2)$$

The magnitude of Y is $\left[(G_1+G_2)^2+(B_1+B_2)^2\right]^{1/2}$

and its phase angle is $\tan^{-1}(B_1 + B_2)/(G_1 + G_2)$.

Hence the current I is $\qquad I = V/Z = VY \qquad (2.41)$

The following example illustrates the solution of a parallel circuit using complexor method as well as conductance susceptance method.

Example 14. *An impedance of 6 ohms resistance and 8 ohms inductive reactance is in parallel with another having 12 ohms resistance and 5 ohms capacitive reactance and the combination takes a current of 10 Amps from the supply. Find the applied voltage, total power consumed and total p.f.*

Solution. Impedance of the first branch

$$Z_1 = R_1 + jX_1 = 6 + j8$$

Impedance of the second branch

$$Z_2 = R_2 + jX_2 = 12 - j5$$

Impedance Z of parallel combination

$$= Z_1Z_2/(Z_1 + Z_2)$$

$$= \frac{(6+j8)(12-j5)}{18+j3} = \frac{72+40+96j-30j}{18+j3}$$

$$= \frac{(112+66j)(18-j3)}{18^2+3^2}$$

$$= (2214 + 852j)/333$$

$$= 6.64 + 2.56j$$

The magnitude and phase angle of Z are

$$\sqrt{6.64^2+2.56^2} \text{ and } \tan^{-1}(2.56/6.64) = 7.13\angle 21°.$$

Supply voltage = ZI = 7.13 × 10 = 71.3 volts

Power factor = cos φ = cos 21° (also 6.64/7.13) = 0.93

Power consumed = 71.3 × 10 × 0.93 = 663 watts.

Susceptance, Conductance Method :

	Branch 1	Branch 2	Total
Conductance R/Z	= 6/100	$12/13^2$ = .0707	.1307
Susceptance $-$ X/Z^2	= $-$ 8/100	$5/13^2$ = .0296	$-$.0504

Note that inductive susceptance is –ve, capacitive susceptance is +ve.

$$\text{Admittance } Y = \left[(0.1307)^2 + (0.0504)^2\right]^{1/2}$$

$$= 0.14 \text{ in magnitude } \angle -21^\circ$$

$$\text{Power factor} = G/Y = 0.1307/0.14 = 0.93$$

$$\text{Applied volts} = I/Y \qquad \text{(since } I = VY\text{)}$$

$$= 10/0.14 = 71.3 \text{ volts.}$$

2.20 IMPORTANCE OF POWER FACTOR

For power engineers, power factor is an important quantity. Since power is given by the relation P = VI × power factor, for the same power, more of current will be drawn (from the source) if the power factor of the circuit is less. Hence alternating current loads should have a good power factor, as near to unity (1) as possible. But alternating current motors can operate only at about 0.7 to 0.8 power factor. Hence, it frequently becomes necessary to improve the power factor. The power factor as seen by the supply source will be improved if a capacitor is connected in parallel with the load. The leading current I_C drawn by the capacitor will improve the power factor.

Example 15. *A 220 V 50 c/s induction motor has an output of 5 H.P. and an efficiency of 80%. It works at a power factor of 0.707. Find a suitable condenser which when connected across the supply, will improve the overall supply p.f. to 0.866.*

Solution. Output of induction motor = 5 × 746 = 3730 W.

Input = Output/efficiency = 3730/0.8 = 4660 W.

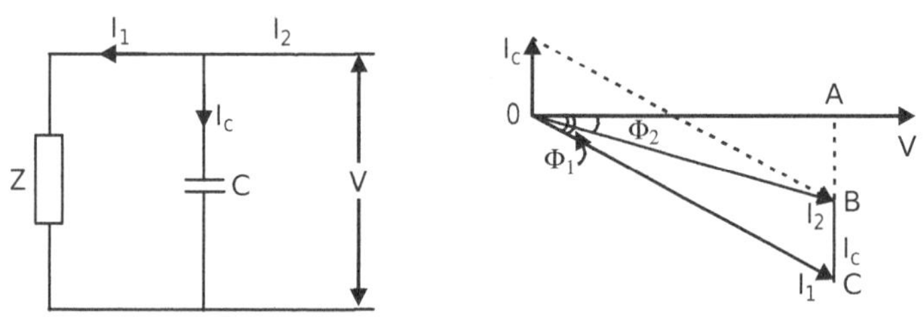

Fig. 2.27 Connecting a capacitor improves p.f.

Hence VI cos φ = 4660

Current drawn = I_1 = 4660/V cos φ = 4660/(220 × .707) = 30 A.

The current phasor is as shown by OC, lagging by φ_1 from the voltage V. When a capacitor is connected in parallel, it takes I_C, leading 90° from V. The total current I_2 is OB, the phasor sum of OC and I_C. The new p.f. angle is φ_2.

The reactive component (AC) of current I_1 = I_1 sin φ = 30 × 0.707

= 21.21 A.

The active component OA = 30 cos φ = 30 × .707 = 21.21.

The new reactive component AB = OA tan φ_2

$$= 21.21\,(1/\sqrt{3}) \qquad (\tan\varphi_2 = 1/\sqrt{3}$$

$$= 12.25 \text{ A} \qquad \text{if } \cos\varphi_2 = 0.866)$$

So $$BC = AC - AB = 21.21 - 12.25 = 9.06.$$

This is also I_C, the current drawn by the capacitor. Hence

$$X_C = V/I = 220/9.06 = 24.3$$

$$C = 1/\omega X = 1/(314 \times 24.3) = 131\ \mu F.$$

Example 16. *Two impedances $R + j\omega L$ and $R - j/\omega C$ are in parallel. Show that the parallel combination will behave as a resistance R and will be independent of ω, the frequency, if $R^2 = L/C$.*

Solution. The admittance of the parallel combination is got as :

$$Y = Y_1 + Y_2$$

$$= \frac{1}{R + j\omega L} + \frac{1}{R - (j/\omega C)}$$

$$= \frac{R - j\omega L}{R^2 + (\omega L)^2} + \frac{R + j/\omega C}{R^2 + (1/\omega C)^2}$$

The admittance will have no imaginary part if the combination is to be purely resistive.

Substituting $$R^2 = L/C, \qquad (a)$$

$$\text{Imaginary part of Y} = \frac{-\omega L}{L/C + (\omega L)^2} + \frac{1/\omega C}{L/C + (1/\omega C)^2}$$

$$= \frac{-\omega LC}{L(1 + \omega^2 LC)} + \frac{\omega C}{1 + \omega^2 LC}$$

$$- 0$$

$$\text{Real part of Y} = \frac{R}{L/C + (\omega L)^2} + \frac{R}{L/C + (1/\omega C)^2}$$

$$= \frac{RC}{L(1 + \omega^2 LC)} + \frac{RC^2\omega^2}{1 + \omega^2 LC}$$

$$= \frac{RC(1 + \omega^2 LC)}{L(1 + \omega^2 LC)} = \frac{RC}{L} = \sqrt{\frac{C}{L}} \text{ using } (a).$$

Hence the circuit behaves resistive, of value $\sqrt{L/C}$ and hence independent of frequency.

2.21 SERIES PARALLEL CIRCUITS

The solution of series parallel circuits is by using Kirchhoff's laws. The polar form of complex numbers can be employed to facilitate the calculations. The following example illustrates the method.

Example 17. *Two impedances 6 + j8 and 5 – j9 are connected in parallel and then in series with another of 3 + j4 ohms and the series parallel circuit is applied a voltage 100 volts a.c. Find the currents in each and the total power and p.f.*

Solution. The currents I_1, I_2 flow through the branches $5 - j9$ and $6 + j8$ as shown and the total current is I. Then, $V = IZ_1 + I_2Z_2$.

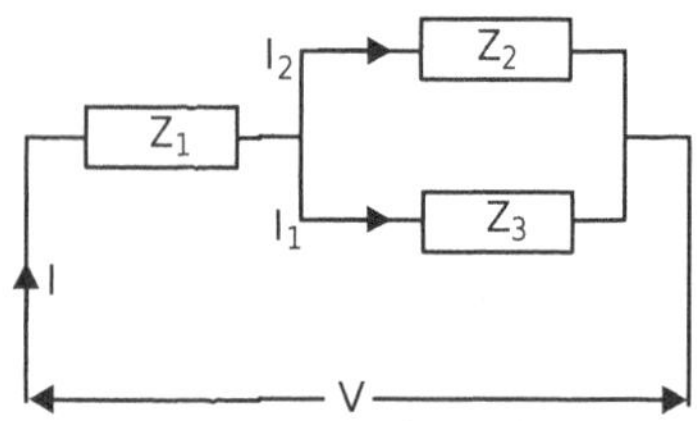

Fig. 2.28

$$I_2Z_2 = I_1Z_3 = (I - I_2)\,Z_3$$

so
$$Z_1I + Z_2I_2 = V$$
$$Z_3I - (Z_3 + Z_2)\,I_2 = 0$$

Hence
$$I = \begin{vmatrix} V & Z_2 \\ 0 & -(Z_2+Z_3) \end{vmatrix} \div \begin{vmatrix} Z_1 & Z_2 \\ Z_3 & -(Z_2+Z_3) \end{vmatrix}$$

$$= V(Z_2+Z_3)/[Z_1(Z_2+Z_3)+(Z_2Z_3)]$$

$$I_2 = -VZ_3/[Z_1(Z_2+Z_3)+Z_2Z_3]$$

Thus
$$[Z_1(Z_2+Z_3)+Z_2Z_3] = [(3+j4)(6+j8+5-j9)+(6+j8)(5-j9)]$$
$$= [(3+j4)(11-j1)+(30+72+40j-54j)]$$
$$= [33 + 4 + 44j - 3j + 102 + 40j - 54j]$$
$$= 139 + 27j$$

$$I = 100(11 - j1)/(139 + 27j)$$
$$= 100 \times 11.05 \angle -5° \, 12'/141.4 \angle 11° = 7.8 \angle -16° \, 12'$$

$$I_2 = -100(5 - j9)/139 + 27j$$
$$= 100 \times 10.32 \angle(-61° + 180°)/141.4 \angle 11° = 7.3 \angle 130°$$

$$I_1 = I - I_2 = 100(11 - j1 + 5 - j9)/139 + 27j$$
$$= 100\,(16 - 10j)/139 + 27j$$
$$= 13.35 \angle -62°$$

$$\text{Total power} = V^*I = 100 \times 7.8 \angle -16° \, 12'$$
$$= 100 \times 7.8 \cos 16° \, 12' = 75 \text{ Watts}$$

$$\text{Power factor} = \cos\,(-16° \, 12') = 0.96 \text{ lag.}$$

Example 18. *In the network of Fig. 2.29, find the value of the reactance of the variable capacitor for which the current through this capacitor I has no component perpendicular to* *the phase of the applied voltage V. Find, for this value of the capacitor reactance, the power consumed in the network in terms of V.*

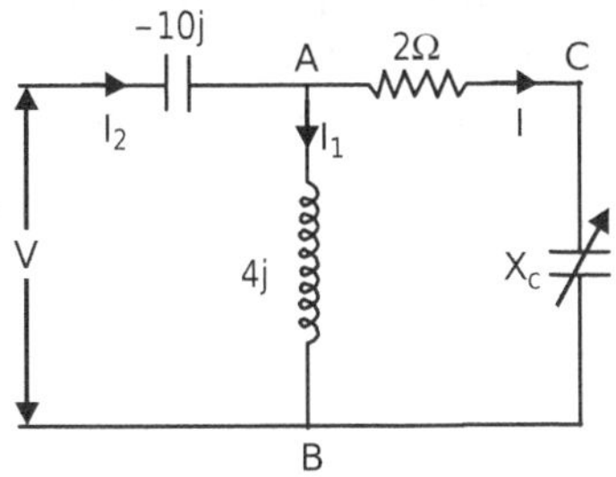

Fig. 2.29

Solution. The current through the capacitor I must not have a component in quadrature with V. Take V as the reference phasor. Then V = V + *j*0. Writing KCL equations,

$$I(2 - jX_C) + (-10j)\, I_2 = V$$

$$4j\, I_1 = (2 - jX_C)\, I \qquad (a)$$

and

$$I + I_1 = I_2$$

$$4j\, I_1 = (2 - jX_C)(I_2 - I_1)$$

$$\therefore \quad (4j + 2 - jX_C)\, I_1 = (2 - jX_C)\, I_2$$

$$I(2 - jX_C) = V + 10j\, I_2$$

$$= V + 10j\frac{(4j + 2 - jX_C)}{2 - jX_C} I_1$$

Using (*a*), we get

$$I(2 - jX_C) = V + 10j\frac{(4j + 2 - jX_C)}{2 - jX_C}\frac{(2 - jX_C)}{4j} I$$

$$= V + 2\frac{1}{2}(4j + 2 - jX_C)\, I$$

$$I\left[-3 + j\left(-10 + 1\frac{1}{2}X_C\right)\right] = V$$

$$I = \frac{V}{-3 + j\left(-10 + 1\frac{1}{2}X_C\right)} \qquad (b)$$

The imaginary part of I must be zero. So $-10 + 1\frac{1}{2}X_C = 0$

$$X_C = 10/1.5 = 6.67\ \Omega.$$

The current is then $\quad I = -\dfrac{V}{3}$

Power in the network = Power in the 2 Ω resistor

$$= I^2 \times 2$$

$$= 2\,(V^2/9) \text{ Watts.}$$

2.22 NODE EQUATIONS FOR NETWORK SOLUTION

In chapter 1, the writing of the node equations of a purely resistive network and the solution of the node voltages were considered. If a network contains inductance and capacitance and is energised by sources which are not constant d.c. sources, the procedure is not in

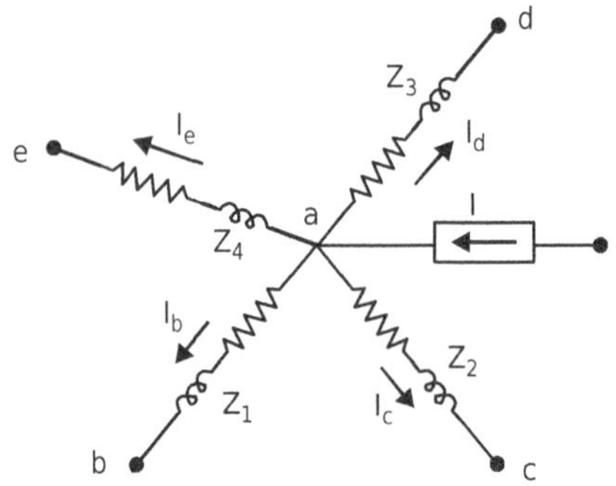

Fig. 2.30

anyway different, except that the equations contain impedance terms (in complex form, of course) in place of resistance terms. The Ohm's law equation, called the terminal equation for impedances, *i.e.*,

$$V = ZI \tag{2.42}$$

is used instead of $v = Ri$ for resistances.

The KCL equation for a node is

$$I_b + I_c + I_d + I_e - I = 0$$

where the I is a current source sending current into node a. Choosing node e as reference and measuring voltages of other nodes w.r.t. it (*i.e.*, $V_e = 0$)

$$\frac{V_a - V_b}{Z_1} + \frac{V_a - V_c}{Z_2} + \frac{V_a - V_d}{Z_3} + \frac{V_a}{Z_4} = I$$

$$V_a\left(\frac{1}{Z_1} + \frac{1}{Z_2} + ... + \frac{1}{Z_4}\right) - \frac{V_b}{Z_1} - \frac{V_c}{Z_2} - \frac{V_d}{Z_3} = I$$

Since 1/Z is called admittance Y.

$$V_a (Y_1 + Y_2 + ... + Y_4) - V_bY_1 - V_cY_2 - V_dY_3 = I \tag{2.43}$$

Similar equations are written for all the nodes and entered in a matrix form. Then, we get

$$\mathbf{YV = I} \tag{2.44}$$

where **Y** is the admittance matrix, **V** is column vector of nodal voltages and **I** is the column vector of currents entering into the nodes.

The matrices **Y**, **V** and **I** can be written down by inspection, without writing equation (2.43) separately for each node. The elements on the main diagonal of the admittance matrix are the sums of admittances connected to the respective nodes. The elements outside the main diagonal are the sums of admittances connected between two nodes, always written with a negative sign. Such a matrix is symmetric to the main diagonal. The current I is a column of nodal currents, the current flowing into the node taken positive.

Let us discuss the applications of the method by a set of typical examples. Example 19 following deals with a simple *a.c. network* containing inductances and capacitances and energised by a.c. current sources. Example 20 describes the method when *voltage sources* are also included in the network. Example 21 describes the procedure if one of the sources is a *dependent source.* Example 22 is a problem similar to 20 but simpler and uses the method of converting voltage sources into current sources by the principle of *node splitting.* The idea of *degenerate source* is introduced in Example 23 which enables the calculation of an open circuit voltage or a short circuit current.

Example 19. *Find the currents I_1, I_2 in the circuit shown. The source is a 40 amperes A.C. source of $i = 40 \sin 10t$.*

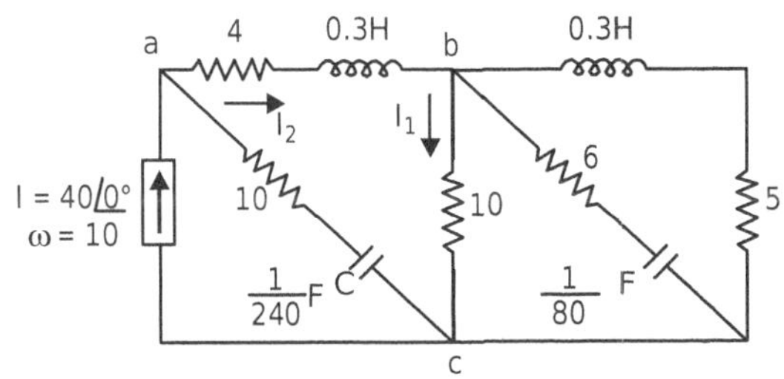

Fig. 2.31

Solution. The current 40 sin 10*t* can be written in complexor form as 40∠0°. The admittances of the branches are first calculated. ω = 10 and hence inductive reactances are 10 L, capacitive reactance are 1/10 C.

$$Z_{ab} = 4 + 0.3 \times 10j = 4 + 3j.$$

$$Y_{ab} = (4 - 3j) / (4^2 + 3^2) = 0.16 - 0.12j.$$

The admittance of the 10 ohm resistor between *b* and *c* is

$$Y_{bc} = 1/10 = 0.1$$

$$Z_{ac} = 10 - j(240/10) = 10 - j24$$

$$Y_{ac} = 10 + j24/(10^2 + 24^2)$$

$$= 0.015 + 0.035j.$$

The impedance of the 6 Ω and 1/80 F between *b* and *c* is 6 – *j*8.

Its admittance is $6 + j8/(6^2 + 8^2) = 0.06 + 0.08j$.

The impedance of the 5 ohm and 1.2 H between *b* and *c* is 5 + 12*j* and its admittance is 0.03 + 0.072*j*.

Taking the node *c* as the reference ($V_c = 0$), the node equations for nodes *a* and *b* are written down.

$$\begin{vmatrix} (0.16-0.12j)+(0.015+0.35j) & -(0.16-0.12j) \\ -(0.16-0.12j) & \begin{matrix}(0.6-.08j)+0.1 \\ +(0.16-.12j) \\ +(.03+.072j)\end{matrix} \end{vmatrix} \begin{vmatrix} V_a \\ V_b \end{vmatrix} = \begin{vmatrix} 40 \\ 0 \end{vmatrix}$$

$$\therefore \quad \begin{vmatrix} 0.175-.085j & -.16+.12j \\ -.16+.12j & -0.35+.322j \end{vmatrix} \begin{vmatrix} V_a \\ V_b \end{vmatrix} = \begin{vmatrix} 40 \\ 0 \end{vmatrix}$$

$$V_a = \begin{vmatrix} 40 & -.16+.12j \\ 0 & 0.350+.322j \end{vmatrix} \div \begin{vmatrix} 0.175-.085j & -0.16+0.12j \\ -0.15+0.12j & 0.350+.322j \end{vmatrix}$$

$$= 40(0.35 + 0.322j) \div (.077 + .064j)$$

$$= (40 \times 4.75 \angle 2° \ 36j).$$

$$V_b = \begin{vmatrix} 0.175-.085j & 40 \\ -0.16+0.12j & 0 \end{vmatrix} \div (.077+.064j)$$

$$= 80\angle -75° \ 36'.$$

Hence current $I_1 = V_b/10 = 8\angle -75° \ 35'$

i.e., in time form, $I_1 = 8 \sin (10t - 75° \ 36'')$

The current $I_2 = (V_a - V_b) \div (4 + 3j)$

$$= [(190+8.6j)-(20-77.5j)]/(4+3j)$$

$$= 191 \angle 26° \ 50'/5 \angle 36° \ 54'$$

$$= 38.2 \angle -10° \ 4'$$

i.e., $i_2 = 38.2 \sin (10t - 10° \ 4')$ from b to a.

Example 20. *For the network shown, write the nodal voltage equations and simplify.*

Solution. In this problem, voltage sources have also included. In such cases, we note the fact that every source (voltage or current) has a current through it and a voltage across it. All the currents flowing from the different sources are added to the nodal currents (including the currents from a voltage source, which is however an unknown quantity), and all the source voltages are transferred into nodal voltages.

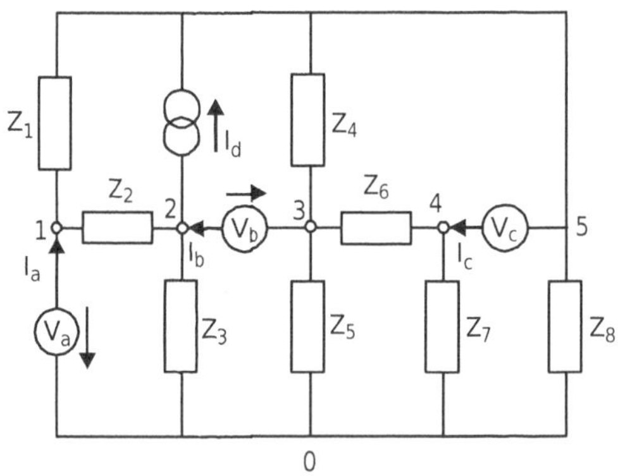

Fig. 2.32

$$V_1 = V_a$$

$$V_2 = V_b + V_3$$

$$V_4 = V_c + V_5$$

and the nodal equations written in the form

$$\begin{vmatrix} Y_1+Y_2 & -Y_2 & 0 & 0 & -Y_1 \\ -Y_2 & Y_2+Y_3 & 0 & 0 & 0 \\ 0 & 0 & Y_4+Y_5+Y_6 & -Y_6 & -Y_4 \\ 0 & 0 & -Y_6 & Y_6+Y_7 & 0 \\ -Y_1 & 0 & -Y_4 & 0 & Y_1+Y_4+Y_8 \end{vmatrix} \times \begin{vmatrix} V_a \\ V_b+V_3 \\ V_3 \\ V_c+V_5 \\ V_5 \end{vmatrix} = \begin{vmatrix} I_a \\ I_b-I_d \\ -I_b \\ I_c \\ I_d-I_c \end{vmatrix}$$

Now, the known variables V_a, V_b, V_c are transferred to the right and the unknown variables I_a, I_b, I_c to the left of the matrix equation giving

$$\begin{vmatrix} -1 & 0 & 0 & -Y_2 & -Y_1 \\ 0 & -1 & 0 & Y_2+Y_3 & 0 \\ 0 & 1 & 0 & Y_4+Y_5+Y_6 & -Y_4-Y_6 \\ 0 & 0 & -1 & -Y_6 & Y_6+Y_7 \\ 0 & 0 & 1 & -Y_4 & Y_1+Y_4+Y_8 \end{vmatrix} \begin{vmatrix} I_a \\ I_b \\ I_c \\ V_3 \\ V_5 \end{vmatrix} = \begin{vmatrix} -V_a(Y_1+Y_2)+V_bY_2 \\ V_aY_2-V_b(Y_2+Y_3)-I_d \\ V_cY_6 \\ -V_c(Y_6+Y_7) \\ V_aY_1+I_d \end{vmatrix}$$

This matrix equation now enables the calculation of all the variables I_a, I_b, I_c, V_3, V_5.

If only the nodal voltages are to be found, the currents I_a, I_b, I_c may be eliminated. For example, the current I_a may be eliminated by omitting the first row and the first column. The current I_b may be eliminated by adding the second and third rows whereas the current I_c will be eliminated by adding the last two rows. The equations, after elimination of I_a, I_b, I_c, become

$$\begin{vmatrix} Y_2+Y_3+Y_4+Y_5+Y_6 & -Y_4-Y_6 \\ -Y_4-Y_6 & Y_1+Y_4+Y_6+Y_7 \end{vmatrix} \begin{vmatrix} V_3 \\ V_5 \end{vmatrix} = \begin{vmatrix} V_aY_1-V_b(Y_2+Y_3)+V_cY_6-I_d \\ V_aY_1-V_c(Y_6+Y_7)+I_d \end{vmatrix}$$

Example 21. *In the circuit Fig. 2.33, a dependent current source depending on the current in ac is included. Find the node voltages.*

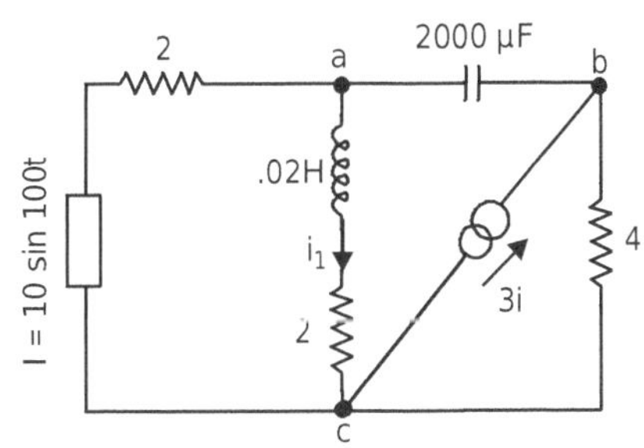

Fig. 2.33

Solution. Choose node *c* as reference. The 0.02 H inductor has reactance of $j\,100 \times 0.02 = j2$.

The admittance of branch *ac* is $1/(2 + 2j) = 0.25 - 0.25j$.

The admittance of branch *ab* is $j\omega C = 0.2j$

The node equations may be written :

$$\begin{vmatrix} 0.25-0.25j+0.2j & -0.2j \\ -0.2j & 0.25+0.2j \end{vmatrix} \begin{vmatrix} V_a \\ V_b \end{vmatrix} = \begin{vmatrix} 10\angle 0^\circ \\ 3i_1 \end{vmatrix}$$

Note that the 2 Ω resistor in series with the current source is to be ignored, since its presence will not alter the current sent by the source.

The dependent current $3i_1$ must be expressed in terms of the nodal voltages.

$3i_1 = 3(V_a - V_c)\, Y_{ac} = 3V_aY_{ac} = 3(0.25 - 0.25j)\, V_a = 0.75(1 - j)\, V_a$.

This is substituted in the second equation

$$\begin{vmatrix} 0.25-0.05j & -0.2j \\ -0.2j & 0.25+0.2j \end{vmatrix} \begin{vmatrix} V_a \\ V_b \end{vmatrix} = \begin{vmatrix} 10 \\ 0.75(1-j)\,V_a \end{vmatrix}$$

The V_a term, a nodal voltage, which is on the right hand side of the second equation is to be transferred to the left.

$$\begin{vmatrix} 0.25-0.05j & -0.2j \\ -.2j-.75(1-j) & 0.25+.2j \end{vmatrix} \begin{vmatrix} V_a \\ V_b \end{vmatrix} = \begin{vmatrix} 10 \\ 0 \end{vmatrix}$$

Solving for V_a, V_b,

$$V_a = \begin{vmatrix} 10 & -0.2j \\ 0 & 0.25+.2j \end{vmatrix} \div \begin{vmatrix} 0.25-0.05j & -0.2j \\ -.2j-.75(1-j) & 0.25+.2j \end{vmatrix}$$

$$= 3.2\angle 38^\circ\, 42' \div .262\angle 45^\circ\, 42'$$

$$= 12.2\angle -7^\circ$$

Similarly, $V_b = 35.7\angle -82^\circ$.

Example 22. *A voltage source V = 100 sin 100t is applied to the circuit shown between a and b and another of 50 sin 100t between c and d. Find the voltages of nodes a and c with respect to node b.*

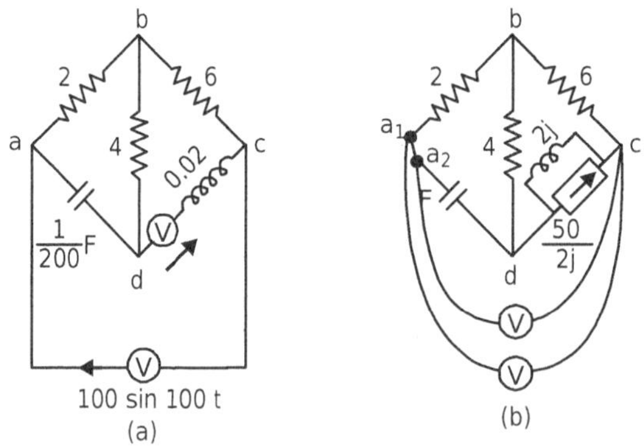

(*a*) Network. (*b*) Node splitting applied to node *a*.

Fig. 2.34

Solution. We have voltage source in this problem. It may be tackled by labelling a current through this source and treating as in Example 20. Alternatively, the network may be solved by first converting the voltage sources into current sources.

The 50 V a.c. source has a series inductance $2j$ ohms (for $\omega = 100$) in series. By the principle of source equivalence, (page 17), this may be represented as a current source $i = 50/2j$ in parallel with an impedance of $2j$ ohms.

The 100 V source has nothing in series. So this principle cannot be employed. So, the node a is split as shown in Fig. 2.35 (b) into 2 nodes a_1, a_2; we apply the same voltage source to each node since the two identical voltage sources in parallel can be replaced by a single voltage source. Now, nodes a_1, a_2 are at the same potential and hence the links a_1, a_2 may be removed, reducing to what is shown in Fig. 2.35 (a). Now each of the 2 voltage sources have series impedances across them. Hence they may be converted into current sources. Thus we get 2 sources which are 100/2 and $100/[-j/\omega C] = 100/-2j$.

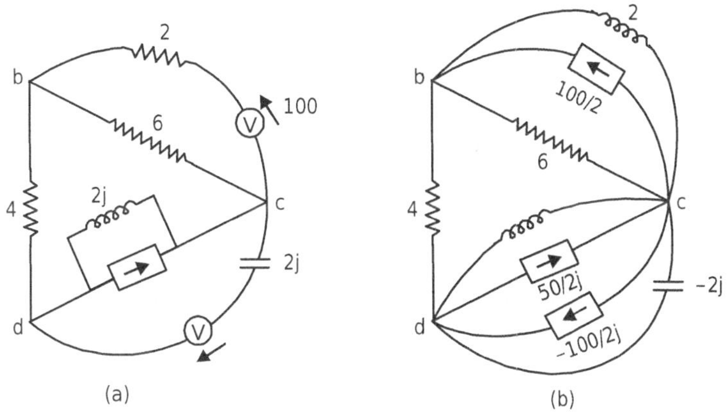

(a) Reduced from 2.35(b). (b) Final network.

Fig. 2.35

Hence the final network appears as in Fig. 2.35(b). The node equations become

$$\begin{vmatrix} 1/6+(1/2j)-(1/2j) & (1/2j)-(1/2j) \\ (1/2j)-(1/2j) & (1/4)+(1/2j)-(1/2j) \end{vmatrix} \begin{vmatrix} V_c \\ V_d \end{vmatrix} = \begin{vmatrix} (50/2j)-j100/2 \\ (100/-2j)-50/2j \end{vmatrix}$$

giving
$$\begin{vmatrix} 1/6 & 0 \\ 0 & 1/4 \end{vmatrix} \begin{vmatrix} V_c \\ V_d \end{vmatrix} = \begin{vmatrix} -50j-25j \\ 75j \end{vmatrix}$$

$\therefore$
$$V_c = -6(50j + 25j) \text{ volts}$$
$$V_d = 75j \times 4 \text{ volts.}$$

Example 23. *Find the open circuit voltage and the short circuit current between nodes b and c of Fig. 2.36 (a).*

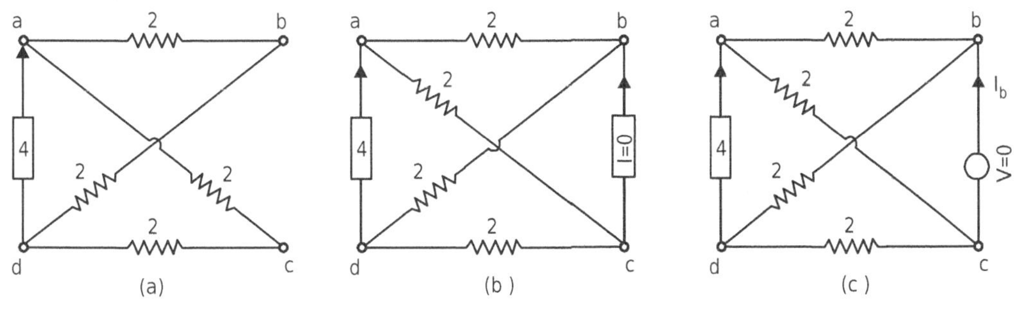

Fig. 2.36 (a) Network, (b) Finding o.c. volts, (c) Finding s.c. current.

Solution. To find the open circuit voltage, a current source I can be connected between *b* and *c* and if later, I is made equal to zero, we will get the equivalent of an open circuit. Such a source I = 0 is called a *degenerate source.*

Similarly, to find the short circuit current, a degenerate voltage source E = 0 is introduced between the node *b* and *c*.

We write the nodal equations as usual choosing the node c as the reference [Fig. 2.36(*b*)].

$$\begin{vmatrix} 1 & -\frac{1}{2} & 0 \\ -\frac{1}{2} & 1 & -\frac{1}{2} \\ 0 & -\frac{1}{2} & 1 \end{vmatrix} \begin{vmatrix} V_a \\ V_b \\ V_d \end{vmatrix} = \begin{vmatrix} 4 \\ I \\ -4 \end{vmatrix}$$

Multiplying the second equation by 2 and adding with first and third, we get

$$V_b = 2I + 4 + (-4) = 2I$$

Hence open circuit voltage $= V_{bc} = V_b = 2 \times 0 = 0$ volts.

The short circuit current is found from Fig. 2.36 (*c*). Choosing node *c* as reference, the node voltage $V_b = V$.

$$\begin{vmatrix} 1 & -\frac{1}{2} & 0 \\ -\frac{1}{2} & 1 & -\frac{1}{2} \\ 0 & -\frac{1}{2} & 1 \end{vmatrix} \begin{vmatrix} V_a \\ V \\ V_d \end{vmatrix} = \begin{vmatrix} 4 \\ I_b \\ -4 \end{vmatrix}$$

Transferring the known voltage to the right and the unknown I_b to the left,

$$\begin{vmatrix} 1 & 0 & 0 \\ -\frac{1}{2} & -\frac{1}{2} & -1 \\ 0 & 1 & 0 \end{vmatrix} \begin{vmatrix} V_a \\ V_d \\ I_b \end{vmatrix} = \begin{vmatrix} 4+\frac{1}{2}V \\ -V \\ \frac{1}{2}V-4 \end{vmatrix}$$

Elimination of V_a, V_d gives I_b. So, multiplying the second equation by 2 and adding to the first and third,

$$-2I_b = 4+\frac{1}{2}V - 2V + \frac{1}{2}V - 4 = -V$$

$\therefore$ $I_b = V/2$. The required short circuit current (put V = 0) is zero.

Example 24. *Find the current in the branch ab.*

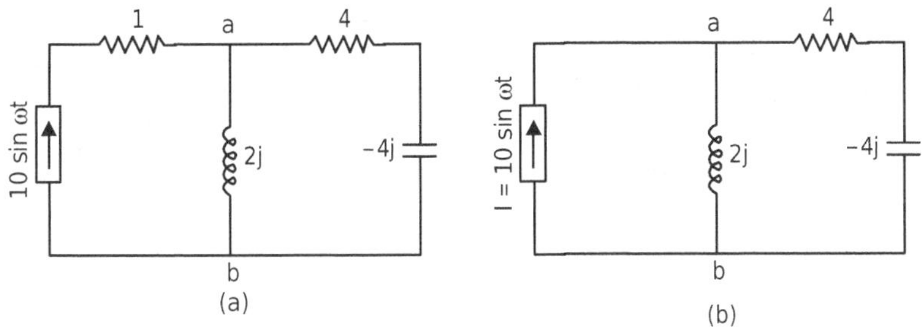

Fig. 2.37

Solution. In this problem, a resistance 1 Ω is a series with the current source. In such cases, when there is an impedance in *series* with the current source, that impedance will not come into the nodal equations. This is because its presence will not alter the current flowing into the node *a* and hence the circuit may be redrawn as in Fig. 2.37 (*b*), omitting the series resistance.

Choose node *b* as the reference

$$\left(\frac{1}{2j}+\frac{1}{4-4j}\right)V_a = 10\ \angle 0°$$

$$V_a = 10 \div 0.125\ (1 - 3j)$$

$$\text{Current in } ab = V_a/2j = \frac{10/2j}{0.125\,(1-3j)} = 4(3-j)$$

$$= 12.6\angle -18\frac{1}{2}°$$

$$I_{ab} = 1.42\sin\left(10t - 18\frac{1}{2}°\right).$$

2.23 METHOD OF LOOP EQUATIONS

The method of loop equations has already been discussed in some detail in pages 14–18. To choose a set of independent loops, we follow the procedure outlined in 6.3, namely, to choose a tree and draw the co-tree branches one by one. The number of loop equations will be given by the formula $b - n + 1$, while the node equations were $(n - 1)$ in number. The choice between the loop and node methods is sometimes made for minimum number of equations. For example if a network has 7 branches and 4 nodes, the number of loop equations will be $7 - 4 + 1 = 4$, the number of node equations will be $4 - 1 = 3$ only. In this case, the node method may be preferred. But when mutual inductances are involved, the loop method is thc simpler of the two, even otherwise.

Example 25. *Write the loop equations for the network of Fig. 2.38 and solve for the current.*

Solution. The basic loops are chosen be drawing the graph, choosing a tree (Fig. 2.38 (*b*)) and finding the loops (*c*). There are 3 loops, which are then marked in the network of (Fig. 2.38 (*a*)).

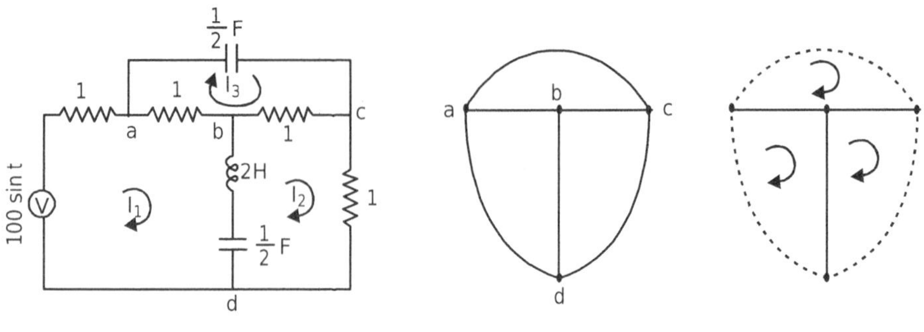

(a) (b) (c)

Fig. 2.38

The impedances of 2H, $\frac{1}{2}$F are (since $\omega = 1$ for the voltage 100 sin t) found as $2j$ and $-2j$. The loop equations are written in matrix form.

$$\begin{vmatrix} 2+2j-2j & -(2j-2j) & -1 \\ (-2j-2j) & 2+2j-2j & -1 \\ -1 & -1 & 2-2j \end{vmatrix} \begin{vmatrix} I_1 \\ I_2 \\ I_3 \end{vmatrix} = \begin{vmatrix} 100\angle 0° \\ 0 \\ 0 \end{vmatrix}$$

$$\therefore \qquad \begin{vmatrix} 2 & 0 & -1 \\ 0 & 2 & -1 \\ -1 & -1 & 2-2j \end{vmatrix} \begin{vmatrix} I_1 \\ I_2 \\ I_3 \end{vmatrix} = \begin{vmatrix} 100 \\ 0 \\ 0 \end{vmatrix}$$

Solving, $I_1 = 55 + 10j$, $I_2 = 5(1 + 2j)$, $I_3 = 10(1 + 2j)$

The branch currents may now be found. For example, the current in branch *ab* from *a* to *b* is $I_1 - I_3$ is $45 - 10j$. The current in branch *bd* from *b* to *d* is $I_1 - I_2$, which is 50.

Example 26. *Find the voltage V across the current source and that across the capacitor in the following network.*

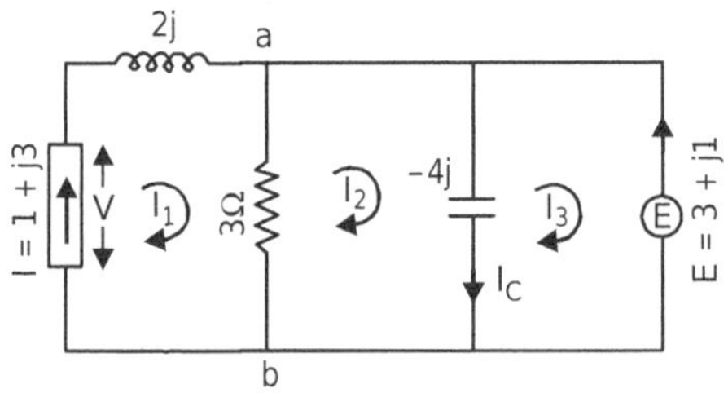

Fig. 2.39

Solution. Let V be the voltage across the current source. Then we write the three loop equations in matrix form

$$\begin{vmatrix} 3+2j & -3 & 0 \\ -3 & 3-4j & 4j \\ 0 & 4j & -4j \end{vmatrix} \begin{vmatrix} I_1 = 1+j3 \\ I_2 \\ I_3 \end{vmatrix} = \begin{vmatrix} V \\ 0 \\ -(3+j1) \end{vmatrix}$$

The variable I_1 is known to be $1 + j3$ and the voltage V across the current source is unknown. Hence, transferring the unknown to the left,

$$\begin{vmatrix} -1 & -3 & 0 \\ 0 & 3-4j & 4j \\ 0 & 4j & -4j \end{vmatrix} \begin{vmatrix} V \\ I_2 \\ I_3 \end{vmatrix} = \begin{vmatrix} -(1+3j)(3+2j) \\ 3(1+j3) \\ -(3+j1) \end{vmatrix}$$

Adding second and the third equations,

$$3I_2 = 3(1 + j3) - (3 + j1) = 8j;\ I_2 = 8j/3$$

$$I_3 = \left[-(3+j) - 4jI_2\right] \div (-4j)$$

$$= \frac{(23j+3)}{12}.$$

$$V = -3I_2 + (1 + 3j)(3 + 2j)$$

$$= -3 + 3j$$

Hence the current $I_c = I_2 - I_3 = (3j - 1)/4$ and the voltage across the current source $-3 + 3j$.

Example 27. *Find the output voltage V_o in terms of V_i in the network. Find the frequency and the value of r for which the voltage V_o is zero.*

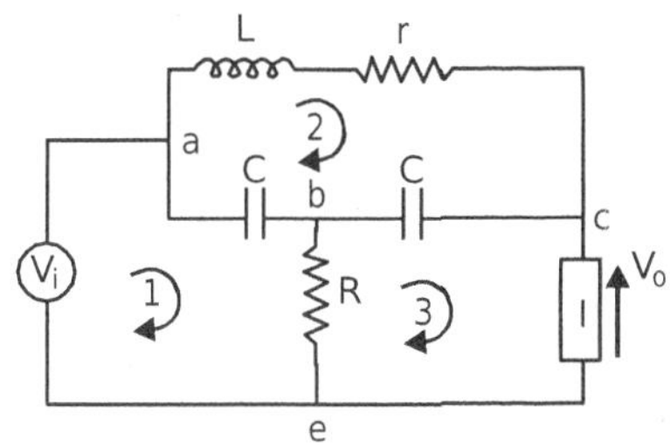

Fig. 2.40

Solution. The loops chosen are shown. A current source which is having I = 0 is inserted at the output terminals *ce* so that voltage V_o may be found by putting I = 0 finally. It is a degenerate current source.

The loop equations are :

$$\begin{vmatrix} R - j/\omega C & -j/\omega C & -R \\ -j/\omega C & r + j\omega L - j2/\omega C & -j/\omega C \\ -R & -j/\omega C & R - j/\omega C \end{vmatrix} \begin{vmatrix} I_1 \\ I_2 \\ -I \end{vmatrix} = \begin{vmatrix} V_i \\ 0 \\ -V_o \end{vmatrix}$$

Note that in the 3rd loop equation, the current – I is the current of the inserted current source *ce* and the voltage across it is V_o from *e* to *c*. But we know I and V_o; so we transfer V_o to the left and I to the right of the equations.

$$\begin{vmatrix} R - j/\omega C & -j/\omega C & 0 \\ -j/\omega C & r + j\omega L - 2j/\omega C & 0 \\ -R & -j/\omega C & 1 \end{vmatrix} \begin{vmatrix} I_1 \\ I_2 \\ V_o \end{vmatrix} = \begin{vmatrix} V_i - RI \\ -jI/\omega C \\ I(R - j/\omega C) \end{vmatrix}$$

Solving, we get,

$$I_1 = V_i(r + j\omega L - 2j\omega C) \div \Delta \qquad \text{(putting I = 0)}$$

$$I_2 = V_i\, j/\omega C \div \Delta$$

where $\Delta = (R - j/\omega C)(r + j\omega L - 2j/\omega C) - 1/(\omega C)^2$

And

$$V_o/V_i = RI_1 + (j/\omega C)I_2 \qquad \text{(putting I = 0)}$$

$$= R(r + j\omega L - 2j/\omega C) + (j/\omega C)^2 \div \Delta$$

For zero V_o, from the above, equating to zero,

$$R_r - 1/(\omega C)^2 = 0$$

and

$$R(\omega L - 2/\omega C) = 0$$

i.e.,

$$\omega = \sqrt{2/LC} \text{ and } r = \frac{1}{R}\frac{1}{C^2}\frac{LC}{2} = \frac{L}{2CR}.$$

Example 28. *A current source of 10 A is applied between a and b. Find the branch current using the loop method.*

Solution. Here again we have a current source problem and also combining a dependent source in branch *bd* of voltage $2V_{cb}$ acting from *d* to *b*.

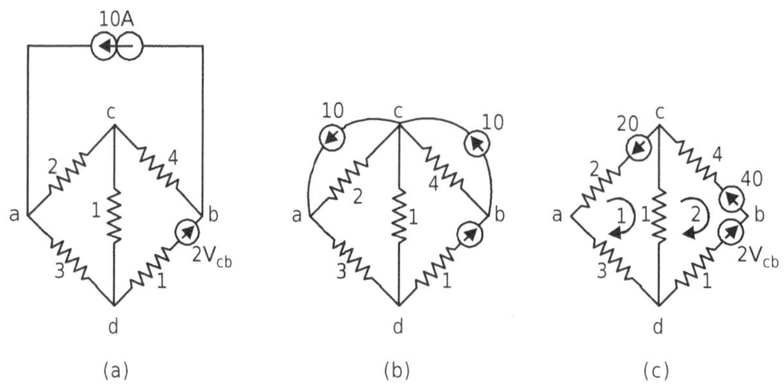

Fig. 2.41

In this example, we illustrate a different procedure for dealing with a current source, namely, the principle of *loop splitting*. We split the current source in two parts in series, with a connection to node *c*. The current enters and leaves *c* and so the conditions are not altered by redrawing the network like this. Now each of the two current sources has a parallel resistance and hence they may be replaced by voltage source with series resistors as in Fig. 2.41 (*c*). The voltage source equivalent is IR so that the voltages in *ac* and *bc* are 20 and 40 volts. The two loops chosen are shown and the loop equations become

$$(2 + 1 + 3)\, I_2 - 1I_2 = -20$$

$$-1I_1 + (1 + 4 + 1)\, I_2 = -40 - 2V_{cb}$$

But

$$V_{cb} = 4I_2 - 40. \text{ (on branch cb)}$$

$\therefore$

$$6I_1 - I_2 = -20 \qquad (1)$$

$$-I_1 + 6I_2 = 40 - 8I_2$$

$$-I_1 + 14I_2 = 40 \qquad (2)$$

Solving equations (1) and (2), $I_1 = 2.64$ A, $I_2 = 3.04$ A. The branch currents may then be found.

EXERCISES

1. The waveform of voltage in an electronic circuit is as shown. Find the r.m.s. value of the waveform. [**Ans.** 5]

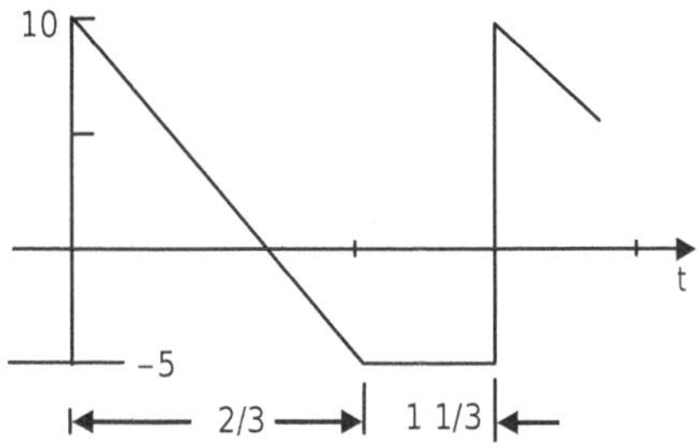

Fig. 1

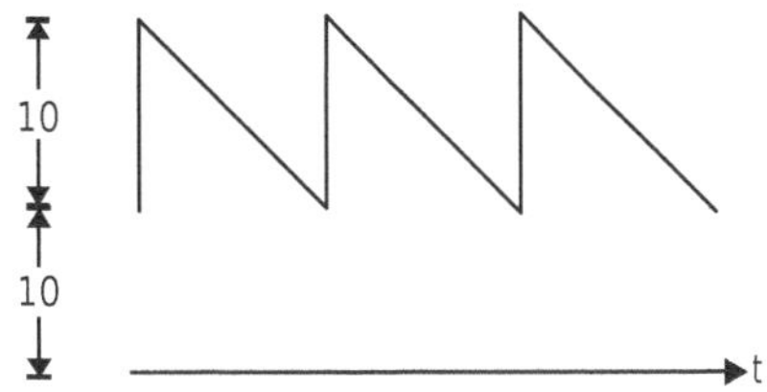

Fig. 2

2. Find the average value and form factor of the wave of Fig. 2. [**Ans.** 15, 1.027]

3. The following are the values at different instants of a periodic wave. The values can be assumed to change suddenly from one value to the next.

Time	0	1	2	3	4	5	6	7	8	9	10	millisecs
Amps	0	3	7	8	9	6	5	3	0	– 3	– 7	...etc.

Find (1) its average value (2) R.M.S. value (3) Form factor (4) Peak factor (5) frequency.

$\left[\textbf{Ans. } 5.125, 5.84, 1.14, 1.54, 62\frac{1}{2}\right]$

4. A certain sine wave has a value of 4 V at t = 0 and 2 millisecs later, it takes its max. value of 12 V. Express this voltage in phasor (r.m.s.) notation. If this voltage supplies current to an impedance of (3 + j4) ohms, find the phasor of current and power supplied.

[**Ans.** (8 + 2.83j), 1.41 – j0.94, 8.64 W

Phasor of voltage in polar form : $6\sqrt{2}\, e^{j(614t+.339)}$

Phasor of current : $1.2\sqrt{2}\, e^{j(614t-0.59)}$]

5. A sinusoidal 50 c/s voltage has the phasor of (4 + j3) volts. If the count of time were to start 0.009 secs prior to an instant, when the reference passes through its positive maximum, find the complex number representation of the above voltage with respect to the new count of time. [**Ans.** – 1.6 + j4.725]

6. A voltage of 3 + j4 is applied to a circuit in which it sends (– 1 + j2) amperes. Find (*i*) the magnitude and angle of impedance of the circuit (*ii*) the active power and reactive volt-amperes.

$\left[\textbf{Ans. } \sqrt{5}, -63.5°, 5\text{ W}, 10\text{ V Ar lead}\right]$

7. The following is an R-L-C series circuit to which a voltage of V volts r.m.s. and frequency f is applied. In each of the following cases, find the current, power factor and power.

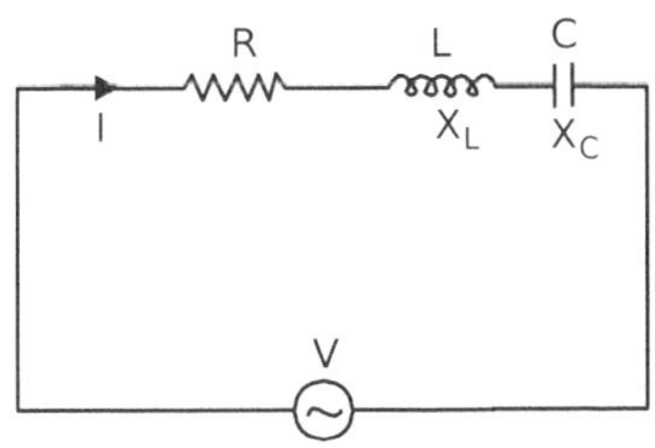

Fig. 3

	V	R	L(H)	C(μF)	*f*
(*a*)	200 V	100	0.01	100	50
(*b*)	200	1000	0.01	100	50
(*c*)	220	25	0.15	50	50
(*d*)	210	10	1.00	10	60
(*e*)	100	100	0.10	100	50
(*f*)	500	100	0.01	10	25

[**Ans.** 1.887, 0.9439, 356.4; 0.1998, 0.9993, 39.95; 1.937, 0.2201, 93.80; 0.3580, 0.0155, 1.282; 0.845, 0.845, 71.42; 0.0774, 0.1547, 59.90]

8. In Fig. 4, $Z_1 = 3 + j5$, $Z_2 = 50 + j50$, $Z_3 = 60 - j80$. The voltage E_1 is 440 V. Using complex notation method, determine (*i*) the total current I_1, (*ii*) the currents I_2 and I_3 (*iii*) the voltage E_2 and p.f. of the total circuit.

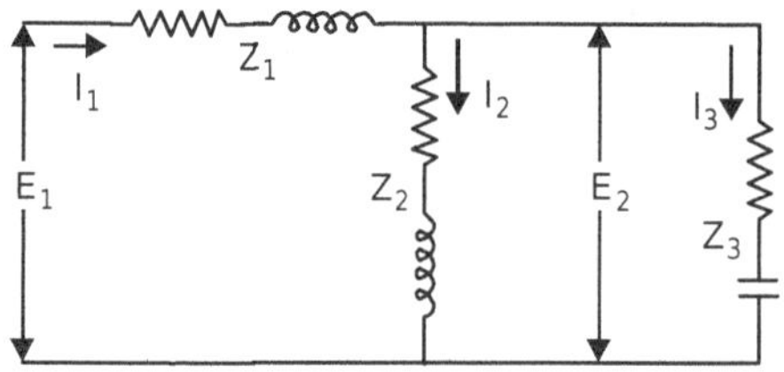

Fig. 4

[**Ans.** 6.7 $\angle -11.1°$, 5.68 $\angle 21.2°$, 4.01 $\angle 13°$
401 $\angle -40°$, 0.981 lag]

9. In Fig. 5, if V = 200 volts, find the current supplied and the power factor if

(*a*) $Z_1 = 2 + j3$, $Z_2 = 3 + j4$, $Z_3 = 5 - j6$.

(*b*) $3 + j4$, $6 - j2$, $3 - j2$.

(*c*) $5 - j2$, $3 - j3$, $6 - j9$. [**Ans.** (*a*) 25.43 A (*b*) 34.31 A (*c*) 21.01 A]

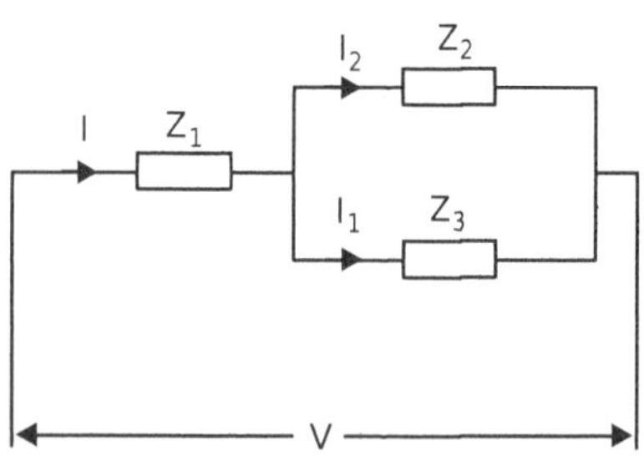

Fig. 5

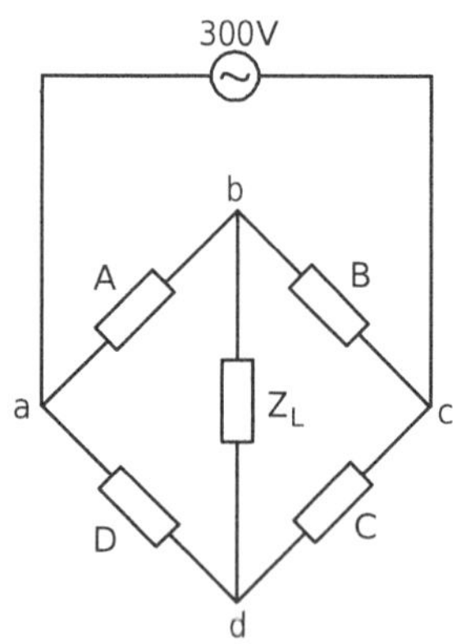

Fig. 6

10. In the bridge circuit (Fig. 6), with a 300 V supply, the impedances are

$Z_A = 2 + j4$, $Z_B = 1 - j4$, $Z_C = j4$, $Z_D = 3 - j4$, $Z_L = (2 - 8j)/3$. Find the current in Z_L.

[**Ans.** 67.1 $\angle 172.9°$ Amp.]

11. The magnitudes of V_{AB}, V_{BC}, V_{CA}, the voltage drops in the circuit (Fig. 7) are found by measurement to be 10, 20 and $10\sqrt{7}$ respectively. Find (*a*) the phases of V_{AB}, V_{BC} with V_{AC} as reference (*b*) the ratio $R_1 : R_2 : X_C$. [**Ans.** 10 $\angle 40.9°$ and 20 $\angle -19.13°$, $1:4:4/\sqrt{3}$]

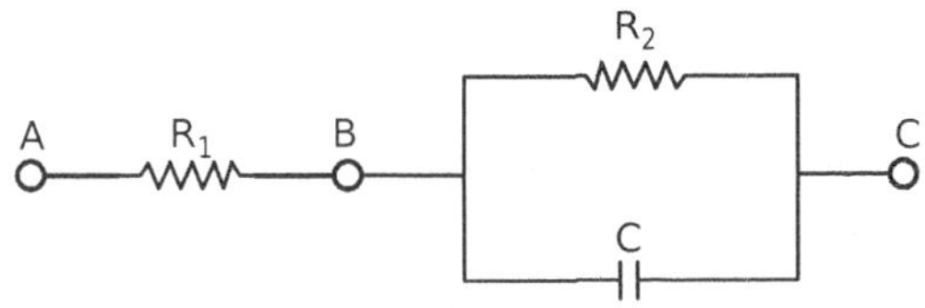

Fig. 7

12. Show that the parallel circuit of Fig. 8 will behave as a pure resistance at the frequency ω such that $(L\omega)^2 + R^2 = L/C$. Find the value of the resistance. [**Ans.** L/CR]

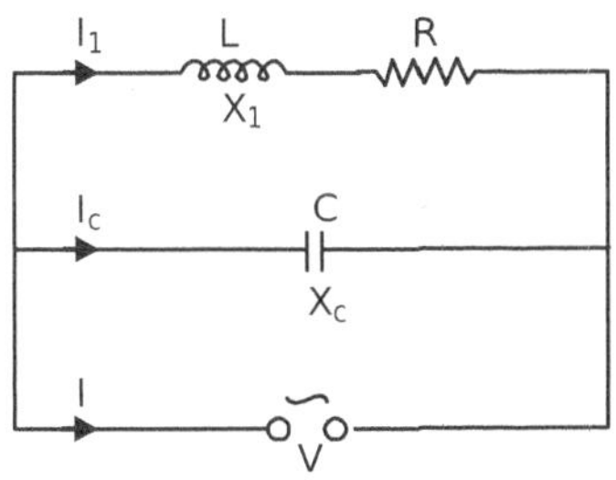

Fig. 8

SIMPLE EXERCISES

1. The *Frequency* of an alternating current is (*a*) the speed at which the current flows in the wire; (*b*) the number of times the current sign (from positive to negative) per second (*c*) the number of times it changes from positive to negative or negative to positive in every second (*d*) the number of times the current is switched off every minute. ( )

2. The waveform of an alternating current specifies (*a*) shape of the wire in which the current flows (*b*) The plot of current versus time in a graph paper (*c*) the radiation it produces as electromagnetic waves. ( )

3. The effective value of an alternating current is (*a*) that value of direct current which is as effective as the latter in producing power (*b*) the usefulness of the current being reckoned in terms of its ability to run a motor (*c*) the bad effect is produces in appliances connected to it (*d*) the shock effect that is produced when come into physical contact. ( )

4. The R.M.S. Value of an a.c. of value $i(t)$ is

(*a*) $\sum i(t)$ (*b*) $\int i^2(t)\,dt$

(*c*) $\int_0^T i^2(t)\,dt \Big/ T$ (*d*) $\left[\int_0^T i^2(t)\,dt \div T\right]^{1/2}$ ( )

5. The form factor of an alternating waveform is

(*a*) average value/maximum value (*b*) peak value/r.m.s. value

(*c*) maximum value/r.m.s. value (*d*) r.m.s. value/average value. ( )

6. A current of sinusoidal waveform $i = 10 \sin 20t$ flows in a wire. Its maximum value. r.m.s. value, frequency are respectively :

(*a*) 10, 7.07, 10/π (*b*) 20, 10, 20/π (*c*) 10, $10\sqrt{2}$, 20 (*d*) 10, 20, $20\sqrt{2}$. ( )

7. Two currents are expressed as 10 sin (*t* + 20°) and 20 cos (*t* – 20°). The phase difference between them is :

(*a*) –10 (*b*) –40° (*c*) 50° (*d*) –10°. ( )

8. Current in a resistance and the voltage across it are

(*a*) equal (*b*) in phase (*c*) independent (*d*) unequal. ( )

9. Current in a pure inductance .... the applied voltage.

(*a*) leads (*b*) equals (*c*) lags (*d*) differs in phase. ( )

10. The voltage across a capacitor .... the current passed through it.

(*a*) lags (*b*) leads (*c*) equals (*d*) in phase with. ( )

11. Inductive reactance = L × (....)

(*a*) 2π/C (*b*) *f* (*c*) ω/C (*d*) 2π*f* (*e*) ω/2π ( )

12. Capacitive reactance = ...

(*a*) C/ω (*b*) 1/ωC (*c*) ω/C (*d*) $1/\sqrt{LC}$ ( )

13. The impedance of a resistance of 3 ohms in series with an inductive reactance of 4 ohms is .... ohms.

(*a*) 13 (*b*) 5 (*c*) 7 (*d*) –1 ( )

14. The impedance of a capacitive reactance of 12 ohms and a resistor of 5 ohms in series is .... ohm.

(*a*) 11 (*b*) 25 (*c*) 13 (*d*) –1 ( )

15. The *Power factor* of a series circuit having 6 ohms resistance and 8 ohms capacitive reactance is

(*a*) 0.8 (*b*) 0.6 (*c*) 0.75 (*d*) 1.33 ( )

16. A current of 10 amps supplied to an appliance at 250 V consumes a power of 2 KW. The p.f. of the appliance is

(*a*) 0.8 (*b*) 0.6 (*c*) 1.0 (*d*) 0.5 ( )

17. A voltage of E = 100 sin (314*t* + 0.6) is passed in series with a rectifier to a 100 ohms load R (Fig. 9). What is the Power supplied?

(*a*) 100 watts (*b*) 50 W (*c*) 25 W (*d*) 314 W ( )

18. The formula for finding the p.f. of an impedance Z using the method in Fig. 10 (3-voltmeter method) is.

(*a*) $\dfrac{V_1^2 - V_2^2 + V^2}{2V_1V_2}$ (*b*) $\dfrac{V^2 + V_1^2 - V_2^2}{2V_1V_2}$ (*c*) $\dfrac{V^2 - V_1^2 - V_2^2}{2V_1V_2}$ ( )

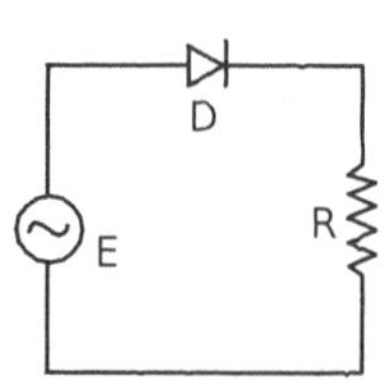

Fig. 9

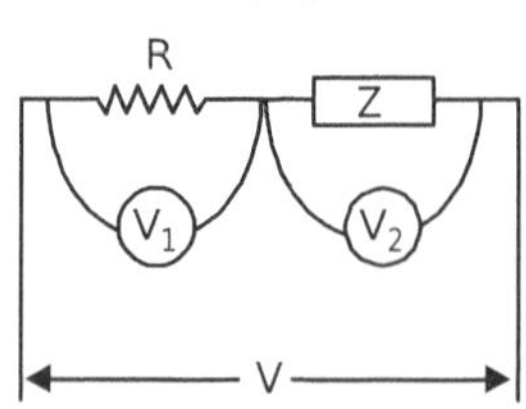

Fig. 10

19. At series *resonance*, the current in a R-L-C circuit is

(*a*) maximum (*b*) Oscillating (*c*) low (*d*) very high. ( )

20. The resonant frequency of a capacitor of 100/μF and a 1 henry inductor in series is .... Hz. (Hertz is the unit for frequency :—Cycles/sec.).

(*a*) 10 (*b*) $100\sqrt{2\pi}$ (*c*) $100/2\pi$ (*d*) $2\pi \times 100$ ( )

21. The p.f. of a resonant circuit is

(*a*) Unity (*b*) Low (*c*) Leading (*d*) exactly 0.5 ( )

22. The resonant circuit consisting of a 10 ohms resistor in series with 10/μF and 0.01 henry is supplied with 100 V. The current in the circuit is

(*a*) 0.01 (*b*) 10 (*c*) 100 (*d*) 20π ( )

23. The voltage across the capacitor and the inductor in a series circuit are

(*a*) Unbalanced (*b*) equal and antiphase(*c*) very low (*d*) very high (*e*) zero. ( )

A.C. CIRCUITS

1. An a.c. wave has the following values which repeat themselves.

Time :	0	1	2	3	4	5	6	7	8	9 millisecs.
Volts :	1	2	2	1	0	1	– 2	– 2	– 1	0

Its frequency is ....

(*a*) 100 (*b*) 50 (*c*) 10 (*d*) 900 ( )

2. The R.M.S. Value of the above waveform is ....

(*a*) 1.414 (*b*) 6 (*c*) 2 (*d*) $\sqrt{10}$ ( )

3. An alternator driven at 375 R.P.M. generates a frequency 50 (Hz). It has .... poles.

(*a*) 12 (*b*) 16 (*c*) 4 (*d*) 8 ( )

4. The form factor of a square wave a.c. is

(*a*) 1.1 (*b*) 1.11 (*c*) 1.414 (*d*) 1 ( )

5. The speed at which an 8-pole alternator should be driven in order to generate 60 Hz. e.m.f. is ....

(*a*) 1000 (*b*) 900 (*c*) 750 (*d*) 1200 ( )

6. A resistance and reactance are connected in series to a 100 V supply. The voltage across the resistance is 60 V and the current is 1 Ampere. The value of the reactance is ....

(*a*) 60 ohms (*b*) 40 ohms (*c*) 100 ohms (*d*) 80 ohms. ( )

7. A carbon lamp rated 100 V 100 watts is to be operated in series with a choke on a 200 V supply. The reactance of the choke required would be .... ohms.

(*a*) 100 (*b*) 173 (*c*) 150 (*d*) 300 ( )

8. If instead of a choke in (7), we used a capacitor for the same purpose, its reactance would be .... ohms.

(*a*) 100 (*b*) 0.001 (*c*) 0.066 (*d*) 173 ( )

9. Two identical impedances each of 10 ohms are connected in series. Total impedance would be .... ohms.

(*a*) 10 ohms (*b*) 14.14 (*c*) 20 (*d*) 15 ( )

10. A resistance of 5 ohms is in series with an inductive reactance of 16 ohms and a capacitive reactance of 4 ohms.

The total impedance is .... ohms.

(*a*) 13 (*b*) 21 (*c*) 25 (*d*) 17 ( )

11. A series circuit consumes a power of 500 watts at p.f. of 0.8 lagging from 250 V supply. Its reactive component of current is .... A and active component of current is .... A.

(*a*) 1.5, 2 (*b*) 1.2, 1.6 (*c*) 8, 6 (*d*) 1.6, 0.8 ( )

12. Find the reactance of a suitable condenser which will draw a current from 250 V supply, of value equal to the reactive component of current in problem 11. This condensers current will .... the applied voltage.

(*a*) 156.25 ohms (*b*) 166.67. (*c*) 12.500 ohms (*d*) 208.33 ohms ( )

(*a*) lag (*b*) lead

13. When the series circuit of problem 11 and the condenser of problem 12 are connected in parallel across the same 250 V supply, the total current drawn will have active component of .... Amps. and a reactive component of .... Amps.

(*a*) 1.2, 1 (*b*) 2, 0 (*c*) 1.0, 1.2 (*d*) 1.2, 1.2 ( )

14. The power factor overall in Problem 13 is ....

(*a*) 0.8 lag (*b*) 0.6 lead (*c*) Unity (*d*) data insufficient. ( )

15. A resistance of 5 ohms and an inductive reactance of 12 ohms are in series. The conductance of this impedance is ....

(*a*) 1/5 (*b*) 1/12 (*c*) 5/12 (*d*) 5/169 (*e*) 12/169 ( )

16. An admittance of an impedance is the .... of the impedance.

(*a*) Square root (*b*) Square (*c*) ratio (*d*) reciprocal ( )

17. What is meant by resonance ? [Anna University, May 2007]

18. Find the power, apparent power, reactive power for the circuit below [Anna University, May 2007]

(*a*) VI product (Apparent power) = 11.18 VA

(*b*) Power = 5W (*c*) Reactive power = 10 VAR, Power factor = 0.447.

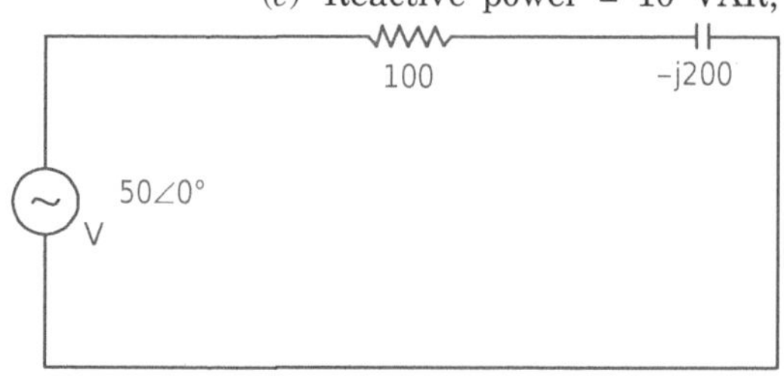

Fig. 11

19. Find the current $i(t)$ in Fig. 12 [Anna University, May 2007]

$I = 55.4\ \angle 180° - \tan^{-1}(55/4)$

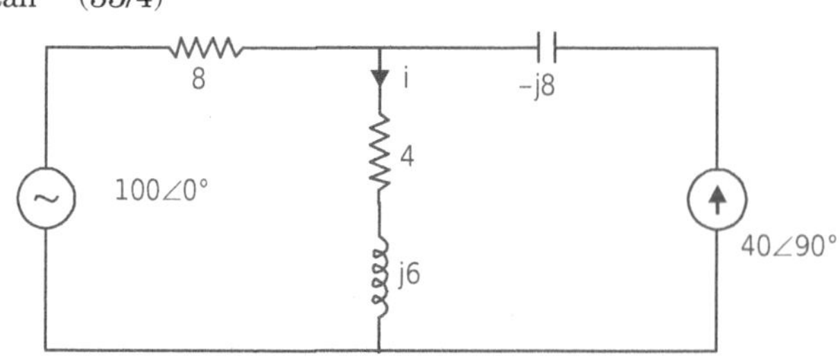

Fig. 12

ANSWERS

SIMPLE EXERCISES

1. (*b*)	**2.** (*b*)	**3.** (*a*)	**4.** (*d*)	**5.** (*d*)	**6.** (*a*)	**7.** (*c*)	**8.** (*b*)
9. (*c*)	**10.** (*a*)	**11.** (*d*)	**12.** (*b*)	**13.** (*b*)	**14.** (*c*)	**15.** (*b*)	**16.** (*a*)
17. (*c*)	**18.** (*c*)	**19.** (*a*)	**20.** (*c*)	**21.** (*a*)	**22.** (*b*)	**23.** (*b*)	

A.C. CIRCUITS

1. (*a*)	**2.** (*a*)	**3.** (*b*)	**4.** (*d*)	**5.** (*b*)	**6.** (*d*)	**7.** (*b*)	**8.** (*d*)
9. (*c*)	**10.** (*a*)	**11.** (*a*)	**12.** (*b, b*)	**13.** (*b*)	**14.** (*c*)	**15.** (*d*)	**16.** (*d*)

3

NETWORK THEOREMS

3.1 SUPERPOSITION PRINCIPLE

Electrical networks containing resistors and capacitors obey this principle which states:

1. The response to a number of *independent* sources (of voltage or current) may be found by adding up the responses due to each of them found separately.
2. If the sources are multiplied by a constant, the response is also multiplied by it.

In the above, we mean by response, the flow of current in a branch or the voltage between any two nodes, obtained due to source(s) of voltages or current.

Example 1. *Find the currents in the branches A, B, C, of the following 2 source network.*

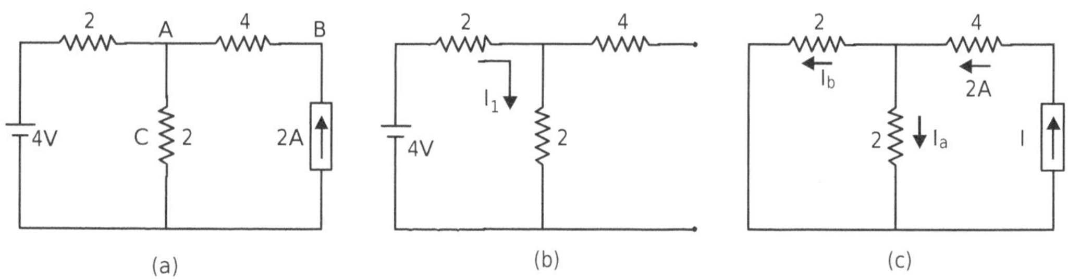

Fig. 3.1 (*a*) Network (*b*) With V source only (*c*) With I source only.

Solution. By superposition, we can solve this problem easily. In Fig. 3.1 (*b*), the response due to the voltage source is found and hence the *current source* does not send any current, which means it *is to be open circuited.* We find the current I_1 shown as 4/4 = 1 ampere and no current flows in the 4 Ω resistor.

The response due to the current source is found in Fig. 3.1 (*c*) in which the *emf source* is to be omitted, *i.e.,* it supplies no emf between it ends, and hence it *is to short circuited.*

Now the 2A current divides in the manner.

$$I_a = \frac{2}{2+2} I = 1 \text{ Amp}$$

$$I_b = 1 \text{ Amp.}$$

Hence the currents in the branches A, B, C are seen to be

$$I_1 - I_b,\ I_1 + I_c \text{ and } 2 \text{ Amps.}$$

giving 0, 2 Amps, 2 Amps, respectively.

The principle of superposition is the only convenient method when a network contains both D.C. and A.C. sources, as seen in the following example.

Example 2. *Solve the network shown in Fig. 3.2 (a).*

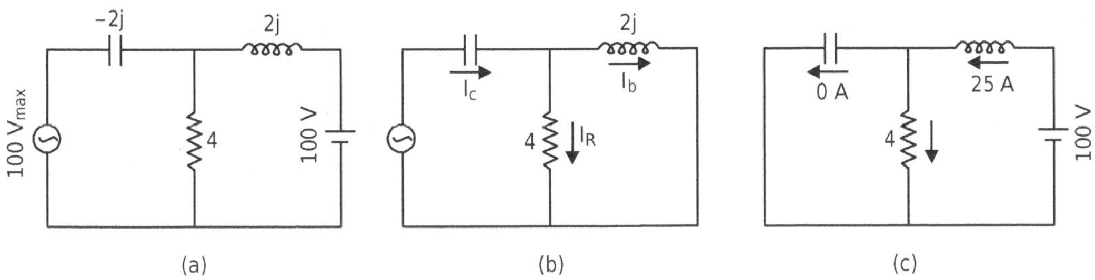

Fig. 3.2

Solution. Fig. 3.2 (*b*) applies the A.C. source only and shorts the D.C. e.m.f. source. The current, for a voltage 100 sin ω*t*, is found by the usual method.

$$Z = -2j + 4(2j)/4 + 2j = 0.8 - 0.4j$$

$$I_C = V/Z = 100/(0.8 - 0.4j) = 50(2 + j)$$

$$i_C = 50\sqrt{2^2 + 1^2}\sin(\omega t + \varphi) \text{ where } \varphi = \tan^{-1}\left(\frac{1}{2}\right).$$

This flows through the capacitor. The current in the resistor is

$$I_R = I\frac{2j}{4+2j} = 50j; i_R = 50\cos\omega t$$

$$I_L = I_C - I_R = 100;\ i_L = 100 \sin \omega t$$

The currents due to the D.C. voltage may be found from Fig. 3.2 (*c*). The inductance offers no voltage drop to D.C. and hence the resistance of 4 ohms sends a current 25 A as shown. No D.C. current can pass through the capacitor.

Hence $i_C = 0,\ i_R = 25,\ i_L = -25$

the minus in i_L because the direction of I_L is different from Fig. 3.2 (*b*).

Hence by superposition, the currents are

$i_C = 50\sqrt{5}\sin(\omega t + \varphi),\ i_R = 50 \cos \omega t + 25,\ i_L = 100 \sin \omega t - 25.$

Example 3. *In the above example, if the A.C. voltage is made only 50 volts (max.) and the D.C. voltage is made 200 V, find the new currents.*

Solution. The A.C. source currents becomes halved while the D.C. source currents are doubled. Hence

$$i_C = 25\sqrt{5}\sin(\omega t + \varphi)$$

$$i_R = 25 \cos \omega t + 50$$

$$i_L = 50 \sin \omega t - 50.$$

3.2 THE RECIPROCITY PRINCIPLE

Let an excitation applied in any branch of a network result in a response in another branch of the same network. Then if the same excitation is applied to the second branch, it should give the same response, now in the first branch, according to the reciprocity principle.

Here by 'excitation' we mean the application of a voltage or current source. This principle is true only for a *passive* network, a network containing no sources. Let the two branches under consideration be separated out from the rest of the network and let us apply voltage sources V_1 and V_2 in these two branches as shown. We write the loop equations for the network, of which I_1, I_2 are two of the loop currents. The other loop currents are inside the network, and to make it simple, we take it that I_3 is the only loop current inside. More loop currents inside is not going to affect the argument. Writing the loop equations,

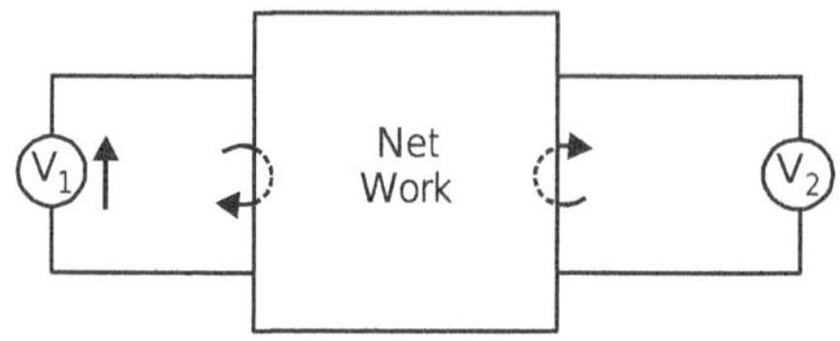

Fig. 3.3

$$Z_1I_1 + Z_{12}I_2 + Z_{13}I_3 = V_1$$

$$Z_{21}I_1 + Z_2I_2 + Z_{23}I_3 = V_2$$

$$Z_{31}I_1 + Z_{32}I_2 + Z_3I_3 = 0 \qquad \text{(as there are no sources inside)}$$

we know that Z_1, Z_2, Z_3 represent the total series impedances in the loop paths 1, 2, 3; and Z_{12}, Z_{13} represent the negative of the impedances common between loops 1-2, 2-3 etc. So, $Z_{12} = Z_{21}$, $Z_{13} = Z_{31}$ etc. and so the matrix Z is symmetrical. The solution for I_1 and I_2 give

$$I_1 = \frac{\Delta_{11}}{\Delta}V_1 + \frac{\Delta_{21}}{\Delta}V_2$$

$$I_2 = \frac{\Delta_{12}}{\Delta}V_1 + \frac{\Delta_{22}}{\Delta}V_2 \qquad (3.1)$$

where Δ is the determinant

$$\begin{vmatrix} Z_1 & Z_{12} & Z_{13} \\ Z_{21} & Z_2 & Z_{23} \\ Z_{31} & Z_{32} & Z_3 \end{vmatrix}$$

and Δ_{11} is the co-factor of Z_1, namely $\begin{vmatrix} Z_2 & Z_{23} \\ Z_{32} & Z_3 \end{vmatrix}$

and Δ_{12} is the co-factor of Z_{12}; Δ_{21} is the co-factor of Z_{21}.

$$\Delta_{12} = -\begin{vmatrix} Z_{21} & Z_{23} \\ Z_{31} & Z_3 \end{vmatrix} \qquad \Delta_{21} = -\begin{vmatrix} Z_{12} & Z_{13} \\ Z_{32} & Z_3 \end{vmatrix}$$

Since $Z_{21} = Z_{12}$, these two determinants are equal. So, $\Delta_{12} = \Delta_{21}$.

Let us apply a voltage $V = V_2$ only to the port 2 and note the current I_1. It is got from (3.1), putting $V_1 = 0$.

$$I_1 = (\Delta_{21}/\Delta)\ V_2 = (\Delta_{21}/\Delta)\ V$$

Now if we apply the same voltage V to port 1 (*i.e.*, $V = V_1$) and find the current I_2 that flows *via.* port 1, we get from (3.1), putting $V_2 = 0$ in the second equation,

$$I_2 = (\Delta_{12}/\Delta)\ V_1 = (\Delta_{12}/\Delta)\ V$$

$$= (\Delta_{21}/\Delta)\ V \qquad (\text{Since } \Delta_{21} = \Delta_{12})$$

$$= I_1$$

thereby proving the principle of reciprocity.

Example 4. *Find the open circuit voltage and the short circuit current at the terminals of the network shown below.*

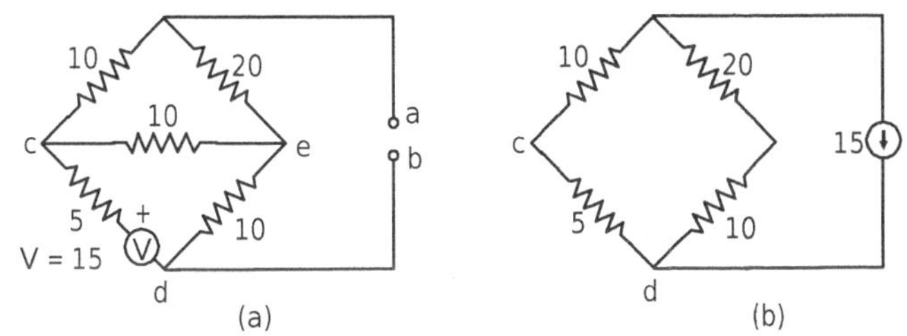

Fig. 3.4

Solution. To obtain the short circuit current between *a* and *b* upon joining *ab*, we can use the reciprocity principle. The *Voltage* of 15 V between *c* and *d* will send as much *Current* through *ab* (when joined) as the voltage applied between *ab* would send through *cd* (when joined). So the current can be found from Fig. 3.4 (*b*). Since 20/10 = 10/5, the bridge is balanced and no current flows through the 10 Ω between *c* and *e*. Hence the current

$$I = V/(10 + 5) = 15/15 = 1 \text{ A}.$$

This is also the short circuit current between *ab* of Fig. 3.4 (*a*).

To find the open circuit voltage we have to apply the dual of the previous principle : a *current* source applied between *c* and *d* will produce as much *voltage* across *ab* as the same current between *ab* would produce a voltage across *cd*. So we first transform branch *cd* using source equivalence principle as in Fig. 3.5 (*a*) below and then apply the above principle as in Fig. 3.5 (*b*). The voltage across *cd* is calculated to be $\frac{3(10+20)}{30+15} \times 5 = 10$ volts, and this is the o.c. voltage in Fig. 3.4 (*a*) across *ab*.

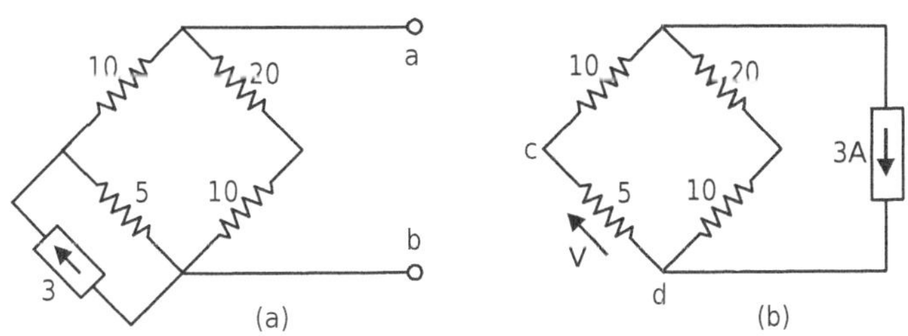

Fig. 3.5

Note however that when reciprocity principle is applied, the currents in the other branches of the network will change. For example, in the above Fig. 3.5, the currents in the 10, 20 Ω resistors will differ from Figs. 3.5 (*a*) to (*b*).

3.3 SUBSTITUTION PRINCIPLE

A branch of a network carrying a current I can be replaced by a voltage source of value IZ where Z is its branch impedance, without affecting the other parts of the network. The understanding of this is simple, because after all, Kirchhoff's voltage law is to be obeyed and replacing by a voltage source IZ does not affect in any way.

As a corollary, we have the *compensation theorem*, which states : when the impedance Z of branch carrying current I is increased to Z + *d*Z, the changes that take place can be found by replacing *d*Z by a series voltage source of I*d*Z opposing the current flowing in the branch.

Example 5*. *In a parallel tuned circuit supplied by a voltage V as shown, the tuning of the condenser is out so as to increase its capacitive reactance by 10%. Find the current then supplied by the voltage source.*

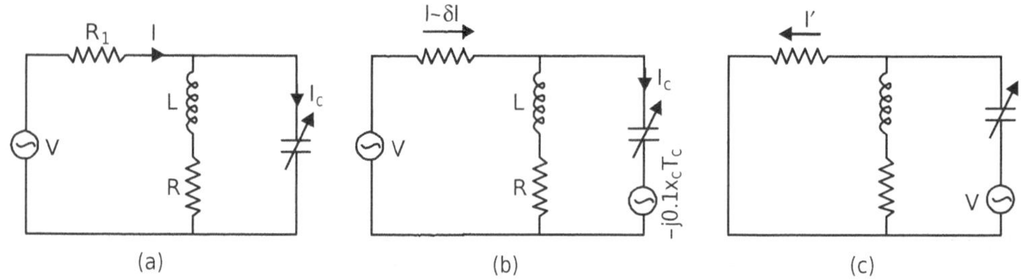

Fig. 3.6

Solution. The change in capacitive reactance of $-j0.1\ X_C$ alters the current I which can be calculated by putting an emf $-\ 0.1jX_CI_C$ in series with C.

The current I *via.* the source in Fig. 3.6 (*a*) is $\dfrac{V}{R_1 + (L/RC)}$ since L/CR is the resistance of the tuned parallel circuit of L, R and C.

The current I_C *via.* the capacitor C is Q times the current I and leads it by 90° at resonance. Hence I_C in Fig. 3.6 (*a*) is $jIQ = jI(\omega L/R)$.

When a voltage V is introduced in, series with C and the input port is a short circuited Fig. 3.6 (*c*), the current I′ applying the reciprocity principle, must also be jIQ. This is for a voltage V but for a voltage $-\ j0.1\ X_CI_C$ of Fig. 3.6 (*b*), the current will be

$$\delta I = -\ j0.1\,X_C I_C \frac{jQ}{R_1 + (L/CR)}$$

so that the actual current supplied when the capacitor is mistuned is by superposition of Fig. 3.6 (*a*) and (*c*), which is Fig. 3.6 (*b*),

$$I - \delta I = \frac{V}{R_1 + (L/CR)} - \frac{0.1jX_C VQ^2}{(L/CR + R_1)^2}$$

using $$I_C = jIQ \text{ and } I = \frac{V}{[R + (L/CR)]}.$$

*Read this problem after reading Chapter 6

3.4 STAR-DELTA TRANSFORMATION OR DELTA-WYE TRANSFORMATION

Between any four nodes (1, 2, 3, 4) in a network, if we have 3 impedance Z_A, Z_B, Z_C connected in the form of a star of Fig. 3.7 (*a*), the middle node can be replaced by three other Z_{AB}, Z_{BC}, Z_{CA} as in Fig. 3.7 (*b*), without altering the currents in the network external to these nodes. Thus it transforms a star type of network into a delta (Δ).

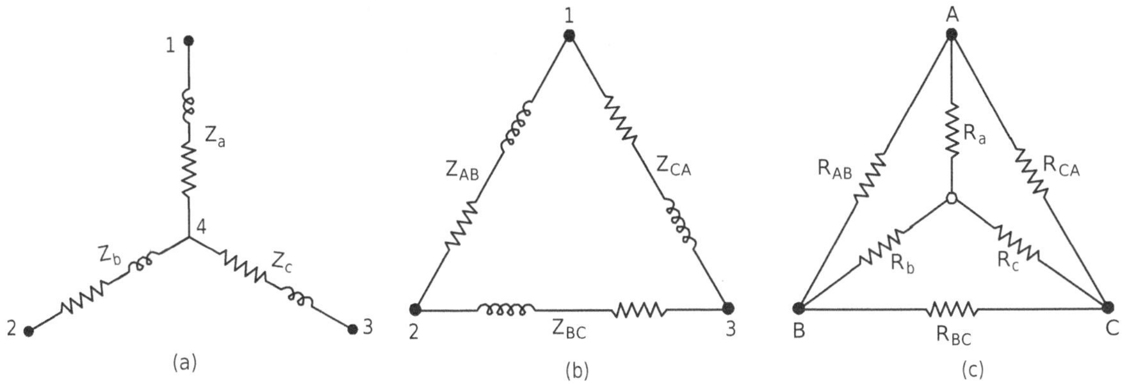

Fig. 3.7 (*a*) Star (*b*) Delta (*c*) The transformation.

The extreme advantage in its use is because of the elimination of one node. But the current in the branches Z_A, Z_B, Z_C themselves will not be those in Z_{AB}, Z_{AC}, Z_{CA} and these will have to be found out by other methods.

The converse is also true in this case, which transforms a Δ into a star. The relationships between the delta and star impedances can be derived easily. Let us assume only resistances R_A, R_B, R_C instead of impedances, but the method is the same for Z also if complex number Z is used in place of R.

The input resistance between any pair of nodes 1-2, 2-3, 3-1 must be the same in both. So, we equate these between each node pair. The resistance 1 and 2 in star is $R_a + R_b$, and in delta, it is R_{ab} in parallel with $R_{bc} + R_{ac}$ Fig. 3.7 (*c*).

$$R_a + R_b = \frac{R_{ab}(R_{ca} + R_{bc})}{R_{ab} + R_{bc} + R_{ca}} = \frac{R_{ab}R_{ca}}{\Sigma R} + \frac{R_{ab}R_{bc}}{\Sigma R}$$

Similarly,

$$R_b + R_c = \frac{R_{ab}R_{bc}}{\Sigma R} + \frac{R_{bc}R_{ca}}{\Sigma R}$$

$$R_c + R_a = \frac{R_{bc}R_{ca}}{\Sigma R} + \frac{R_{ca}R_{ab}}{\Sigma R_{ab}}$$

where $\Sigma R = R_{ab} + R_{bc} + R_{ca}$. By inspection of the above equation,

$$R_a = \frac{R_{ab}R_{ca}}{\Sigma R_{ab}}; R_b = \frac{R_{bc}R_{ab}}{\Sigma R_{ab}}; R_c = \frac{R_{ca}R_{bc}}{\Sigma R_{ab}} \quad (3.2)$$

These relations may be remembered easily by looking at Fig. 3.7 (*c*) and noting that the star resistance R_a is equal to the product of the adjoining delta resistors divided by the sum of all the three delta resistors. Relation (3.2) holds good for impedances as well.

To find the delta resistors R_{ab}, R_{bc}, R_{ca} from the star resistors, (the converse), we use (3.2). Dividing the first relation of (3.2) by the second,

$$\frac{R_a}{R_b} = \frac{R_{ca}}{R_{bc}}$$

$\therefore$ $$R_{ca} = R_{bc}(R_a/R_b)$$

Similarly, $$R_{ab} = R_{bc}(R_a/R_c)$$

Substituting, we get $$R_a = \frac{R_{ab}R_{ca}}{R_{ab}+R_{bc}+R_{ca}} \quad \text{(from 3.2)}$$

$$= \frac{(R_{bc}\ R_a/R_b)(R_{bc}\ R_a/R_c)}{R_{bc}\left[1+R_a/R_b+R_a/R_c\right]}$$

$$= \frac{R_a^2R_{bc}}{R_bR_a+R_aR_c+R_bR_c}$$

from which $$R_{bc} = \frac{R_bR_c+R_cR_a+R_aR_b}{R_a} = R_b+R_c+\frac{R_bR_c}{R_a}$$

and likewise $$R_{ca} = \frac{R_bR_c+R_cR_a+R_aR_b}{R_b} = R_c+R_a+\frac{R_cR_a}{R_b} \text{ etc.}$$

The delta resistor between two nodes is the sum of the star resistances connected to the nodes plus their product divided by the resistor to the third node.

This star-delta transformation has no connection to the motor starters using Star-Delta winding re-arrangement.

Example 6. *Solve the bridge network of Fig. 3.8 (a) using delta star transformation.*

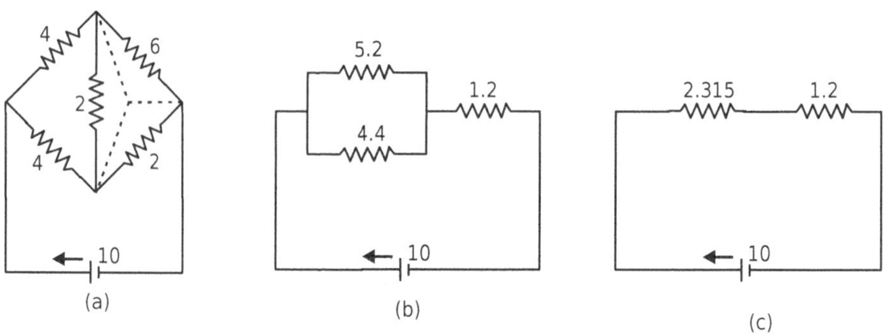

Fig. 3.8

Solution. The delta of 2, 6 and 2 ohms can be replaced by a star, each arm of which can be found by equation (3.2) as

$$\frac{6\times2}{6+2+2}, \frac{2\times2}{6+2+2}, \frac{2\times6}{6+2+2}$$

giving 1.2, 0.4, 1.2 ohms. The circuit now becomes a combination of 2 resistors in parallel together with a 1.2 ohm series resistor giving

$$1.2+\frac{5.2\times4.4}{5.2+4.4} = 1.2 + 2.315 = 3.585$$

and hence the battery current is 10/3.585 = 2.787 amperes.

Example 7. *Find the capacitance of the connected capacitors between the terminals in Fig. 3.9 (c) below. The values are in μF.*

Solution. Suppose an a.c. voltage of frequency ω is applied; then we can compute the reactances of each of them by $X_c = 1/\omega C$, and hence the reactances are proportional to 1/C.

Using Y-Δ transformation for the star at node *b*, (noting that the capacitive reactance X_c is to be used in the formula and not C directly),

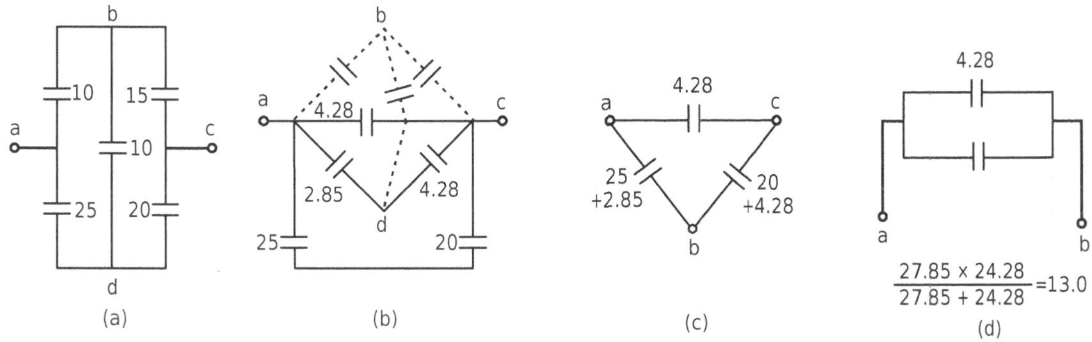

Fig. 3.9

$$\omega X_{ac} = (1/10) + (1/15) + [(1/10)(1/15) \div (1/10)]$$
$$= 0.1 + .0667 + .0667 = 0.233$$
$$\omega X_{ad} = (1/10) + (1/10) + [(1/10)(1/10) \div (1/15)]$$
$$= 0.1 + 0.1 + 0.15 = 0.35$$
$$\omega X_{cd} = (1/10) + (1/15) + [(1/10)(1/15) \div (1/10)] = 0.233$$

The respective capacitances are thus

$$(1/.233),\ (1/.35),\ (1.233) \text{ or } 4.28,\ 2.85,\ 4.28\ \mu\text{F}. \tag{3.3}$$

The parallel branches across *ad* and *cd* may be reduced by the formula $C = C_1 + C_2$ for two capacitors in parallel as in Fig. 3.9 (*c*).

Ultimately, we get Fig. 3.9 (*d*) and the capacitance between *a* and *c* is 13.0 + 4.28 = 17.28 μF.

3.5 ROSEN'S THEOREM

Rosen's theorem is a generalisation of the star delta transformation. A star containing a set of branches, can be replaced by a mesh of different branches, thereby excluding the star node. The transformed mesh will have one branch between any two nodes. Let Y_1, Y_2, ... be the branch admittances. Let *n* be the star node. Let us consider all node voltages above an external O node in the circuit. I_1, I_2, I_3, ... are currents entering the branches from the external part of the network. The currents I_1, I_2, ... may be expressed in terms of the node voltages V_1, V_2, ... and the admittance Y_1, Y_2, etc. V_n is the voltage of the star node.

$$(V_1 - V_n)Y_1 = I_1$$
$$(V_2 - V_n)Y_2 = I_2$$
$$\cdots \quad \cdots \cdots$$

$$(V_1 - V_n)Y_1 + (V_2 - V_n)Y_2 + \ldots. = I_1 + I_2 + \ldots.$$

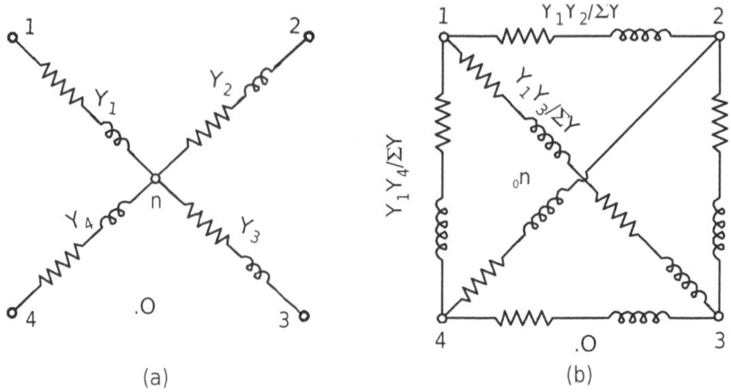

Fig. 3.10

But since the sum of the currents entering a node is zero,

we get $$(V_1 - V_n)Y_1 + (V_2 - V_n)Y_2 + ... = 0$$

from which $$V_n = \frac{V_1Y_1 + V_2Y_2 + ...}{Y_1 + Y_2 + ...} \tag{3.4}$$

So, $$I_1 = (V_1 - V_n)Y_1$$

$$= V_1Y_1 - \left(\frac{V_1Y_1 + V_1Y_2 + ...}{Y_1 + Y_2 + ...}\right)Y_1$$

$$= \left(\frac{(V_1 - V_2)Y_2 + (V_1 - V_3)Y_3 + ...}{\Sigma Y}\right)Y_1$$

$\therefore$ $$I_1 = \left[(V_1 - V_2)Y_1Y_2 + (V_1 - V_3)Y_1Y_3 + ... + ...\right] \div \Sigma Y$$

This is the current which would flow into terminal I if it were connected by admittances $Y_1Y_2/\Sigma Y$, $Y_1Y_3/\Sigma Y$ ... etc. to the terminals 2, 3, ... respectively.

Similarly, the current entering terminal 2 is the same if it were connected by admittances

$$\frac{Y_1Y_2}{\Sigma Y}, \frac{Y_2Y_3}{\Sigma Y}, ...$$

to the nodes 1, 3, ... etc.

Hence the star network may be replaced by the mesh of Fig. 3.10 (*b*) where

$$Y_{12} = \frac{Y_1Y_2}{\Sigma Y}, Y_{13} = \frac{Y_1Y_3}{\Sigma Y}, ..., Y_{pq} = \frac{Y_pY_q}{\Sigma Y} \text{ etc.} \tag{3.5}$$

Note that this theorem transforms a star into a mesh. There is no converse to this general theorem *i.e.,* from mesh to star, except for the case of 3 nodes which have the delta star transformation also. For the above, for the case of the 3 branch star,

$$Y_{12} = Y_1Y_2/(Y_1 + Y_2 + Y_3)$$

$$Z_{12} = \frac{1}{Y_1} + \frac{1}{Y_2} + \frac{Y_3}{Y_1Y_2} = Z_1 + Z_2 + \frac{Z_1Z_2}{Z_3}$$ as before if *a*, *b*, *c* are used in place of 1, 2, 3.

Example 8. *Using Rosen's theorem, determine the stray capacitances of the bridge network below in terms of the bridge arm reactances.*

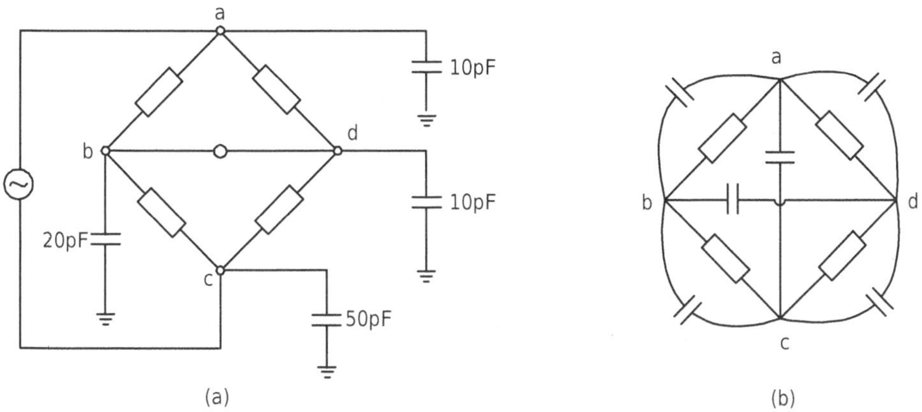

Fig. 3.11

Solution. The stray capacitances at the four nodes to earth may be taken as a star of four capacitances with earth as the star node. In Fig. 3.11 (*a*), the effect of the capacitances on the bridge balance equations cannot be determined, but when the same is as in Fig. 3.11 (*b*), the reactances of the capacitances may be included in the bridge arm values.

$$Y_{ab} = \frac{Y_a Y_b}{\Sigma Y} = \frac{\omega C_a \omega C_b}{\omega(C_a + C_b + C_c + C_d)}$$

$$Y_{ab} = \omega C_{ab}$$

$$\therefore \quad C_{ab} = \frac{C_a C_b}{C_a + C_b + C_c + C_d} = \frac{10 \times 20}{10 + 20 + 50 + 10} = 2.225 \text{ pF}$$

$$C_{ac} = C_a C_c / \Sigma C = 10 \times 50/90 = 5.56 \text{ pF}$$

$$C_{ad} = 10 \times 10/90 = 1.115 \text{ pF}$$

$$C_{bc} = 20 \times 50/90 = 11.15 \text{ pF}$$

$$C_{bd} = 20 \times 10/90 = 2.23 \text{ pF}$$

$$C_{cd} = 50 \times 10/90 = 5.56 \text{ pF}$$

3.6 MILLMAN'S THEOREM

This is also called parallel generator theorem. We have proved this theorem in the course of Rosen's theorem itself (Art. 3.5). This theorem states that a set of parallel branches will have voltage across them equal to

$$V = \frac{E_1 Y_1 + E_2 Y_2 + \ldots + E_n Y_n}{Y_1 + Y_2 + \ldots. + Y_n}$$

where Y_1, Y_2, ..., Y_n are the admittances of the *n* branches (Fig. 3.12 (*a*)) and E_1, E_2, ... are the voltage sources in these *n* branches (some of which may be zero also).

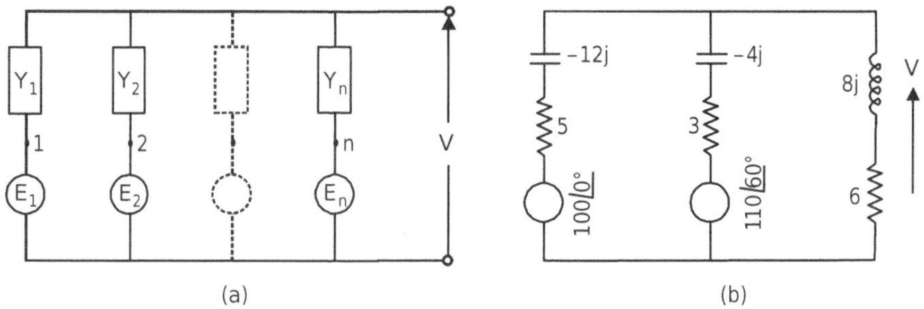

Fig. 3.12

Example 9. *Two a.c. generators of voltage 100 sin ωt and 110 sin (ωt + 60°) are in parallel and supply a load of 6 + 8j ohms. The internal admittances of the generators are 5 – 12j and 3 – 4j. Calculate the voltage of the load.*

Solution. Millman's theorem is easily applied to this problem Fig. 3.11 (*b*). The load voltage is

$$V = \frac{(100\angle 0°)(5-12j)+(110\angle 60°)(3-4j)+(0)(Y_L)}{(5-12j)+(3-4j)+1/[6+8j]}$$

Note that $0 \times Y_L$ is with respect to the load branch, since there is no voltage in that branch.

$$V = \frac{100(5-12j)+110(\cos 60 + j\sin 60)(3-4j)}{5-12j+3-4j+0.06-0.08j}$$

$$= 90 + 19j \text{ on simplification.}$$

so that the magnitude of load voltage = $(90^2 + 19^2)^{1/2}$ = 92.2 Volts.

3.7 THEVENIN'S THEOREM AND NORTON'S THEOREM—EQUIVALENT CIRCUITS

The theorems are of special importance. They are very useful in the analysis of rather complex networks, if currents in certain branches of the network are only required to be found. The Norton's theorem is a dual form of Thevenin's theorem.

Take any terminal pair of a network, marked *a* and *b* in Fig. 3.13 (*a*). The theorem replaces the network by a single source of voltage E and of internal impedance Z, as shown in Fig. 3.13 (*b*). The value of E is the open-circuit voltage across *ab* (*i.e.*, as it is in Fig. 3.13 (*a*)) without any branch between *a* and *b*. The impedance Z is the input impedance between terminals *a and b*, after short circuiting all independent voltage sources inside the network.

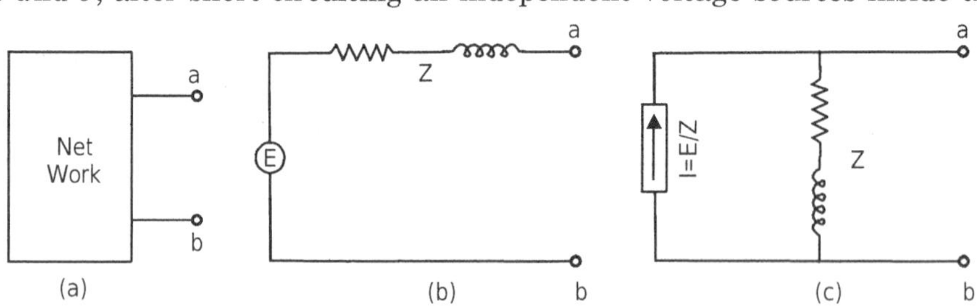

Fig. 3.13 (*a*) Network (*b*) Thevenin equivalent (*c*) Norton's equivalent.

Norton's theorem is a dual of Thevenin's theorem. According to it, the network can be replaced by a current source in parallel with an impedance Z. The current source is the short circuit current which will flow into *ab* when short-circuited. Z is the same as in Thevenin's. The Norton equivalent of the network is shown in Fig. 3.13 (*c*). We see that the Fig. 3.13 (*c*) is the dual of (*b*) and so the current source is E/Z, by source equivalence principle applied to Fig. 3.13 (*b*).

Before we proceed to prove the theorem, let us take an illustrative example.

Example 10. *Find the current which will flow in the detector of the following bridge circuit.*

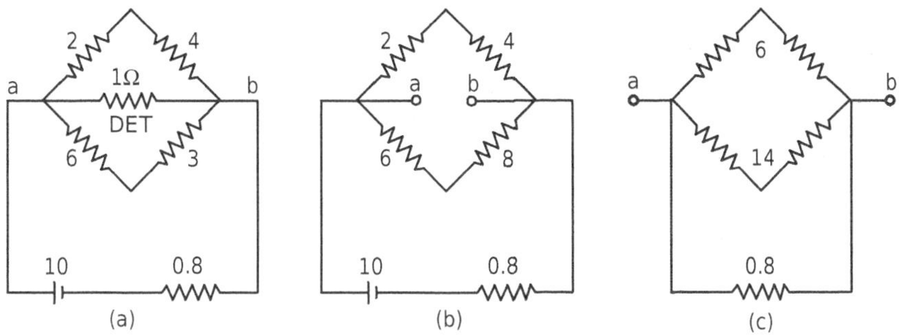

Fig. 3.14

Solution. Mark the detector terminals as *a*, *b*. To find the current through *ab*, we first replace the circuit other than the branch *ab* by its Thevenin equivalent. For that, we remove branch *ab* and find the open circuit voltage across *ab* by Fig. 3.14 (*b*). The two branches of (2 + 4) and (6 + 8) ohms in parallel give 6 × 14/20 or 4.2 ohms and in series with the 0.8 Ω, makes 5 ohms, so that the battery current in Fig. 3.14 (*b*) is 10/5 or 2 amperes. Hence voltage between *ab* in Fig. 3.14 (*b*) is

$$E_{ab} = 10 - 2\,(0.8) = 8.4 \text{ volts}$$

which is open circuit voltage across *ab*. The Thevenin's impedance is found between terminals *ab* after short circuiting the battery, which is an independent voltage source. So from Fig. 3.13 (*c*), the value of Thevenin's circuit resistance is,

$$\frac{1}{R} = \frac{1}{6}+\frac{1}{14}+\frac{1}{0.8} = 0.166 + 0.071 + 1.25 = 1.487$$

giving $\quad R = 1/1.487 = 0.672$ ohms.

Hence Thevenin's equivalent to the network other than the detector branch is a voltage of 8.4 volts in series with 0.672 ohms. So, the current through detector may now be found easily as,

$$I_{ab} = \frac{E}{R+R_E} = \frac{8.6}{0.672+1} = 5.15 \text{ Amps.}$$

where R is the Thevenin's resistance, R_E, the external resistance.

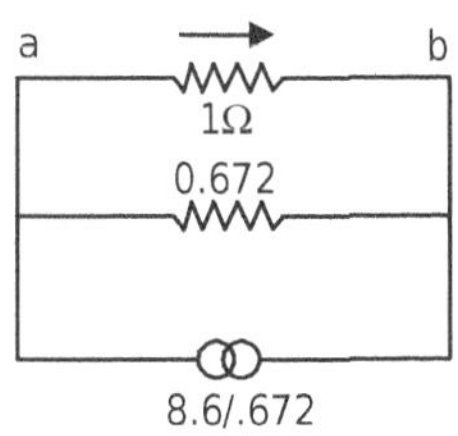

Fig. 3.15

The same circuit has the Norton's equivalent as in Fig. 3.15, because it is simply the dual form of Thevenin's. Using this too, we may find the current through *ab*. The current source sends a current of I = 8.6/0.67 and this divides into the two branches of 0.672 and 1 Ω, respectively. Hence current in *ab* is,

$$I_{ab} = \frac{0.672\,I}{0.672+1.0} = \frac{0.672}{1.672}\times\left(\frac{8.60}{0.672}\right)$$

$$= 5.15 \text{ A as before.}$$

Proof of Thevenin's theorem. Thevenin's theorem can be proved using the Superposition principle. To prove it, we have to show that Fig. 3.13 (*a*) and (*b*) are equivalent. For this, it is enough if we show that equal currents flowing into the network result in equal voltages at the terminals in both Figs. 3.16 (*a*) and (*b*).

Connect any current source of I amperes at the terminals *a* and *b* as shown in the figure below. The voltage V at these terminals can be found by superposition principle as the sum of :

v = voltage at *ab* due to internal source with the current source removed + voltage at *ab* due to the current source only with internal voltages removed.

= Open circuit volts at *ab* + IZ = E + IZ since IZ is the voltage due to the current flowing through an input impedance Z between *a* and *b*.

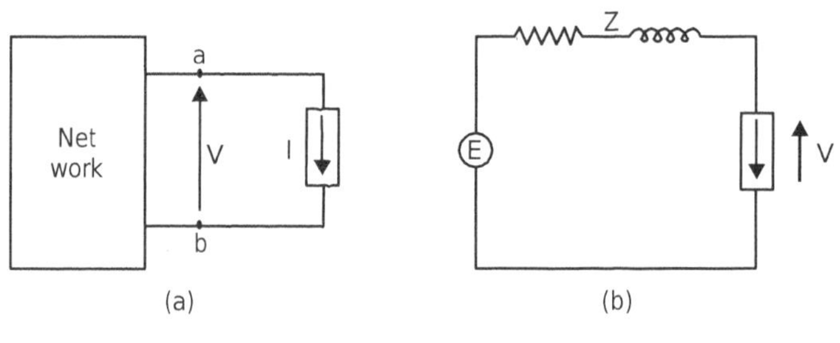

Fig. 3.16

In Fig. 3.16 (*b*), the same current I is applied and then the voltage V will be, by Kirchhoff's voltage law,

$$v = E + IZ$$

and hence Fig. 3.16 (*b*) is the equivalent of Fig. 3.16 (*a*). Thus, the equivalence is for any arbitrary I and the theorem is proved.

Norton's theorem follows by the source equivalence principle from Fig. 3.13 (*b*), and Norton's network is shown in Fig. 3.13 (*c*). The same can be shown by duality as well, the networks of Fig. 3.13 (*b*) and (*c*) being duals. Alternatively it may be proved by connecting an arbitrary voltage source V to *ab* of Fig. 3.13 (*a*) and using the superposition principle as in the above argument.

Example 11. *Determine the current in the resistance of 4 ohms connected between A and B using Norton's theorem.*

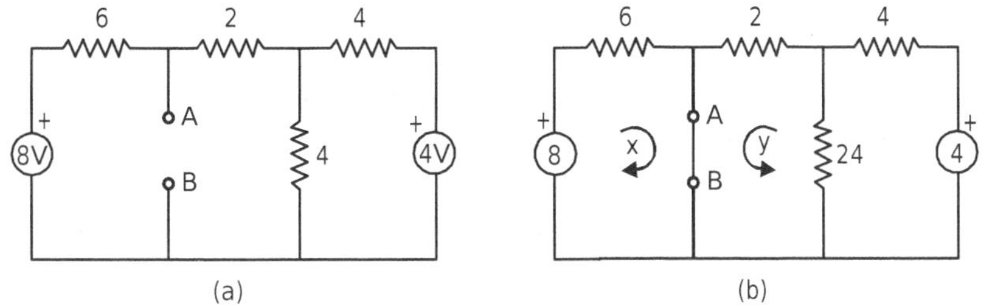

Fig. 3.17

Solution. We have to find the short circuit current between A and B to use Norton's theorem. If x and y are the two currents flowing as shown,

$$x = \frac{8}{6} = 1.33 \text{ and } y = \frac{4 \text{ volts}}{4 + (48/26)} \times \frac{24}{26} = \frac{12}{19}$$

So that $x + y = I_{AB} = 1.33 + \left(\frac{12}{19}\right) = 1.960$ amp.

'The input impedance' with all independent sources short circuited is found from Fig. 3.18 (*a*) as :

The 4 ohm and 24 ohms in parallel give $4 \times \frac{24}{28} = 3.43$ and hence $R_{AB} = \frac{5.43 \times 6}{11.43} = 2.86.$

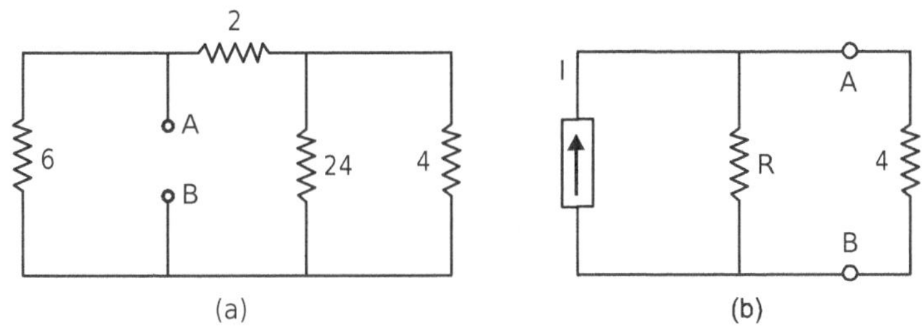

Fig. 3.18

So, by Norton's theorem, the current is found in the 4 ohm resistor between A and B as

$$I \times \frac{2.86}{2.86 + 4} = \frac{1.96 \times 2.86}{6.86} = 0.82 \text{ A.}$$

Open and short circuit method for finding internal Z of a network. By Thevenin's theorem, every network will be replaced by an internal source E in series with an internal impedance Z. For a practical network, the values of E and Z may be found by two experiments.

(1) Keep the terminals open. Then, by Fig. 3.19 (*a*), the *Voltage between the open terminals is E*, which can be measured with a high resistance voltmeter connected to the terminals.

(2) Short circuit the terminals, and connect an ammeter in series with the short circuit to measure the short circuit current I_S Fig. 3.19 (*b*). The ammeter does not affect the conditions of a short, because its resistance is negligible.

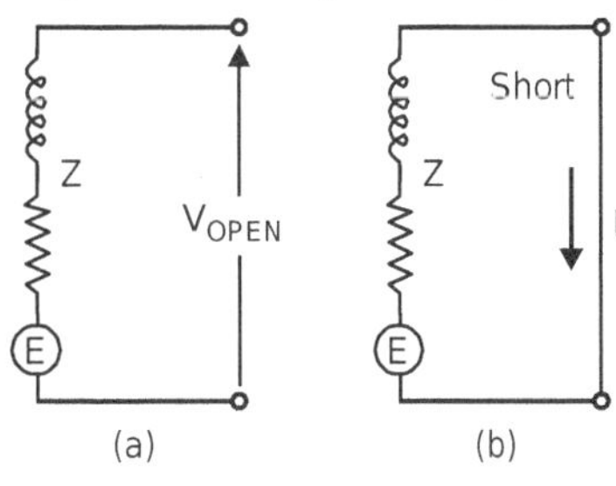

Fig. 3.19

From the Fig., $I_S = E/Z$

$\therefore$ $Z = E/I_S$ = Open circuit volts/Short circuit current

$= V_{open}/I_{short}$ (3.6)

These two tests are the ones that are generally employed to determine the network equivalent of electric machines, such as the transformer, alternator, induction motor and so on.

Example 12. *An alternator field is adjusted to give an open circuit voltage of 400 V. When the terminals of the alternator are shorted by an ammeter with the magnetic field remaining the same, the ammeter reads 100 A. Find the synchronous impedance of the alternator. Find its terminal voltage when delivering 40 amperes to a purely resistive load.*

Solution. The internal 'synchronous' impedance

$$= \frac{V_{open}}{I_S} = \frac{400}{100} = 4$$

and since this is purely reactive, $Z = 4j$.

$$I = \frac{E}{Z + R_L} = \frac{400}{4j + R_L} = \frac{400}{\sqrt{4^2 + R_L^2}} = 40 \text{ A} \qquad \text{(given)}$$

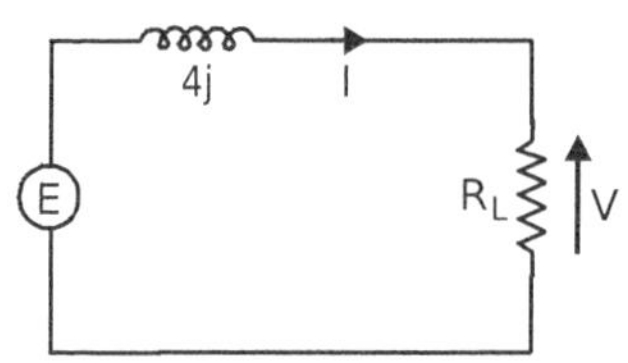

Fig. 3.20

and so $R_L = \sqrt{84} = 9.17$

and the terminal volts = IR_L = 40 × 9.17 = 367 volts.

Treatment of dependent sources. It has been already stated that when finding the Thevenin's (or for that matter Norton's) immittance* all internal sources, *excepting the dependent* sources, will be reduced to zero. (Every voltage source is short circuited and every current source opened). The dependent sources will be left as such. In such cases, to find Thevenin's Z_T, we may follow one of the following two methods.

1. Apply an arbitrary voltage V to the terminals *a*, *b*; reduce to zero any independent internal source. Determine what current *i* flows into the network. The ratio *v*/*i* gives Z. The method is called *applied source method.*
2. Open and short circuit method: Find the open circuit voltage at the terminals and calculate I_s. We have already seen that $Z = V_o/I_s$. In this method, no internal source, dependent or not, need be reduced to zero.

---

* Immittance is a common name given to impedance or admittance. To what it refers, will be clear from the context.

Example 13. *Calculate the current flowing through R_L of the network of Fig. 3.21 (a), if R = I, R' = 2, k = 2, E = 10 V, R_L = 5/7 ohm*

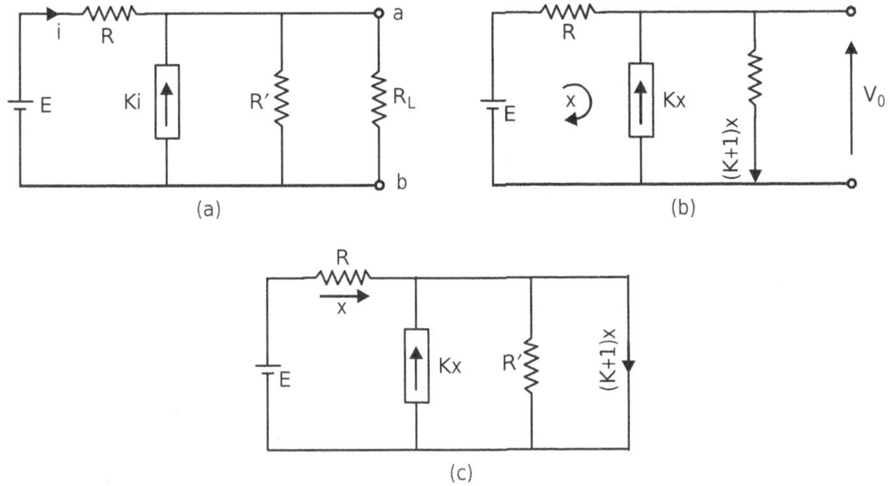

Fig. 3.21

Solution. We first find the open circuit voltage and Z_T of Thevenin at terminals *ab*.

The open circuit voltage is found in Fig. 3.21 (*b*).

(*a*) $$V_{open} = (k + 1)\, x\, R'$$

$$E - Rx = (k + 1)\, x\, R'$$

giving $$x = E/[R + (k+1)R']; \; V_0 = (k + 1)\, ER'/[R + (k + 1)R'] \qquad (a)$$

(*b*) We use the open and short circuit method to find Z_T.

In Fig. 3.21 (*c*), I_s is $(k + 1)\, x$.

Also $$E = Rx, \text{ or } x = E/R.$$

$$I_s = (k + 1)\, E/R$$

∴ $$Z_T = V_o/I_s = RR'/[R + (k+1)R']$$

Hence, in Fig. 3.21 (*a*), current through $R_L = V_o \div (R_T + R_L)$

Substituting given values,

$$V_0 = 3 \times (10/7) \times 2 = 60/7$$

$$R_T = 2 \times 1/[1 + 2 \times 3] = 2/7$$

So, $$I_L = 60/7 \div [(2/7) + (5/7)] = 12 \text{ amps.}$$

Power calculation using Thevenin's theorem. We have seen that the theorem replaces a network (with sources) by a voltage E and a series impedance Z, as far as any external load (circuit) connected to the terminals of the network is concerned. The power dissipated in the load will be $I^2 R_L$ (Fig. 3.20 for instance) and since I is the actual current, the power calculation will also be correct. that means, the theorem can be used to find power in the external circuit.

But what about the power internally dissipated? The Thevenin equivalent shows that when the terminals are open as in Fig. 3.19 (*a*), the internal power is zero because no current flows then. But, in the actual network, for instance in Fig. 3.21 (*b*), even when the terminals are open, currents flow in R, R′ and hence there is power loss. We may choose to call it 'no-load power loss'. The power calculations are true externally but not internally.

The power loss in the actual network and in its Thevenin equivalent are not equal, but the difference between the two losses is constant, (does not vary with changes in the external load). The constant difference is the no load power loss.

The above paragraph may be taken as a theorem by itself but its proof is very involved and will be omitted here.

Example 14. *Find the power supplied to the load of (1 – j) and the power then generated by the two sources.*

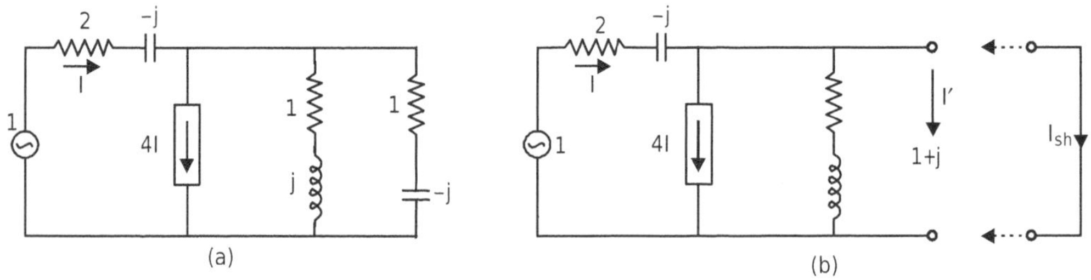

Fig. 3.22 (*a*) Network (*b*) Open and shorting methods.

Solution. The open circuit voltage is found, after removing $(1 - j)$ as

$$V_o = I' (1 + j) = (I - 4I)(1 + j)$$

and

$$1 \text{ volt} = (2 - j)\, I + (1 + j)\, I'$$
$$= (2 - j)\, I + (1 + j)(-3\, I)$$
$$= (-1 - 4j)\, I \qquad (a)$$

and so

$$V_o = -3\,(1 + j)/(-1 - 4j)$$

Shorting the output terminals gives I_{sh} as

$$I_{sh} = I - 4\, I \text{ where I is now } = 1/(2 - j)$$

$$\therefore \quad I_s = -3/(2 - j)$$

Hence

$$Z_T = \frac{V_o}{I_s} = \frac{(1+j)(2-j)}{-1-4j} = -0.412 + 0.646j$$

The current through load of $1 - j$ in Fig. 3.22 (a) is

$$\frac{V_o}{Z_L + Z_T} = \frac{-3(1+j)}{(-1-4j)(1-j-0.412+0.646j)}$$
$$= 1.50 \angle 61°$$

and hence power in load $(1 - j)$ is $= I^2R = 1.50^2 \times 1 = 2.25$ Watts.

The power loss in Thevenin's impedance $= (-0.412) \times 1.5^2$

$$= -0.927 \text{ Watts}$$

The no-load power loss can be found from the currents in the open circuit diagram as

$$2I^2 + 1(I - 4I)^2 = 11\, I^2 = 11[1/(-1-4j)]^2 = 11/17$$
$$= 0.65 \text{ W}.$$

and hence the power generated by the sources when the load is $(1 - j)$ is

$$= 0.65 + 2.25 - 0.927 = 1.973 \text{ Watts}.$$

Incidentally, note that Z_T in this problem has a *negative resistance of* – 0.412. A positive resistance dissipates power when current flows through it, while a negative resistance generates when current flows in it. This is because of the dependent current source in this example, which generates 0.93 W. The 1 V e.m.f. source generates 1.32 Watts.

Example 15. *Find the Thevenin's equivalent of the ideal transformer circuit shown in Fig. 3.23 (a) below, at the load Z_L terminals.*

Solution. The open circuit voltage at the terminals of Z_L is to be found first. In the ideal transformer the inductances of the windings 1 and 2 are high and $e_1/e_2 = N_1/N_2 = I_2/I_1$. On open circuit Z_L is removed, I_2 is zero and hence I_1 also 0. So $e_1 =$ V and,

$$V_{open} = e_1 = V(N_2/N_1)$$

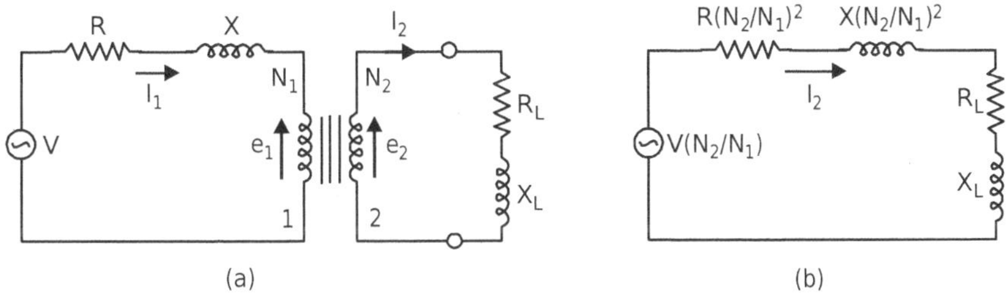

Fig. 3.23

To find the Thevenin's impedance at the Z_L terminals, Z_L is short circuited. Then

$$I_{sh} = I_2 = I_1 (N_1/N_2) \text{ and} \tag{a}$$

$e_2 = 0$ since Z_L is short circuited.

$$\therefore \quad e_1 = 0 \text{ and}$$

$$V = (R + jX) I_1$$

$$\therefore \quad I_{sh} = (N_1/N_2) V/(R + jX) \qquad \text{from } (a)$$

Hence $$Z_t = V_{open}/I_{sh} = (N_2/N_1)^2 (R + jX)$$

and the equivalent circuit becomes as in Fig. 3.23 (*b*).

3.8 MAXIMUM POWER TRANSFER THEOREM

In communication networks, it is frequently required to obtain maximum power from an e.m.f. source having an internal impedance Z = R + *j*X, into a load Z′. Fig. 3.24 shows that the load is variable, so that maximum power is absorbed in it.

The current $$I = \frac{E}{R + jX + R' + jX'}$$

The power in $$Z' = I^2R' = \frac{E^2R'}{(R + R')^2 + (X + X')^2}$$

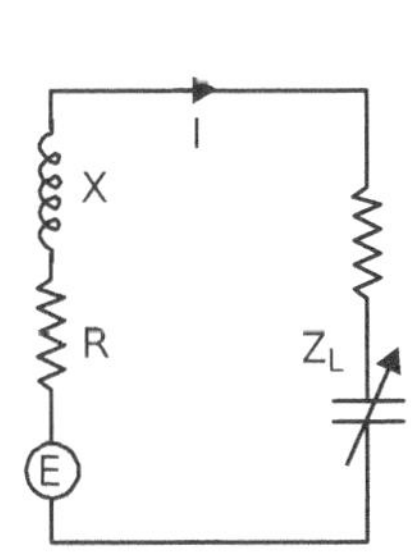

Fig. 3.24

where E is the R.M.S. value of the voltage. This power is a function of the variables R′ and X′ of the load. Since X′ appears only in the denominator, the power will be maximum when X = – X′. Substituting this value for X′, we get

$$P_m = \frac{E^2R'}{(R + R')^2} \tag{3.7}$$

If we further vary R′ also we get a maximum of what we have in equation (3.7) and it is got by differentiating with respect to R′, as

$$E^2\left[(R+R')^2 - 2(R+R')R'\right] = 0$$

$$R = R'$$

Hence maximum power by varying both R′ and X′ is got as

$$E^2R/(2R)^2 = \frac{1}{2}E^2/R \tag{3.8}$$

and this absolute maximum is only when R′ = R and X′ = – X, or

$$Z' = R' + jX' = R - jX = Z^*, \tag{3.9}$$

when the load impedance is the conjugate of the series impedance Z. This relation is called *maximum power transfer theorem*. The power loss in the internal resistance R is also $E^2/2R$.

Suppose a network were not as simple as Fig. 3.24. But it does not matter, because even a general network can be replaced as per Thevenin's theorem (as far as Z' is concerned) by an E and a series Z. However, the internal power will be more by the amount of no load loss in the network, as seen already.

Example 16. *In the bridge network of Fig. 3.25, A = 18 + j30, B = 36 + j60, C = 10 – j30, D = 10 – j30, e = 300 volts. Find the value of Z_L for maximum power in Z_L and its value. Find also the efficiency given by (power in load) ÷ power supplied by sources.*

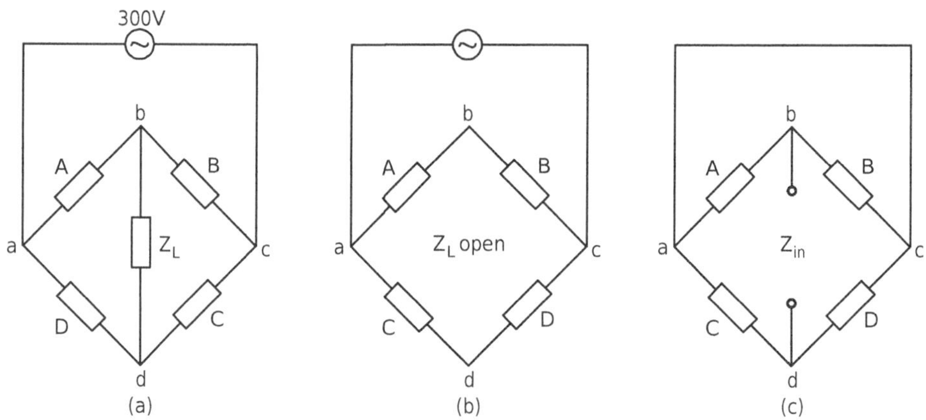

Fig. 3.25

Solution. On open circuiting Z_L, Fig. 3.25 (*b*) currents flow along *abc*, *adc*, their values being

$$\frac{300}{54+j90} \text{ and } \frac{300}{20-j60}$$

and the voltage V_{open} at *bd* is given by

$$V_{ab} + V_{bd} + V_{da} = 0,\ V_{bd} = -V_{ab} + V_{ad}$$

$$V_{bd} = \frac{-300(18+j30)}{54+j90} + \frac{330(10-j30)}{20-j60} = -200+150 = -50$$

and so there is a rise in voltage of 50 V from *b* to *d*. The input impedance Z_{in} (Fig. 3.25(*c*)) at *bd* is found after shorting the 300 V source as that of A, B in parallel in series with C, D in parallel.

$$Z = \frac{A \times B}{A+B} + \frac{C \times D}{C+D} = \frac{2}{3}(18 + j30) + \frac{1}{2}(10 - j30)$$

$$= 17 + j5$$

Hence the value of Z_L for max. power in it is $17 - j5$, by maximum power transfer theorem, and the maximum power is $V_{open}^2 / 2R = 50^2/2 \times 17 = 73.5$ Watts. The internal loss, in Thevenin's equivalent circuit, also is 73.5 watts. But the efficiency of power in the load to the source is not 73.5/(73.5 + 73.5) or 50% because the actual internal network power differs from that in Thevenin's equivalent by the no-load power. The latter is found using the open circuit currents along *abc*, *adc* as

$$\frac{300^2 \times 54}{54^2 + 90^2} + \frac{300^2 \times 20}{20^2 + 60^2} = 907 \text{ watts.}$$

and the required efficiency $= \dfrac{73.5}{907 + 2(73.5)} = 6.97\%$.

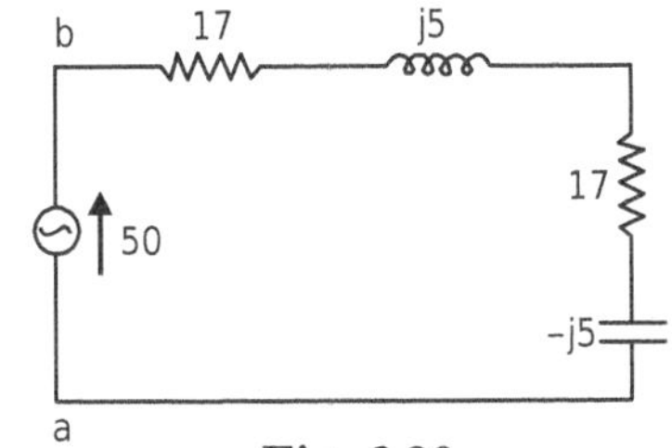

Fig. 3.26

3.9 TELLEGEN'S THEOREM

This is a basic principle. It is based on conservation of power. If we take one loop of a circuit, as we do for solving using loop equations, then some elements may dissipate power, but some other elements (the sources of voltage) may supply power. The total algebraic sum of the power taken by summing up for all elements in a loop is necessarily zero.

$$v_b^T . \bar{i}_b = \bar{o} \qquad (a)$$

v_b = voltage across branch

i_b = current in the branch.

T denotes transpose, treating v and i as vectors. The rule over zero means it is a vector!

That means, the inner product of the branch voltage and current vectors is zero.

A positive value for the product in a branch represents that the branch dissipates power—for example, a resistor. Thus some branches may be sourcing power so that others may dissipate or store power, so that (*a*) is valid.

The node equations have a matrix A and a voltage vector $\bar{V}_n$.

$$A^T \bar{v}_n = \bar{v}_b$$

$$A \bar{i}_b = \bar{o}$$

3.10 MINIMUM ENERGY THEOREM

Instead of using Kirchhoff's voltage law, the following theorem could be used in some cases.

Theorem. In a network carrying steady currents i_1, i_2, ... flowing in the branches containing resistance R_1, R_2, ... and having voltage sources of emf E_1, E_2, ... respectively, the function $F = \sum \left[Ri^2 - 2Ei \right]$ is a minimum, the summation extending over all the branches; provided the currents divide as per KCL.

We accept KCL and so the clause "provided currents divide as per KCL" is always satisfied. Suppose that the branch carrying i is common to m loops (this depends on the choosing of the loops however). Let us assume that additional currents ε_1, ε_2, ..., ε_n are introduced into the n loops. These additional currents satisfy KCL because a circulating current enters and leaves each junction as it enters. Hence due to these extra currents, the change δF in the function F considered is

$$\delta F = F + \delta F - F$$

$$= \Sigma\left[R(i+\varepsilon_1+\varepsilon_2+....+\varepsilon_m)^2 - Ri^2\right] - 2\Sigma\left[E(i+\varepsilon_1+....+\varepsilon_m) - Ei\right]$$

$$= 2\varepsilon_1\Sigma(Ri - E) + 2\varepsilon_2\,\Sigma(Ri - E) + ... + \Sigma R(\varepsilon_1 + ... + \varepsilon_m)^2$$

Since $\Sigma(Ri - E) = 0$ by KVL,

$$\delta F = \Sigma R(\varepsilon_1 + \varepsilon_2 + ... + \varepsilon_m)^2 > 0$$

because the above expression is a perfect square. So F is a minimum.

In a passive network where E's are zero, the theorem : ΣRi^2 is a minimum; *i.e.*, the total energy is a minimum.

The Kirchhoff's voltage law could be derived from this principle as a converse.

Example 17. *A cube with its four diagonals is made up of resistances r for the sides and r' for the diagonals. The diagonals cross at the centre, so these wires are insulated. Find the resistance between opposite corners of the cube.*

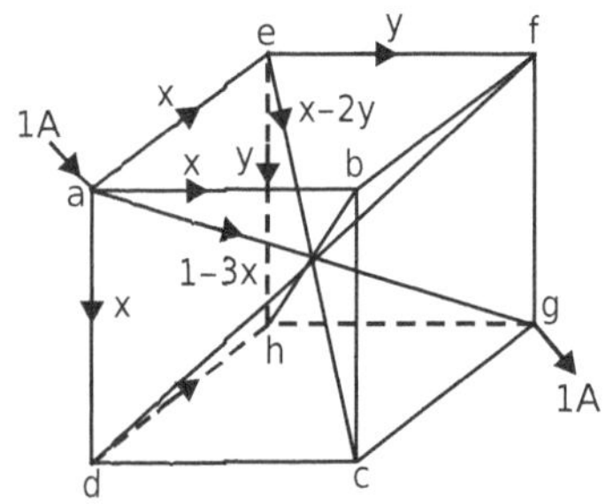

Fig. 3.27

Solution. By symmetry, the current 1 ampere entering A must divide into three equal currents x along ae, ab, ad (the sides of the cube have equal resistances) and $(1 - 3x)$ along the diagonal ag. The current along ae (*i.e.*, x), divides into y amperes along the sides ef, eh; and $(x - 2y)$ along ec. Similarly, the current in hb is $(2y - x)$.

$$F = \Sigma Ri^2 = 6rx^2 + 6ry^2 + r'(1-3x)^2 + 3r'(2y-x)^2$$

For F to be a minimum, the partial derivatives are zero.

$$\frac{\partial F}{\partial x} = 12rx + 2r'(1 - 3x)(-3) + 3r' \times 2(2y - x)(-1) = 0$$

$$2rx - r'(1 - 3x + 2y - x) = 0$$

$$\frac{\partial F}{\partial y} = 12ry + 6r'(2y - x) = 0$$

Solving for x, $$x = \frac{r'(r+2r')}{2r^2 + 8r\,r' + 6r'^2}$$

But voltage drop a to g is $(1 - 3x)r'$

Hence resistance between a and g is

$$= (1 - 3x)r'/(\text{current } a \text{ to } g)$$
$$= (1 - 3x)r'$$
$$= rr'(2r + 5r')/2(r^2 + 4rr' + 3r'^2)$$

on simplification.

EXERCISES

1. In Fig. 1, find the current in the 2 Ω resistor using superposition and reciprocity.

[**Ans.** $(1 + 7j)/8$)]

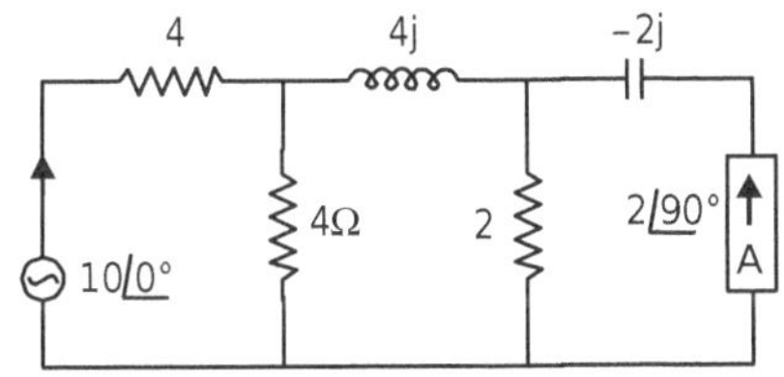

Fig. 1

2. Find the current in the 4 Ω resistor using superposition. [**Ans.** $(3 - 9j)/40$, A to B]

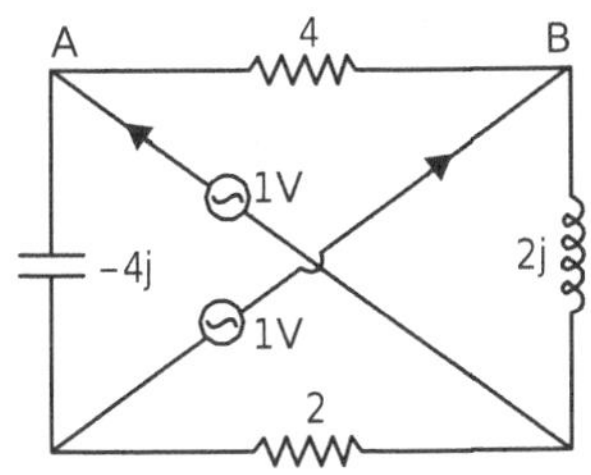

Fig. 2

3. In Fig. 3, N is the network which contains resistors only. Making use of the data in Fig. 3 (a), ($i_1 = 2$, $i_2 = 1$), and suitable network theorems, determine the current i_1 in Fig. 3 (b).

[**Ans.** $i_1 = 0$]

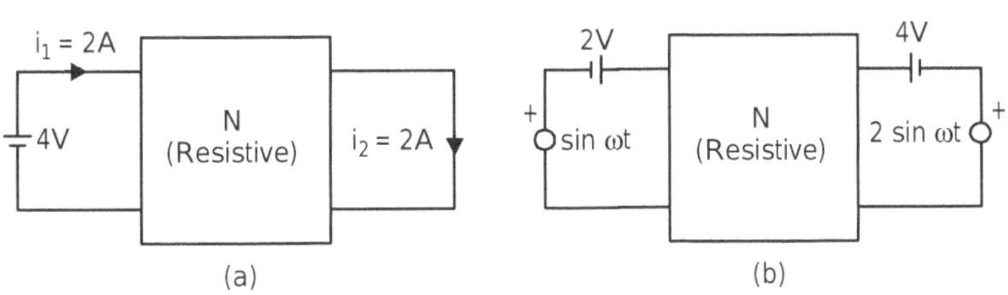

Fig. 3

4. In Fig. 4, if the 2 Ω resistors are both changed to 3 ohms find i_1, i_2. $\left[\textbf{Ans. } \frac{3}{4} \text{ A each}\right]$

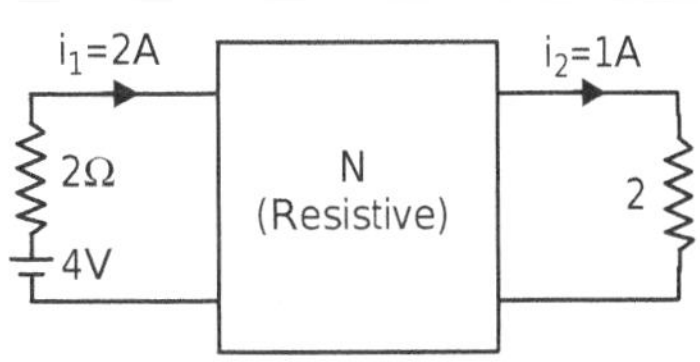

Fig. 4

5. In Fig. 5, show that the resistance between AB is 4/5 if all resistors are one ohm.

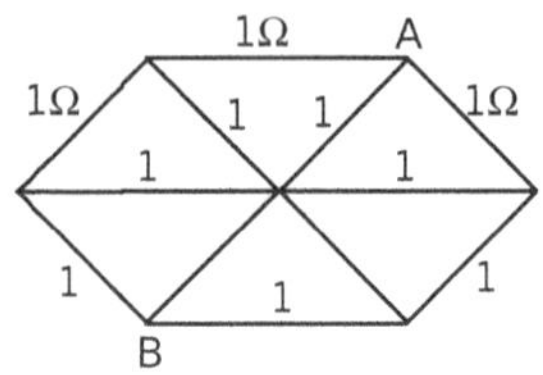

Fig. 5

6. In Fig. 6, find the resistance between A and B by successive application of Rosen's theorem. [**Ans.** 0.4 Ω]

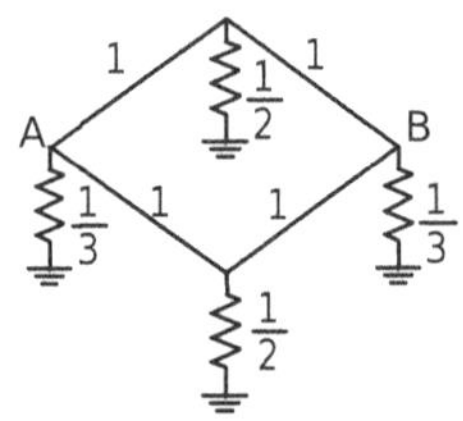

Fig. 6

7. Reduce the network Fig. 7 to a 3 resistance star between points A, B, C if each resistance is one ohm. [**Ans.** 5/9]

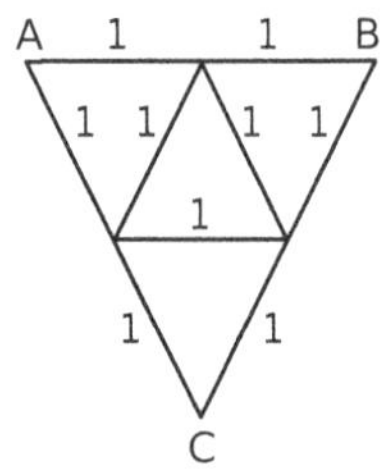

Fig. 7

8. Find the voltage V_{AB} in Fig. 8 using Thevenin's theorem. [**Ans.** $12j - 20$]

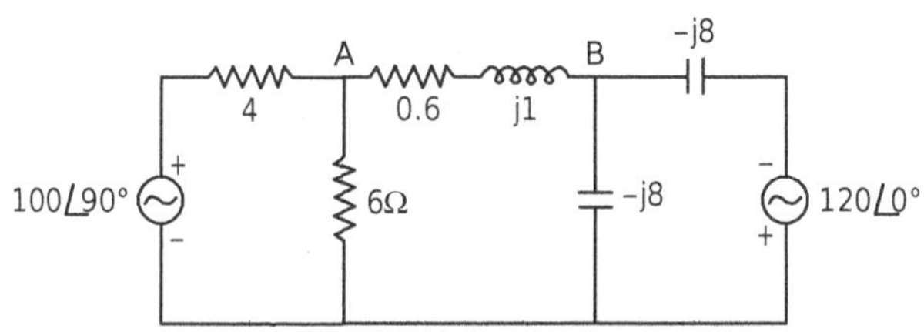

Fig. 8

9. In Fig. 9, find V_{AB} at the terminals. [**Ans.** 1/1 V]

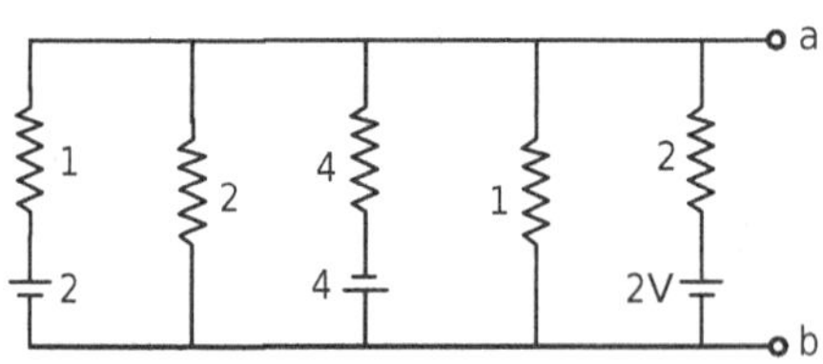

Fig. 9

10. If $X_L = 2X_C$, deduce the Norton equivalent circuit for the circuit shown in Fig. 10.

[**Ans.** A current source $E/2X_C$ in parallel with $2X_C$]

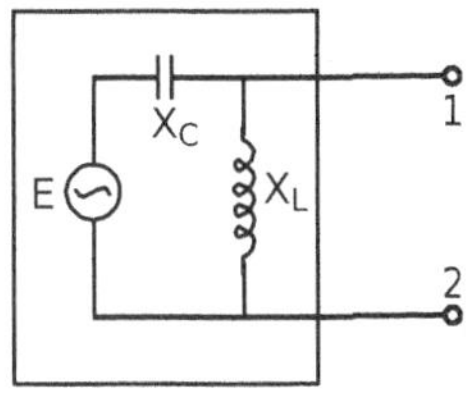

Fig. 10

11. Find the Thevenin and Norton equivalents of the dependent source network of Fig. 11.

$\left[\textbf{Ans.}\ -\frac{15}{4}\text{V in series with } \frac{45}{4}\Omega;\ -\frac{1}{2}\text{A in parallel to } \frac{45}{4}\Omega\right]$

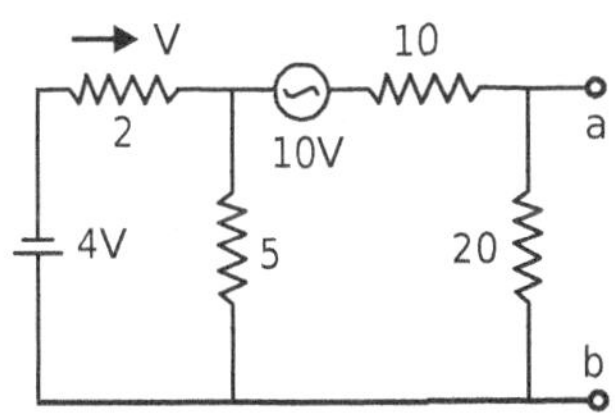

Fig. 11

12. A certain network including a generator has an open circuit voltage of 125 V and on short circuit sends 5.59 A. When a 10 ohm resistive load is connected, the current is 4.41 A. By Thevenin's theorem, find an equivalent voltage source of the network. How can you determine the sign of the reactive term in such a network?

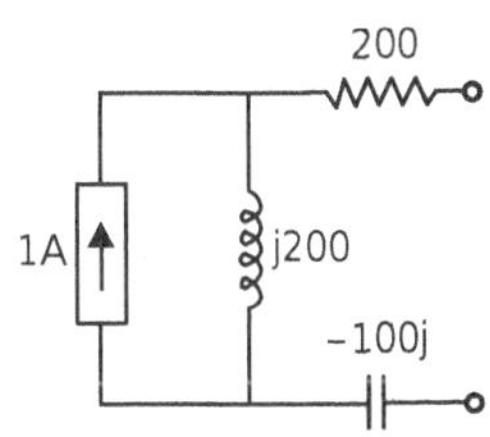

Fig. 12

[**Ans.** 125 V in series with $10 \pm j20$ ohms. Apply a 10 Ω inductive and a 10 Ω capacitance load and find for which the current is greater.]

13. In Fig. 13, what load is needed to give maximum power output? Find its value.

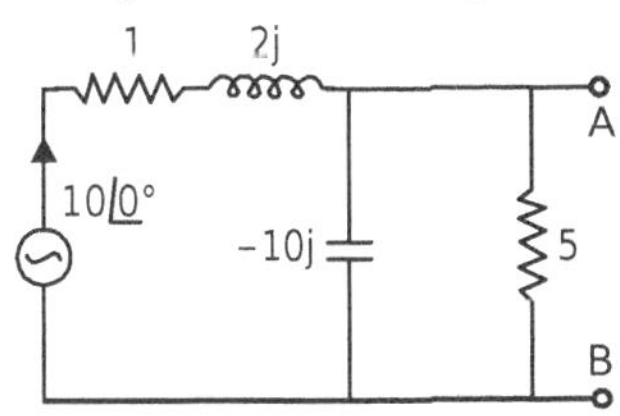

Fig. 13

[**Ans.** 100 $(2 - j)$, 50 W]

14. Find the impedance to be connected across ab in Fig. 13 so that maximum power output is obtained. For the 10 V source, what is the then efficiency?

[**Ans.** $(30 - 20j)/13$. 300 W to load. 48.387%]

15. Find the maximum power across the load impedance in Fig. 14

[Anna University, May 2007]

[**Hint:** The open circuit voltage at AB is 16.6 V. The S.C. resistnace is 10.42. R should be 10.42 (Z is only R here). P_{max} = 0.413 W.

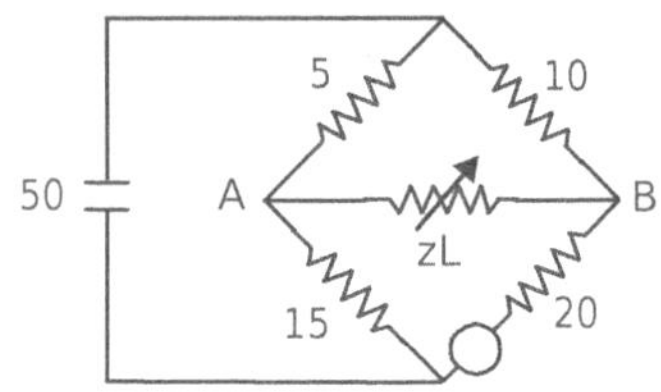

Fig. 14

16. Find the equivalent resistance at AB for the network in Fig. 15 (all resistors equal, R).

[Anna University, May 2004]

[**Hint:** A delta of equal resistors R can be replaced by a star of R/3 each. Inner delta becomes a star or R/3.

Adding to the link resistors, we have a star of $1\frac{1}{3}$ R inside. Make the outer delta to a star of R/3. So,

R_{AB} = 2 × (R/3 | | 4R/3)

= 8R/15.

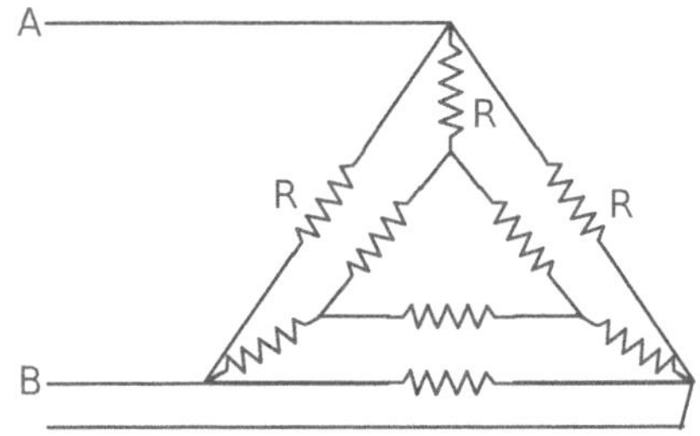

Fig. 15

SIMPLE EXERCISES

1. What is star-delta transformation ?

2. What are the limitations of superposition theorem ?

[Can't use for nonlinear circuit elements difficult in dependant source circuits]

3. State Reciprocity theorem.

4. Three batteries of internal voltage 1.5 v and having internal resistances of 10, 20 and 25 ohms are connected in parallel to a resistor of 100 ohms. Find the voltage across 100 ohms.

[**Hint:** Millman's theorem] [**Ans.** 1.425 V]

5. When a transformer is applied a reduced voltage and its secondary winding was first open, then short-circuited; it gave an open circuit voltage of 10 V and a D.C. current of 10 A.

6. State Thevenin's theorem.

What modification is required for assessing power in the elements ?

7. State one application for maximum power transfer theorem.

[Audio transformer matching ratio for loudspeaker connection.]

8. State Rosen's theorem and give an example of its application.

4

TRANSIENTS AND RESONANCE IN CIRCUITS

4.1 TRANSIENTS

A transient in a circuit is the short-time change in the current (or voltage) in part or whole of a circuit due to some switching action (or a similar disturbance) brought about in the circuit. The transient current has an initial value (before the switching) and a final value (when the transient has elapsed); between these, a current change that occurs is the transient current; because, the current cannot change suddenly to the final value in a circuit which has either an inductance or capacitance or both.

4.2 THE D.C. R-L TRANSIENT

The d.c. voltage switched on to an R-L circuit gives rise to a transient current given by

$$L\frac{di}{dt} + Ri = E \qquad (t > 0)$$

The solution is

$$i(t) = \frac{E}{R}(1 - e^{-Rt/L}) \tag{4.1}$$

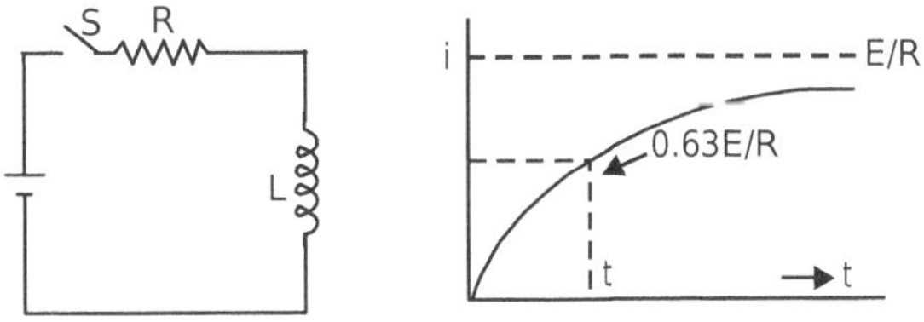

Fig. 4.1 D-C R-L transient.

At $t = 0$, $i = 0$ (initial value); and at $t = \infty$, $i = E/R$ (final value). The quantity L/R is called the *time constant* of the R-L circuit. Time constant is usually denoted by the greek letter τ. In the time $t = \tau$, the current increases to $(1 - e^{-1})$ or $(1 - 0.3679)$ or 0.6321 of E/R. Thus, the time constant may be defined as the time in which the current reaches 63% of its total change.

4.3 PROPERTY OF AN EXPONENTIAL CHANGE—STEP INPUT RESPONSE

(*a*) During every time constant from any instant *t*, the current changes by 63% of the difference between final value and the value at the instant *t*.

$$\left.\begin{matrix}\text{Current at}\\ \text{the instant } t\end{matrix}\right\} i_t = \frac{E}{R}(1 - e^{-t/\tau})$$

$$\left.\begin{matrix}\text{Current at}\\ \text{time } t+\tau\end{matrix}\right\} i(t+\tau) = \frac{E}{R}(1 - e^{-t-\tau/\tau})$$

$$\begin{aligned}\text{The change in time } \tau &= \text{the Difference between the above values}\\ &= [1 - \exp(-t - \tau/\tau) - 1 + \exp(-t/\tau)]\ E/R\\ &= (E\ e^{-t/\tau}/R)(1 - e^{-1})\\ &= (\text{Final value} - i_t) \times 0.63,\end{aligned}$$

thereby proving the statement.

In the above, the sudden application of this voltage source (via Switch S) is called 'Step Input'. Here, we have given a step of E volts to this R-L circuit.

(*b*) The above property can be used to sketch the transient, in cases where only one time constant is involved.

Let the final value be 10, initial value 0, time constant 2 sec.

t	0	2	4	6
i	0	6.3	6.3 + (10 – 6.3) .63 = 8.63	8.63 + (10 – 8.63) .63 = 9.49

and so on.

(*c*) In half a time constant, the change is $(1 - e^{-1/2}) = 0.393$, (39.3%).

In quarter time constant, the change is $(1 - e^{-1/4})$ or 22.2%.

4.4 TRANSIENT IN AN R-C CIRCUIT

In this, a battery is switched on at $t = 0$ and it charges the capacitor via the resistor R. The equation is

$$Ri + (q/C) = E$$

$$R\frac{dq}{dt} + \frac{q}{C} = E$$

Solving this differential equation,

$$q = CE\ (1 - e^{-t/CR)})$$

$$v = q/C = E(1 - e^{-t/CR}) \tag{4.2}$$

At $t = 0$, the charge q as well as voltage v across C are both zero. At $t \to \infty$, $v \to E$. The time constant τ is *CR* in this case.

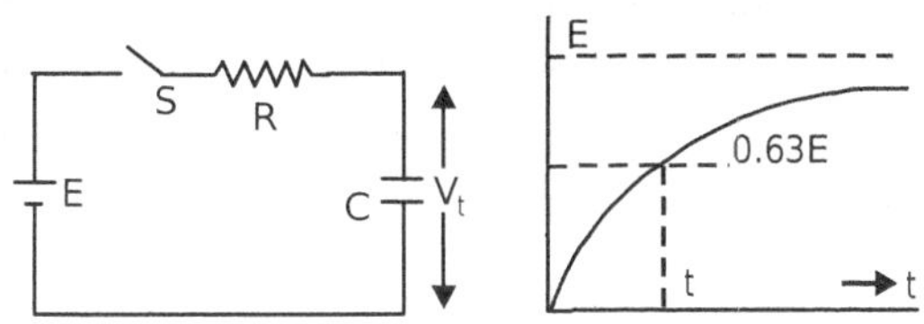

Fig. 4.2 Transient in an R-C circuit.

4.4.1 Discharge of a Charged Capacitor-Source Free R-C Transient

In the above circuit if we have no battery E, (source-free circuit), then the current flow is only due to stored voltage (charge) on the capacitor C.

$$R\frac{dq}{dt}+\frac{q}{C} = 0$$

Here at $t = 0$, $q = CV_S$, where V_S is the stored voltage on C.

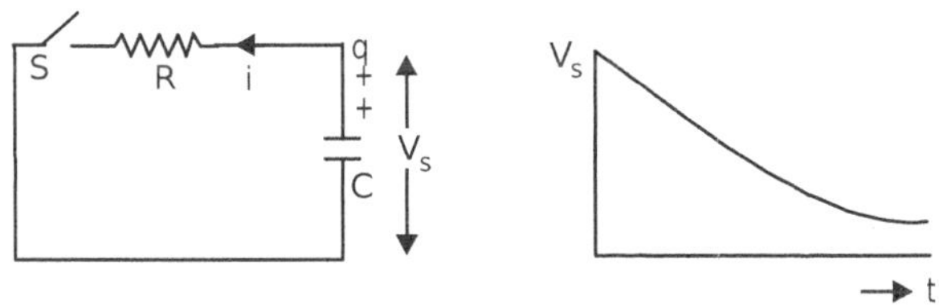

Fig. 4.3 Stored charge on C—discharge.

$$q = CV_S\, e^{-t/RC}$$

or

$$v_C = q/C = V_s e^{-t/RC}$$

The voltage on C, (that was charged prior to S closing) exponentially falls to zero at $t \to \infty$.

The quantity CR = τ, the time constant of the circuit.

Example 1. *A resistor R and a 2 μF capacitor are in series across a 100 v D.C. supply. Across the capacitor is a cold-cathode lamp which strikes at 60 V. Calculate R so that the lamp strikes 5 seconds after the switch is closed. If R = 5 megaohms, after how much time would the lamp glow?*

Solution. The voltage across C $= v_c = E(1 - e^{-t/CR})$

The lamp strikes when $V_c = 60$

$$60 = 100\left[1-\exp(1-t/CR)\right]$$

$$= 100\left[1-\exp(-t\times 10^6/2\,R)\right]$$

From this $0.4 = \exp(-t \times 10^6/2\,R) = \exp(-5 \times 10^6/2\,R)$

giving $R = 2.74$ Megohms.

$R = 5$ Meg., $v_c = 60 = 100\left[1-\exp(-t/10)\right]$ $(\because CR = 10)$

$$\exp(-0.1t) = 0.4$$

$$t = 9.1 \text{ seconds.}$$

The lamp with glow 9.1 seconds later after the D.C. supply in switched. But as soon as the lamp strikes, the capacitor will be discharged by the conducting path of the lamp. Hence

v_c drops to zero at once and the lamp extinguishes immediately. The capacitor again begins to charge up as before. After further 9 secs., the lamp glows again. The process repeats again and again. So the voltage across C rises and falls, giving rise to a sawtooth wave shape shown in Fig. 4.4.

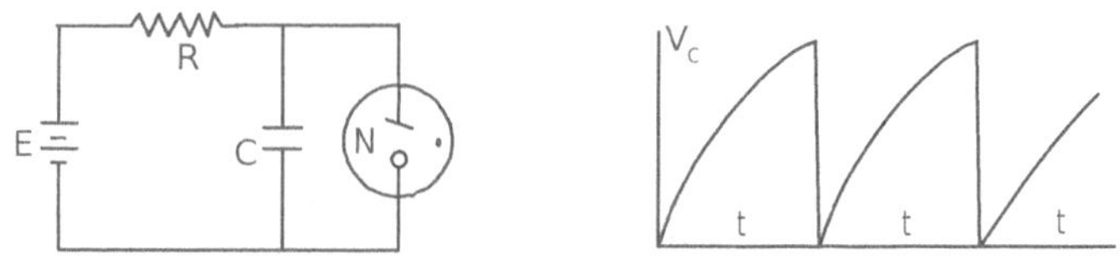

Fig. 4.4 Sawtooth wave generated by R-C circuit N = Neon.

4.5 OSCILLATION IN AN L-C CIRCUIT

Whenever inductance and capacitance happen to be together, there is a likelihood of oscillations.

(The oscillations refer to current and voltage of course).

Consider the capacitor C with initial charge Q discharged via an inductor L after switch S is closed (t = 0). (See Fig. 4.5 (a))

Voltage across C + Voltage across L = 0

$$q/C + L\frac{di}{dt} = 0 \tag{4.3}$$

But $dq/dt = i$; hence differentiate 4.3 to get

$$\frac{i}{C} + L\frac{d^2i}{dt^2} = 0$$

$$\frac{d^2i}{dt^2} + \frac{1}{LC}i = 0 \tag{4.4}$$

The solution is $i = A\sin\left(t/\sqrt{LC}\right) + B\cos\left(t/\sqrt{LC}\right)$

At $t = 0$, $i = 0$; (initial value). Hence

$$0 = A \times 0 + B \times 1 \text{ i.e., } B = 0.$$

$$i = A\sin t/\sqrt{LC} \tag{4.5}$$

Since $i = dq/dt$, we get at $t = 0$, $q = Q$,

$$\therefore \qquad Q = A\sqrt{LC} \text{ or } A = Q/\sqrt{LC}.$$

Hence $$i = Q/\sqrt{LC}\sin t/\sqrt{LC} = I_{max} \times \sin \omega_o t$$

where $$\omega_o = 1/\sqrt{LC} \text{ and } I_{max} = Q/C\left[\sqrt{C/L}\right] = V_o\sqrt{C/L},$$

V_o being the voltage to which C is initially charged.

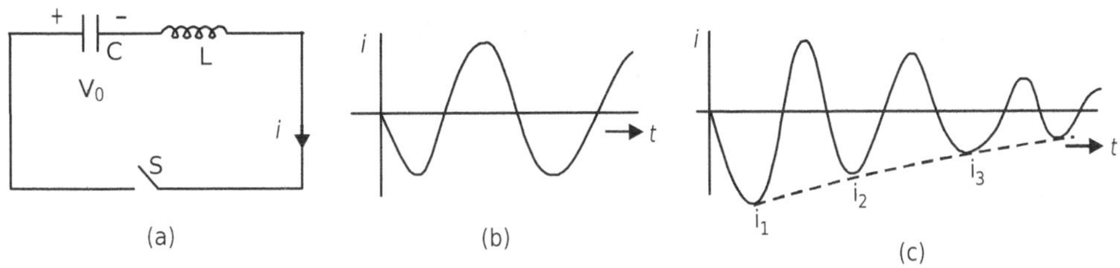

Fig. 4.5 Oscillations in an L-C circuit.

Thus ω_o is the natural frequency of oscillation of the L-C series circuit. The oscillation continues for ever in this case (Fig. 4.5(*b*)). But if there is a resistance in circuit (neglected in the above case), the oscillations decay as in Fig. 4.5(*c*). The resistance *damps* the oscillations by consuming energy.

With resistance, the differential equation (4.3) becomes

$$\frac{q}{C} + L\frac{di}{dt} + Ri = 0 \tag{4.6}$$

Differentiating with respect to t,

$$(d^2i/dt^2) + (R/L)\, di/dt + (i/LC) = 0$$

The characteristic equation is

$$m^2 + (R/L)m + 1/LC = 0$$

$$m = -R/2L \pm \sqrt{(R^2/4L^2) - (1/LC)}$$

For the usual case when R is small.

$$m = -R/2L \pm j\sqrt{(1/LC) - (R^2/4L^2)} = -\alpha \pm j\omega$$

and the solution is

$$i = e^{-at}\left[A\cos\omega t + B\sin\omega t\right]$$

Using the initial condition that $i(0) = 0$, A = 0. So,

$$i = Be^{-at}\sin\omega t \tag{4.7}$$

The values α, ω are given by $\alpha = -R/2L$ and $\omega \simeq 1/\sqrt{LC}$ if R is small.

Equation 4.7 shows a decaying oscillation, shown in Fig. 4.5 (*c*).

The decay is due to the term with the negative exponent. If i_1, i_2, i_3, ... are successive peaks of the oscillations,

$$i_1 = A\exp(-\alpha t_1)$$

$$i_2 = A\exp(-\alpha t_2) = A\exp\left[-\alpha(t_1 + T)\right]$$

$$i_3 = A\exp\left[-\alpha(t_1 + 2T)\right] \text{ etc.,}$$

where T is the *period* of each oscillation, being $2\pi/\omega$.

The ratios i_1/i_2, i_2/i_3, i_3/i_4 etc. are all equal;

$$\delta = \text{Ratio between peaks} = \exp(\alpha T)$$

The quantity δ is called *decrement*; and αT is log δ, known as *logarithmic decrement*.

Example 2. *An inductor of 10 Henry having negligible resistance is carrying 1 Amp steady current. Suddenly an uncharged capacitor of 100 μF is across it. Find the transient voltage across the capacitor.*

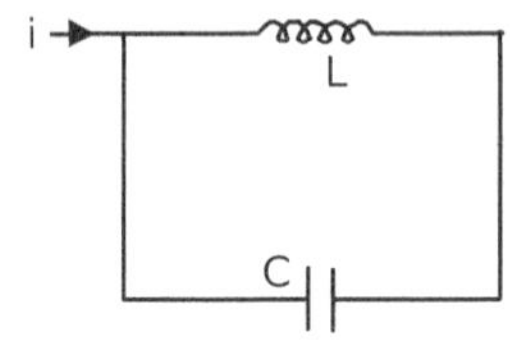

Fig. 4.6

Solution. The solution of the L-C discharge equation is

$$i = A \sin \omega t + B \cos \omega t$$

$$\omega = 1/\sqrt{LC} = 1/\left[10 \times 100 \times 10^{-6}\right]^{1/2} = 31.6$$

At $t = 0$, $i = 1$ and hence we get B = 1.

$$\therefore \quad i = A \sin \omega t + \cos \omega t$$

Integrating,
$$q = \int i\, dt = -(A/\omega) \cos \omega t + (1/\omega) \sin \omega t$$

But since the capacitor is uncharged at $t = 0$; $q(0) = 0$.

$$\therefore \quad 0 = -A/\omega \quad \text{or} \quad A = 0.$$

$$\therefore \quad i = \cos \omega t$$

The voltage across capacitor is

$$v_c = [1/C] \quad i\, dt = (1/\omega C) \sin \omega t = 316 \sin \omega t.$$

Example 3. *The natural oscillations of a series R.L.C. circuit have a frequency of 1000 Hz and a logarithmic decrement of 0.1 (a) Determine the number of cycles in which the amplitude of the oscillations decays to 36.8% of its initial value. (b) Given that the resistance of the circuit is 10 Ω, determine L and C.*

Solution. It is given that $0.368 = \exp(-\alpha t)$

Where t is the time taken for decaying upto 36.8%

$$\log_s \delta = \alpha T = 0.1. \text{ Since } T = 1/1000, \alpha = 100.$$

$$\exp(-100t) = 0.368$$

giving $t = 10^{-2}$ seconds. *i.e.*, $1000 \times 10^{-2} = 10$ cycles.

Also
$$\omega = 1000 \times 2\pi = 1/\sqrt{LC} \qquad (1)$$

$$\alpha = 100 = R/2L \qquad (2)$$

Hence, from (2), L = 0.05 H and C is got from (1) as 800 μF.

4.6 INITIAL AND FINAL VALUES

Consider the relations

$$v = L\frac{di}{dt} \quad i = C\frac{dv}{dt}$$

In the steady state, no current can pass through a capacitor and is therefore similar to an open circuit.

In the steady state, there is no change of current in an inductor and hence L di/dt is zero; or, the voltage across an inductor is zero; it is equivalent to a short-circuit.

In the initial state, no current can flow through an inductance suddenly, because it will develop a large voltage to oppose it. So, it is similar to an open circuit initially. **Likewise, no voltage can suddenly be applied to a capacitor initially.** It will send a large current to oppose it.

In the initial state, for inductance :

$$i = 0 \quad v = \text{large} \rightarrow \text{open circuit}$$

For capacitor :

$$v = 0 \quad i = \text{large} \rightarrow \text{short circuit}$$

In the final state, for inductance:

$$i = \text{Constant}, \; di/dt = 0 \text{ or } v = 0 \rightarrow \text{short circuit,}$$

For capacitor :

$$v = \text{Constant}, \; dv/dt = 0, \; i = 0 \rightarrow \text{Open circuit.}$$

4.7 SOLUTION OF FIRST ORDER TRANSIENTS

Transient problems involving only one time constant can be solved easily. For example, the C-R circuit has a single time constant, of value CR. Let v_{co} be the initial voltage to which C is charged. The final value of the voltage will be E itself as it will ultimately get charged to the battery voltage. In the transient condition, the voltage changes as exp $(-t/CR)$. The voltage v_c at any instant can be got from

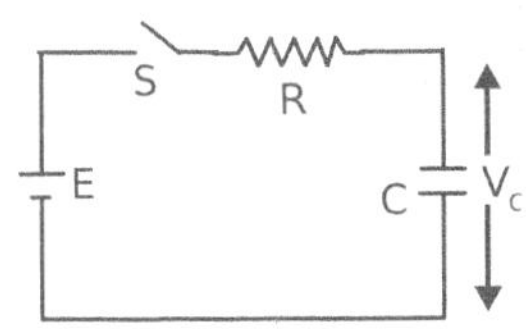

Fig. 4.7

$$v_c = \text{Final value} - (\text{Final value} - \text{initial value}) \, e^{-t/\tau}$$

So, $$v_c = E - (E - v_{co}) \, e^{-t/\tau}) \qquad (4.8)$$

If $$v_{co} = 0, \quad v_c = E\,(1 - e^{-t/\tau})$$

The formula (4.8) enables us to determine the transient solution for single time-constant networks.

But how to recognise a network has a single time constant? For that, reduce all sources (voltage and current) to zero and try to combine the capacitors (or inductors) into one. If this is possible, the network has a single time-constant.

For example, network of Fig. 4.8 has 3 capacitors but they can be combined into an equivalent capacitor,

$$C_c = C_1(C_2 + C_3)/[C_1 + (C_2 + C_3)]$$

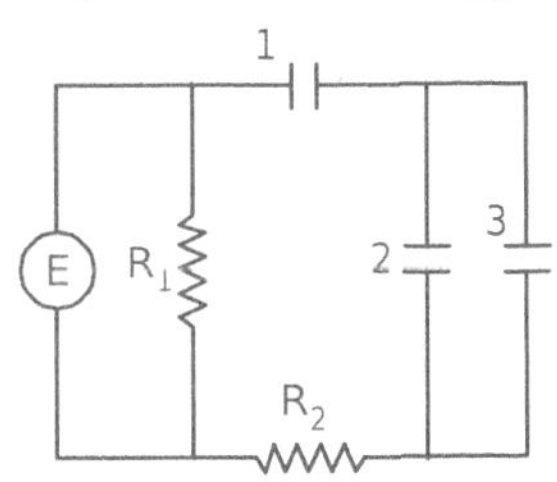

Fig. 4.8

Hence it is a single τ network. In this network R_1 is directly across the source and hence there is no transient current in it. So, R_2 alone decides the time constant, which is thus C_c R_2.

After separating the capacitor or inductor from the rest of the network, Fig. 4.9 (*a*), we have to reduce it to the Thevenin or Norton equivalents (Figs. 4.9 (*b*) & (*c*)). The time constant will be got as L/R_e, for instance, when R_e is the Thevenin's equivalent resistance.

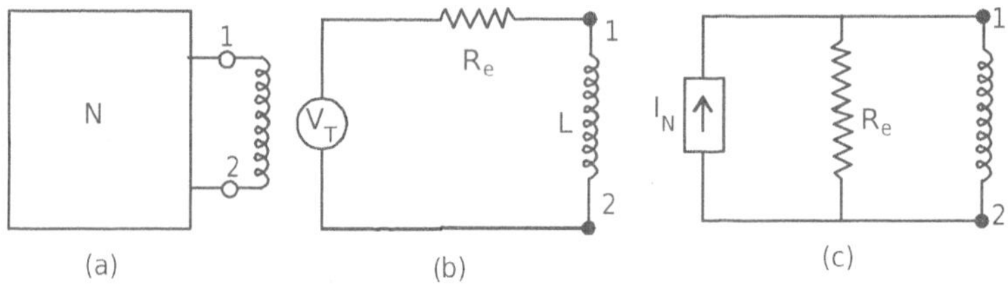

Fig. 4.9 (*a*) Network, after separating L or C is reduced as in (*b*) (Thevenin equivalent) or as in (*c*) (Norton eqv.).

Example 4. *In the network shown, find the voltage v(t) after the switch S is opened.*

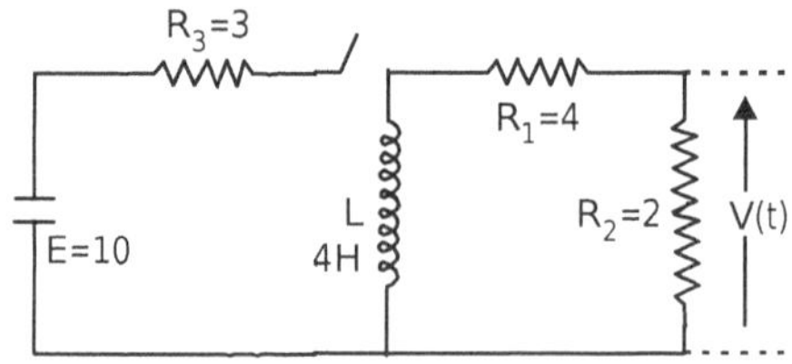

Fig. 4.10

Solution. This is a single time-constant network as it has only one inductance. Initially, the switch remains closed. The inductor short circuits the 4 Ω and 2 Ω resistor and the current in it is 10/3 amps, via R_3. Once the switch is opened, the current in the inductor cannot die down immediately and hence it now passes through the 4 ohm and 2 ohm resistors, thereby the voltage across R_2 is $(10/3) \times 2 = 20/3$ volts at $t = 0$.

The final value of current and voltages are zero in L and R_2 etc., as the stored energy in L will ultimately be dissipated away in the resistors. The time-constant is $L/(R_1 + R_2) = 4/(2 + 4) = 2/3$ secs. Hence the voltage $v(t)$ is

Final value – (Final – Initial) exp (– 3/2 *t*).

$$v(t) = 0-[0-(20/3)]\,e^{-3t/2} = 20/3\,e^{-3t/2}.$$

Example 5. *The switch S is closed in the adjoining network. Find the transient voltage v(t).*

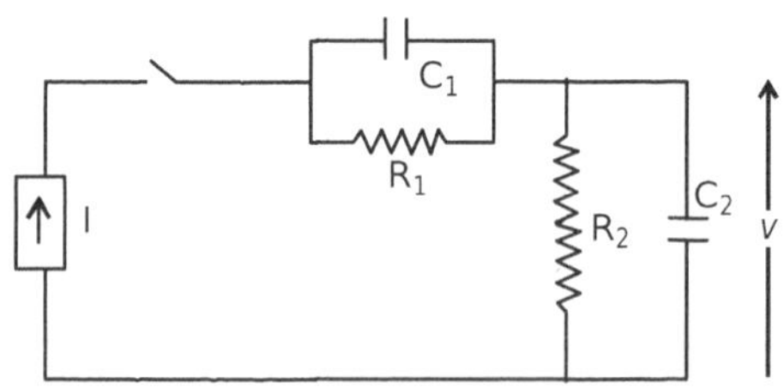

Fig. 4.11

Solution. The initial value for $v(t) = 0$ since C_2 is uncharged.

The final value will be R_2I (the capacitor then carries no current)

Applying Norton's theorem to the left of terminals of R_2, C_2;

(1) Open the current source, get the equivalent (Fig. 4.12 (*a*))

(2) Find the short-circuit current. It is I itself. (Fig. 4.12 (*b*))

(3) Draw the network using Norton's equivalent in parallel. (Fig. 4.12 (*c*))

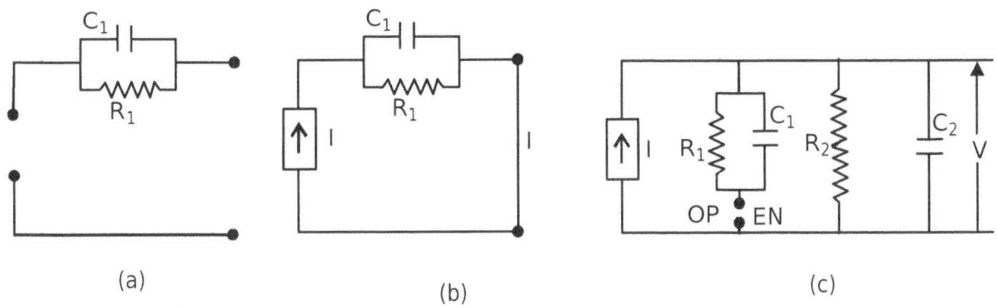

Fig. 4.12

The reduced network has R_2C_2 in parallel. The R_1C_1 is open and does not affect $v(t)$. Hence $T = R_2C_2$.

$$v(t) = \text{Final value} - (\text{Final} - \text{Initial}) \exp(-t/\tau)$$

$$= I R_2 - (I R_2 - 0) \exp(-t/R_2C_2) = IR_2 (1 - e^{-t/CR}).$$

Example 6. *If the capacitor stores a voltage V_0 at t = 0 when switch is closed, find v(t).*

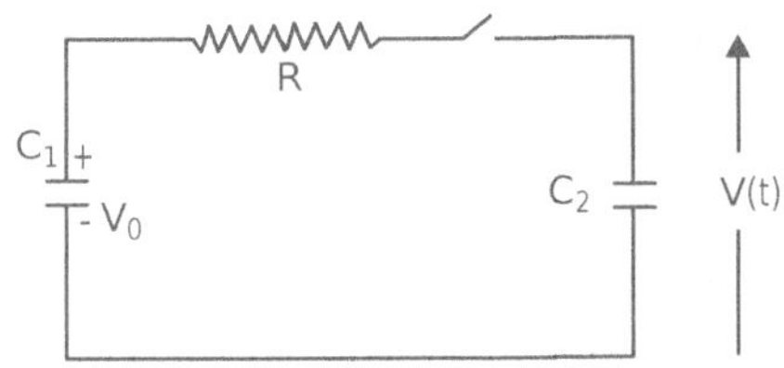

Fig. 4.13

Solution. The initial value of current is got by noting that the capacitor is initially a short-circuit; $i(0) = V_0/R$.

The final value is evidently zero, because when C_1 and C_2 get equal potentials, no more current flows.

The time constant is $R \times C_1 \dfrac{C_2}{(C_1 + C_2)}$.

$$\therefore \qquad i(t) = 0 - (0 - v_0/R) \exp(-t/\tau)$$

4.8 RESONANCE IN A.C. CIRCUITS—ACCEPTOR RESONANCE

In the R-L-C series circuit, consider the special case when $X_L = X_C$. Then the voltages V_L and V_C will be equal and in phase opposition (Fig. 4.14). So the supply voltage to the series circuit will be equal to just $V_R = IR$.

$$V = IR$$

$$I = V/R$$

So the circuit behaves as a simple resistance. The power factor of R-L-C circuit under this condition (*i.e.*, $X_L = X_C$) is unity. The impedance

$$Z = \sqrt{R^2 + (X_L - X_C)^2}$$

becomes minimum, when $X_L = X_C$. The minimum value is R.

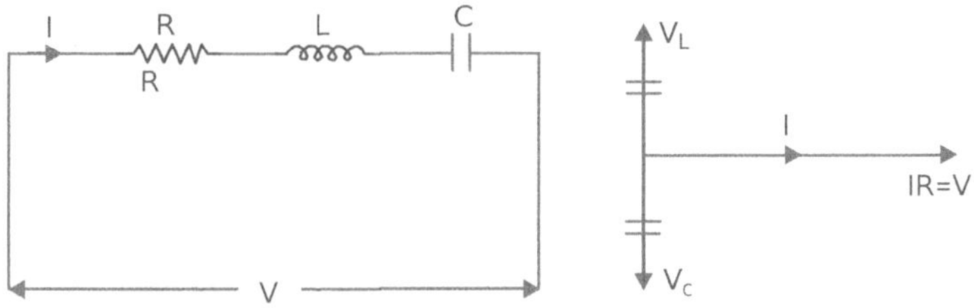

Fig. 4.14 Phasor diagram of series circuit at resonance.

Let us now examine under what condition X_L becomes equal to X_C.

$$X_L = \omega L;\ X_C = 1/\omega C.$$

$$\therefore \quad \omega L = 1/\omega C$$

$$\omega^2 = \frac{1}{LC}$$

$$\omega = \frac{1}{\sqrt{LC}} \tag{4.9}$$

Thus if the supply voltage to the R-L-C series circuit has a frequency given by

$$f_r = \frac{1}{2\pi\sqrt{LC}} \tag{4.10}$$

then the condition is satisfied. For other frequencies, $X_L \neq X_C$ and hence Z will be greater than R. At this frequency, f_r, Z is minimum and so the current is a maximum. If we vary the frequency of the voltage applied to an R-L-C circuit, and measure the current at any frequency and make a plot of f versus I, we get the graph as in Fig. 4.15. The frequency at which the circuit passes maximum current is called the *Resonant Frequency* f_r.

4.8.1 Frequency Response

We know that

$$X_L = \omega L = 2\pi f L$$

$$\therefore \quad X_L \propto f$$

$$X_C = 1/\omega C = 1/2\pi f C$$

$$\therefore \quad X_C \propto 1/f$$

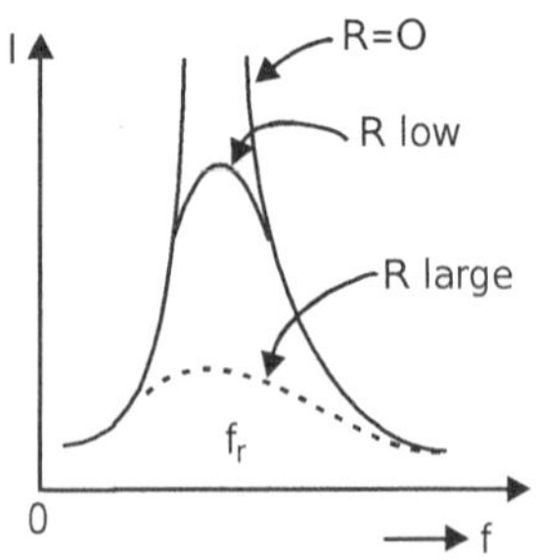

Fig. 4.15 Frequency response—LC series circuit.

The inductive reactance varies with f, and so X_L versus f is a straight-line Fig. 4.16. X_C varies inversely and so it is a hyperbola. In Fig. 4.16, $- X_C$ is drawn. The net reactance of the series circuit is $X_L - X_C$ which is shown as the dotted line. The reactance becomes positive for frequency greater than f_r. The current is given by E/Z. The variation of Z with frequency is plotted. The shape is in the form of a U, the minimum point is f_r. Hence the current is in the form of an inverted U. For R = 0, the impedance is :

$$Z = \sqrt{R^2 + (X_L - X_C)^2}$$

$$= X_L - X_C \text{ if } R = 0$$

Hence I = V/Z, and the current is infinite at f_r for R = 0. For low values of R and large values of R, the variation of current with frequency is shown in Fig. 4.15.

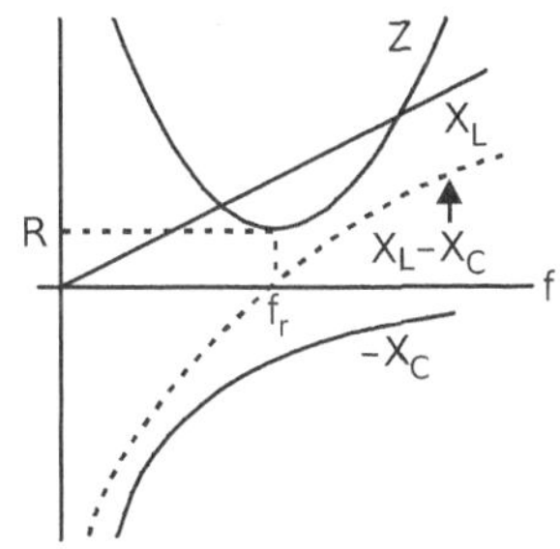

Fig. 4.16

The voltages V_L and V_C at resonance frequency are in phase opposition Fig. 4.14.

$$V_L = \omega LI = \omega L\ (V/R) \quad \text{at resonance.}$$

$$V_C = (1/\omega C)\ I = V/R\omega C \quad \text{at resonance.}$$

Since $\omega L = 1/\omega C$,

$$V_L = V_C = (\omega L/R)\ V$$

$$V_L/V = \omega L/R \tag{4.11}$$

The ratio of the voltage across L (or C) to the applied voltage V is called the 'magnification factor' Q.

$$Q = \omega L/R$$

Hence, summarising :

(1) At resonance, the inductive and capacitive reactances cancel.

(2) The current is a maximum. If R = 0, it becomes infinitely large.

(3) The circuit is purely resistive at resonance and hence the power factor is unity.

(4) The impedance is a minimum value (= R).

(5) The resonant frequency is $f_r = \dfrac{1}{2\pi\sqrt{LC}}$.

(6) The voltages across L and C are generally more than the applied voltage V, the ratio is called 'magnification factor' Q. Q = ωL/R.

Example 7. *A current of 4 amps flows through a coil of resistance 20 ohms and inductance 0.5 H, and a condenser 10 μF in series. Find the voltage across the condenser and that across the whole circuit at the resonant frequency. What is the Q of the coil?*

Solution. The resonant frequency for a series circuit

$$f_r = \frac{1}{2\pi}\frac{1}{\sqrt{LC}}$$

$$\sqrt{LC} = \left[0.5\times10\times10^{-6}\right]^{1/2} = 0.002236$$

$$f_r = \frac{1}{2\times3.14}\times\frac{10^6}{2236} = 71.4 \text{ cycles/sec (Hz)}$$

The Q or 'magnification factor' of the coil = $\omega L/R$

$= 2\pi \times 71.4 \times 0.5/20 = 11.2$

At resonance, the inductive reactance and capacitive reactance cancel each other and the circuit is resistive. So

$$V = RI = 4 \times 20 = 80 \text{ Volts.}$$

is the voltage across the circuit. The voltage across C is I X_C.

$$X_C = 1/\omega C = 10^6/2\pi \times 71.4 \times 10 = 224$$

So the voltage across condenser = $224 \times 4 = 896$ V

The magnification is also got from 896/80 = 11.2.

The value of Q, the magnification factor, is of interest at high frequencies, because, at these frequencies, the resistance increases with frequency but the ratio $Q = \omega L/R$ is fairly constant. So, radio frequency coils are specified not by their resistance and inductance but by their Q value, a high Q being preferred.

The name *acceptor resonance* is given to the series circuit resonance. This is because, at or near resonance, the circuit accepts a large current and has a high magnification of the voltage V_L. The current I rises as a peak near resonance. In radio work, the sharpness of the peak is important. The difference in frequencies $f_2 - f_1$ in Fig. 4.17 is used as a measure of the sharpness of the peak. Here f_2 and f_1 are frequencies for which the current is 0.707 times the maximum current. Since power is I^2R, at these frequencies the power is $(0.707)^2$ times the power at the resonant frequency. Hence these points are called *half power points*. Let us find $f_2 - f_1$ in terms of L, R and ω_r.

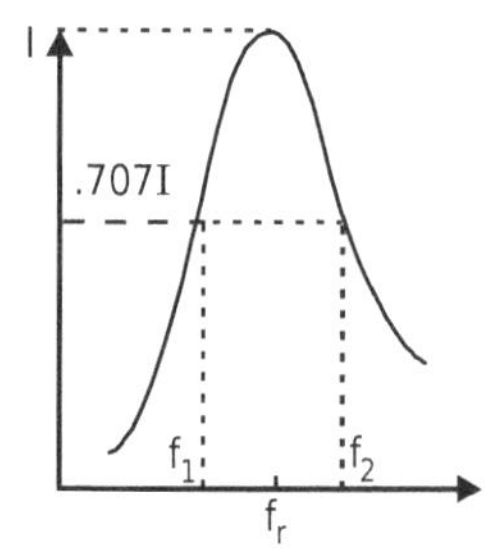

Fig. 4.17

At f_1 and f_2, the current is 0.707 V/R = $V/\sqrt{2}R$.

$$\therefore \qquad \frac{V}{\sqrt{2}R} = V \div \sqrt{R^2 + [\omega L - (1/\omega C)]^2}$$

Squaring and multiplying with R,

$$2R^2 = R^2 + [\omega L - (1/\omega C)]^2$$

$$\therefore \qquad R = \pm[\omega L - (1/\omega C)]$$

$$\therefore \qquad \omega^2 LC \pm \omega RC - 1 = 0$$

$$\omega = \pm \frac{R}{2L} \pm \sqrt{\left(\frac{R}{2L}\right)^2 + \frac{1}{LC}}$$

Usually R/2L may be small compared to 1/LC so that it may be neglected. Also, $1/\sqrt{LC} = \omega_r$.

$$\therefore \qquad \omega = \pm (R/2L) \pm \omega_r$$

The negative sign for ω_r is not significant. Hence the two frequencies f_1 and f_2 are

$$2\pi f_1 = -(R/2L) + \omega_r$$

$$2\pi f_2 = +(R/2L) + \omega_r$$

so that

$$(f_2 - f_1) = [1/2\pi][R/L]$$

$$\frac{f_2 - f_1}{f_r} = \frac{R}{2\pi f_r L} = \frac{R}{\omega_r L} = \frac{1}{Q} \qquad (4.12)$$

Hence, the sharpness or *selectivity* of the circuit varies inversely as the Q of the coil. Hence a high Q coil gives a small width $f_2 - f_1$, *i.e.*, a good selectivity.

For the previous problem, the selectivity is

$$(f_2 - f_1)/f_r = 1/Q = 1/11.2$$

$$f_2 - f_1 = f_r/11.2 = 71.4/11.2 = 6.34 \text{ c/s (Hz)}.$$

4.9 PARALLEL RESONANCE

When the two branches of a parallel circuit contain an inductance and a capacitance, resonance can occur in the parallel circuit. The total current drawn from the source will be at unity power factor at a particular frequency, called the resonant frequency. The current in branch 1 lags behind the voltage but the branch 2 current leads it. Both the currents will vary with frequency. It is possible that at a certain frequency, the total current which is the phasor sum of the two branch currents will be in phase with the voltage. In such a case, the imaginary components of I_1 and I_2 cancel each other. Then,

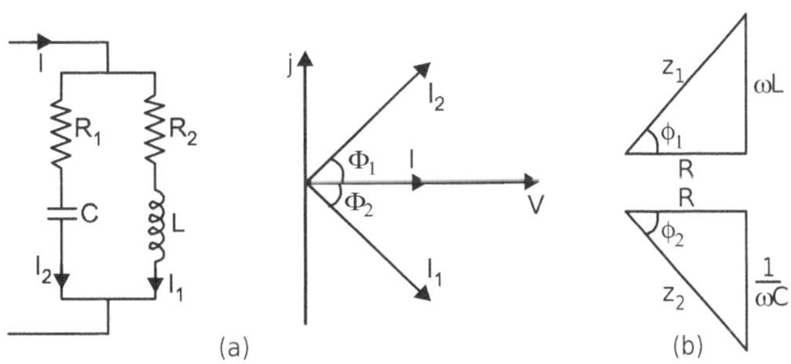

Fig. 4.18 (*a*) Current in a parallel resonant circuit. (*b*) Impedance triangles.

$$I_1 \sin \varphi_1 = I_2 \sin \varphi_2$$

From the Impedance triangle,

$$\frac{V\,\omega L}{R_1^{\,2} + (\omega L)^2} = \frac{V/\omega C}{R_2^{\,2} + (1/\omega C)^2}$$

$$\frac{\omega L}{R_1^2 + (\omega L)^2} = \frac{1/\omega C}{R_2^2 + (1/\omega C)^2} \tag{4.13}$$

This reduces to

$$\omega^2 LC\ (L - CR_2^2) = L - CR_1^2$$

from which, the value, ω_r, the frequency of resonance, is

$$\omega_r = \frac{1}{\sqrt{LC}}\sqrt{\frac{L - CR_1^2}{L - CR_2^2}} \tag{4.14}$$

and $$f_r = \omega_r/2\pi.$$

Usually, for parallel resonant circuits, no resistance is connected in the capacitor branch. (*i.e.*, $R_2 = 0$). In such a case, putting $R_2 = 0$ in the above, we get

$$\boldsymbol{f_r = \frac{1}{2\pi}\sqrt{\frac{1}{LC} - \frac{R^2}{L^2}}} \tag{4.15}$$

We note some interesting points from (4.14).

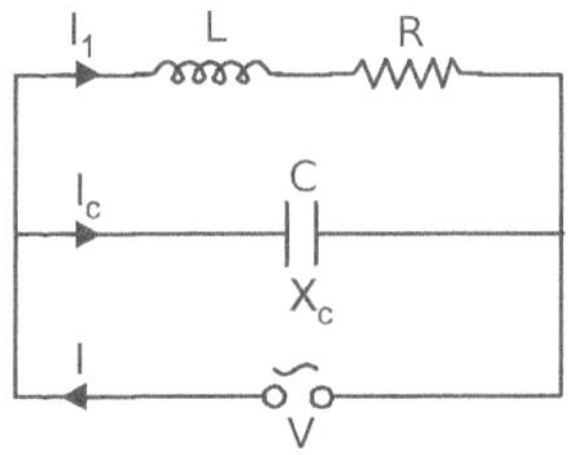

Fig. 4.19

1. When $L < CR_1^2$ or $L < CR_2^2$, the quantity within the big square root is imaginary. Then, there is no resonant frequency.
2. But, if $L < CR_1^2$ as well as CR_2^2, then resonance occurs at a particular frequency.
3. If $R_1 = R_2 = \sqrt{L/C}$, then the value of ω_r is indeterminate. That is, resonance occurs at all frequencies, *i.e.*, the circuit is purely resistive.

Let us find the magnitude of the total current at resonance. From Fig. 4.19, the total current is the phasor sum of the two branch currents.

$$\begin{aligned} I_t &= I_1 + I_2 \\ &= I_1 \cos\varphi_1 - jI_1 \sin\varphi_1 + I_2 \cos\varphi_2 + jI_2 \sin\varphi_2 \\ &= I_1 \cos\varphi_1 + I_2 \cos\varphi_2 \quad (\text{since } I_1 \sin\phi_1 = I_2 \sin\phi_2) \\ &= \frac{VR_1}{R_1^2 + (\omega L)^2} + \frac{VR_2}{R_2^2 + (1/\omega C)^2} \end{aligned}$$

If $$R_2 = 0,$$

$$I_t = \frac{VR_1}{\left[R_1^2 + (\omega L)^2\right]} \tag{4.16}$$

Hence, the ratio V/I_t, which represents only a resistance since I_t is in phase with V, becomes

$$\frac{V}{I_t} = \frac{R_1^{\,2} + (\omega L)^2}{R_1}$$

From (4.13), we get, putting $R_2 = 0$,

$$R_1^{\,2} + (\omega L)^2 = \omega L\,(1/\omega C) = L/C, \text{ so that}$$

$$\frac{V}{I_t} = \frac{L}{CR}$$

This resistance is called the '*Dynamic Resistance*' of the parallel resonant circuit.

Example 8. *If a coil of 50 ohms and 0.5 H and a condenser of 100 μF are connected in parallel, find the resonant frequency and the current for an applied voltage of 100 V to the parallel combination.*

Solution. For a parallel circuit having R, L in one branch and C in the other, the resonant frequency is

$$\frac{1}{2\pi}\sqrt{\frac{1}{LC} - \frac{R^2}{L^2}}$$

Hence in this case,

$$(1/LC) - R^2/L^2 = (10^6/0.5 \times 100) - (50^2/0.5^2) = 10{,}000.$$

$$f = 100/2\pi = 15.9 \text{ c/s.}$$

At resonance the impedance is resistive and has the value of L/CR (dynamic impedance).

$$L/CR = 0.5 \times 10^6/100 \times 50 = 100 \text{ ohms.}$$

Parallel resonance curves showing the variation of current and p.f. with frequency are shown in Fig. 4.20. Note that the current is not a minimum at the resonant frequency f_r, but at a near frequency f_r'. But, if $R_1 = R_2 = 0$, (*i.e.*, if both the parallel branches have no resistance), then f_r and f_r' will coincide and the current I will be zero. This can be shown as follows.

If $R_1 = R_2 = 0$, then

$$\omega_r = 1/\sqrt{LC}$$

Hence, the current, from (4.14), putting $R_1 = 0$,

$$I_t = 0$$

which is evidently the minimum value.

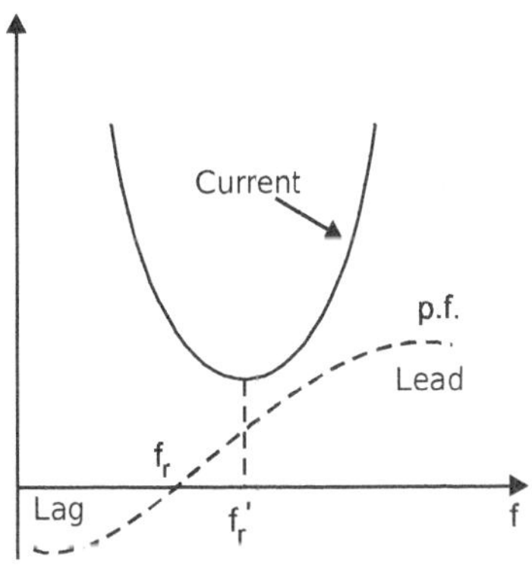

Fig. 4.20

A parallel resonant circuit therefore admits a very minimum current at resonant frequency. Therefore it can be employed as a "wave-trap" to block as completely as possible any unwanted frequency by *tuning* the circuit to be resonant at that frequency. Hence the name '*rejector resonance*' to the parallel resonance.

The parallel resonant circuit has a minimum current near resonance and hence the maximum impedance at that frequency. Hence the circuit will absorb maximum power from a high impedance voltage source at the resonant frequency. This finds another application of the parallel resonant circuit.

EXERCISES

1. A direct voltage of 10 V is applied to a 1 Ω resister in series with an 1 H inductor. What is the current 0.1 sec after switching on and what is the time taken for the current to reach one half of its final value? [**Ans.** 0.95 A, 0.69 sec.]

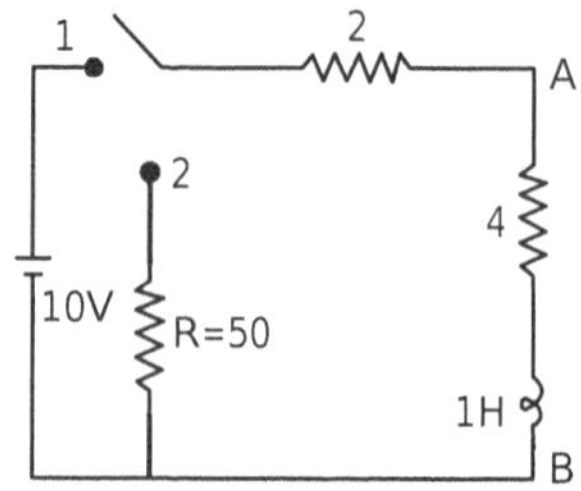

Fig. 1

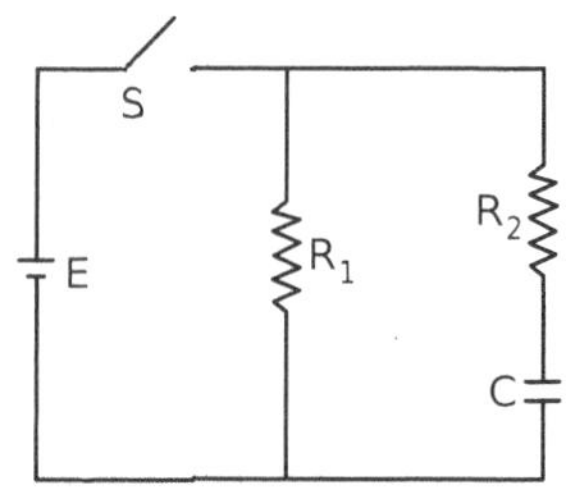

Fig. 2

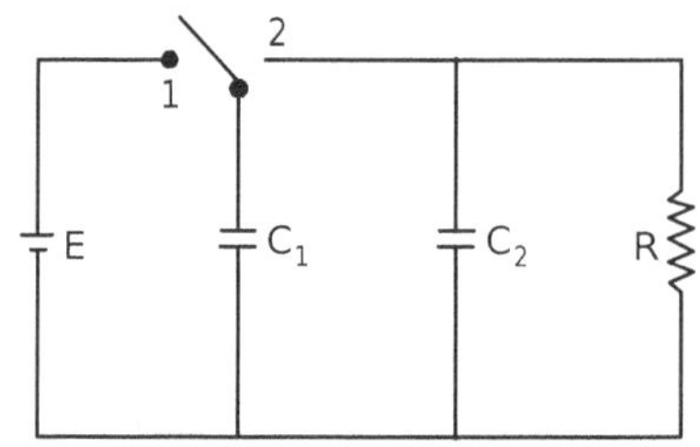

Fig. 3

2. The switch (Fig. 1) is initially in position 1 and then switched to position 2. Find V_{AB} at the instant of switching, if R = 50 Ω. If this voltage is to be reduced to 10 V, what will be the new R? [**Ans.** 83.67 V, 4 Ω]

3. After S is closed (Fig. 2), calculate the time at which the current is 1 A. E = 40 V, R_1 = 50, R_2 = 70 and C = 10 μF. [**Ans.** 73.5 milliseconds]

4. The switch in Fig. 3 is rotating with a uniform speed of n rotations per second alternatively making brief contacts with terminals 1 and 2. Find the maximum voltage across C_2 and the voltage fluctuation.

$$\textbf{Ans. } V_{max} \approx \frac{C_1 E}{(C_1 + C_2 - C_2 x)} \text{ where } x = \exp.(-t/n\, C_2 R) \text{ and } V_{min} = x\, V_{max.})$$

[**Hint.** When C_1 meets the battery, it gets charged to C_1E coulombs. When it touches C_2, it imparts part of its charge to C_2. After a number of rotations, the max. and min. values of the voltage across C_2 would have settled. If V_1 is the max. voltage across C_2, it as would discharge, during one rotation, to V_2, the minimum value. V_2 is got

$$V_2 = V_1 \exp(-1/nC_2R), \tag{a}$$

as the time of discharge is $1/n$ and C_2R is the time-constant.

Let us put $x = \exp.(-1/nC_2R)$. When the switch touches C_2, the charges on C_1 and C_2 are C_1E and C_2xV_1 respectively. The charges distribute immediately so that

$$(C_1 + C_2) V_1 = C_1E + C_2xV_1 \tag{b}$$

thereby changing C_2 to the max. voltage V_1. Then C_1 separates from C_2 leaving it to discharge. V_1 can be found from (*b*) and V_2 from (*a*)].

5. A resistor R and a 10 μF capacitor are in series with a switch across a 150 V d.c. supply. A neon lamp whose striking potential is 90 V, is across the capacitor. Find the value of R if the neon strikes 5 sec. after the switch is closed. [**Ans.** 540 Kilo-ohm]

6. A coil has 1 H inductance but negligible resistance and carries a d.c. of 1 ampere. A capacitor of 100 μF is suddenly connected across the capacitor. Find the maximum value of the voltage across the capacitor and the instant at which it occurs.

[**Ans.** 100 V, 15.9 ms]

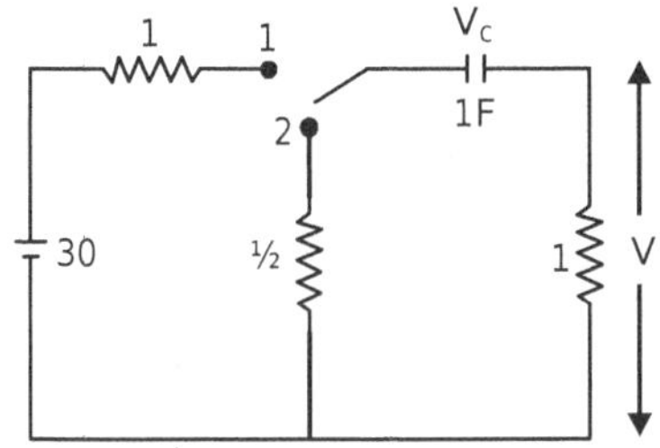

Fig. 4

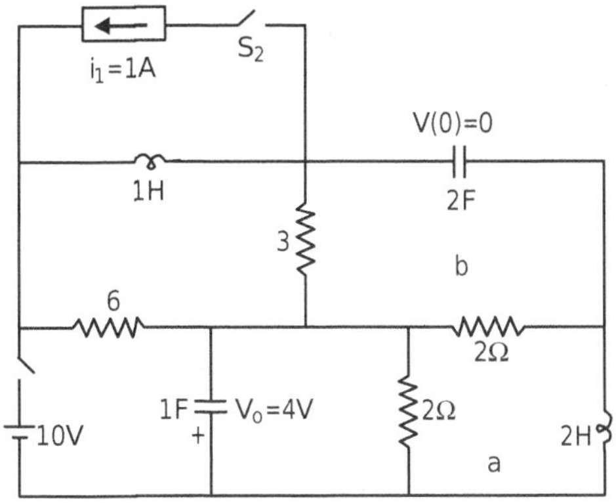

Fig. 5

7. In Fig. 4, switch is held in position 1 so long until v_c reaches 50% of the battery value and then to position 2. Find when $v(t)$ reaches – 5 Volts. [**Ans.** Total time = 2.43 secs.]

8. In Fig. 5, find all the currents and voltages at $t = 0^+$ and $t \to \infty$. S_1 is closed at $t = 0$ and S_2 opened at $t = 0$.

[**Ans.**		1 F	2 F	1H	2 H	6 Ω	3 Ω	2_a Ω	2_b Ω
$t = 0$	i	5.33	+0.6	1	0	2.38	0.4	2	0.6
	v	– 4	0	12.8	– 2.8	14	1.2	4	1.2
$t = \infty$	i	0	0	20/9	5/3	10/9	20/9	5/3	5/3
	v	10/3	10	0	0	—	—	—	—]

[**Hint.** The inductor carrying current I at $t = 0$ is equivalent to a source of value I in parallel with the inductor. A capacitor with initial charge is equivalent to a battery in series with the capacitor. For finding values at $t = 0$, open circuit the inductor and short circuit the capacitor, leaving the current and voltage sources only in place. For $t = \infty$, inductor is short circuited, capacitor is open circuited.]

9. A 1/9 Farad capacitor, initially charged to 10 V is switched to an inductor of 1 H in series with a resistor of 10 Ω. Find the voltage across the resistor after switching. What is the time taken to reach the maximum voltage and what is the rate of change of this voltage at $t = 0$? [**Ans.** $10\,(e^{-t} - e^{-9t})$, 0.55 sec; –80 V/sec.]

10. A 1 F capacitor, initially charged to 10 volts, is connected across an L-R series circuit (L = 2 H, R = 2 Ω) at $t = 0$. Find the maximum current in the circuit, the instant at which it occurs and the charge lost by the capacitor by that time.

$$\left[\textbf{Ans. } i = -10^{\,0.5t} \sin\frac{1}{2}t,\ i_m = 3.23 \text{ at } \pi/2 \text{ secs. } 3.54 \text{ Coulomb.}\right]$$

11. In a series RLC circuit, as the frequency is varied keeping the supply voltage constant at 10 volts, it is observed that the power consumed by the circuit is maximum at a frequency of f_0 cycles/sec and is one watt. (*i*) Determine the R (*ii*) Show that if equal values of power are consumed by the circuit at two frequencies f_1 and f_2 then $f_1 f_2 = f_0^2$. (*iii*) If the voltage across C is 100 V at f_0, determine the Q factor.

[**Ans.** (*i*) 100 (*iii*) 10]

12. The bandwidth of a series resonant circuit is 0.05 times the resonant frequency. Find the resonant frequency if L/R = 10 milliseconds. [**Ans.** ω = 2000]

13. Determine the angular frequencies for which the following circuit is in resonance (Fig. 6).

[**Ans.** rejector at $\omega = 1/\sqrt{2}$, accepter at $\omega = \sqrt{\frac{1}{2}(2+\sqrt{3})}$]

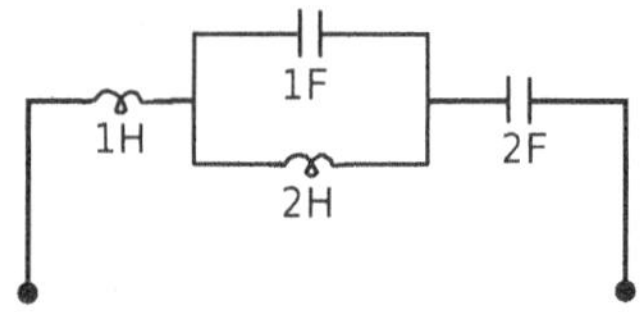

Fig. 6

SIMPLE EXERCISES

1. The time constant of an R-L circuit is

(*a*) L/R (*b*) R/L (*c*) L^2/R (*d*) LR [**Ans.** (*a*)]

2. The time constant of a C-R circuit is

(*a*) CR (*b*) C/R (*c*) $\frac{1}{CR}$ (*d*) R + L [**Ans.** (*a*)]

3. If C is in microfarads and R is a in Kilo-ohm, the time constant is in ....

[**Ans.** millisec]

4. The formula for a series resonant frequency is .... [**Ans.** $f_r = \frac{1}{2\pi\sqrt{LC}}$]

5. How does the resonant frequency of an R-C-L series circuit change with R ?

[**Ans.** No Change]

6. What is the resonant frequency of a parallel resonant circuit with R-L and C parallel branches ?

[**Ans.** $\omega_r = \sqrt{\frac{1}{LC} - \frac{R^2}{L^2}}$]

7. The Q of a resonant circuit is .... [**Ans.** $\left(\frac{\omega_r L}{R}\right) = \left(\frac{1}{\omega_r CR}\right)$]

8. The dynamic resistance of a parallel resonant circuit is .... [**Ans.** L/CR]

5

COUPLED CIRCUITS

5.1 MUTUAL INDUCTANCE

Two coils kept side by side such that the magnetic flux due to a current in one links the other (either partly or wholly) have a mutual effect between them.

Let φ_1 be the flux produced by coil 1 due to a current i_1 in it. A part of this flux links coil 2 also and let us call it ϕ_{1m}, the mutual flux of φ_1. We know that

$$N_1\varphi_1 = Li_1$$

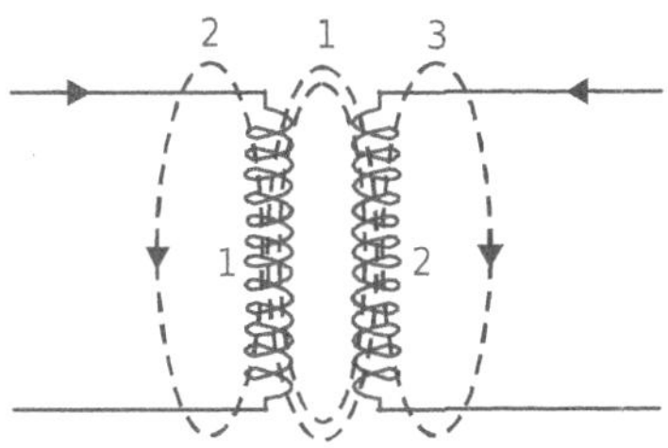

Fig. 5.1

If i_1 changes, φ_1 also changes and therefore φ_{1m}. The flux φ_{1m} links coil 2 and so induces and e.m.f. in coil 2, $e_2 = -\, Nd\varphi_{1m}/dt$. (5.1)

Since φ_{1m} is proportional to the current producing it, we may write

$$e_2 = M_{12}\frac{di_1}{dt} \tag{5.2}$$

where M_{12} is called the *mutual inductance* due to flux in coil 1 inducing e.m.f. in coil 2.

Similarly, if coil 2 carries current i_2, it will produce a flux φ_2, part of which namely φ_{2m} will link coil 1. It will induce e.m.f. in coil 1 if i_2 is changing.

$$e_1 = -\, N_1 d\varphi_{2m}/dt \tag{5.3}$$

which may be written $\qquad e_1 = M_{21}\, di_2/dt$ (5.4)

M_{21} is the mutual inductance due to flux of coil 2 inducing e.m.f. in coil 1. So, from (5.1 to 5.4)

$$M_{12} = N_2\varphi_{1m}/i_1 \text{ and } M_{21} = N_1\varphi_{2m}/i_2 \tag{5.5}$$

Example 1. *Two coils A and B of 500 and 800 turns respectively are wound with their axes parallel to each other. 50% of the flux produced links coil B. A current of 1 amp in coil A produces 0.04 milliweber of flux, but in coil B, the same current produces 0.02 mWb only. Find the self inductances of A and B, and the mutual inductance.*

Solution.

$$L_1 = \frac{N_1\varphi_1}{i_1} = \frac{500\times 0.04\times 10^{-3}}{1} = 20 \text{ mH}.$$

$$L_2 = \frac{N_2\varphi_2}{i_2} = \frac{800\times 0.02\times 10^{-3}}{1} = 16 \text{ mH}.$$

$$M_{12} = \frac{N_2\varphi_{1m}}{i_1} = \frac{800\times(0.5\times 0.04)\times 10^{-3}}{1} = 16 \text{ mH}.$$

Co-efficient of coupling. The ratio of φ_{1m}/φ_1 represents the fraction of the flux produced by coil 1 and linking coil 2 and the ratio φ_{2m}/φ_2 that fraction of the flux produced by coil 2 which links coil 1. The geometric mean of these ratios is defined as the co-efficient of coupling K.

$$K = \left(\frac{\varphi_{1m}}{\varphi_1}\frac{\varphi_{2m}}{\varphi_2}\right)^{1/2} \tag{5.6}$$

$$= \left(\frac{M_{12}/N_2}{L_1/N_1}\frac{M_{21}/N_1}{L_2/N_2}\right)^{1/2}$$

$$\therefore \quad K^2 = \frac{M_{12}\,M_{21}}{L_1\,L_2}$$

It is proved later that $M_{21} = M_{12}$.

Hence $K = M/\sqrt{L_1L_2}$

In the above problem, the coupling co-efficient

$$K = 16/\sqrt{20\times 16} = 0.9. \tag{5.7}$$

Case of Both Coils Carrying Current

In this case, induced voltages appear across both the coils. The induced e.m.f. in coil 1 consists of self induced e.m.f. and the mutually induced e.m.f. and they may either add up or oppose each other so that the M di/dt term has either a + or a – sign.

$$e_1 = -L_1\frac{di_1}{dt} \pm M_{21}\frac{di_2}{dt}$$

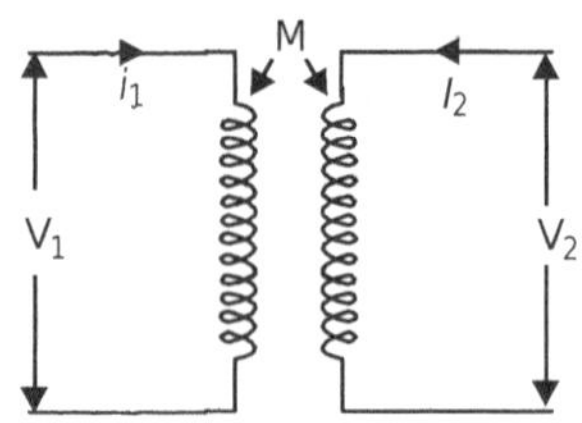

Fig. 5.2

Similarly, $$e_2 = \pm M_{12}\frac{di_1}{dt} - L_2\frac{di_2}{dt}$$

The sign of the mutually induced voltage must be found by experiment . The applied voltage v_1 and v_2 to the coils must be equal and opposite to the induced e.m.f.'s so that

$$v_1 = L_1\frac{di_1}{dt} \pm M_{21}\frac{di_2}{dt}$$

$$v_2 = \pm M_{12}\frac{di_1}{dt} + L_2\frac{di_2}{dt} \quad (5.8)$$

The sign of the mutually induced voltage can be known if the winding of the coils in the core is known. In Fig. 5.3, the winding sense of the coils is clearly shown. Assume a current i_1 entering A and increasing. It produces a magnetic flux φ which mutually links coil 2 as shown. Now a current must flow in coil 2 in such direction that its flux opposes the flux φ, *i.e.*, from C to D *via* the external resistor shown. The polarity is thus + at C and – at D. So, we find the polarity of the mutually induced voltage.

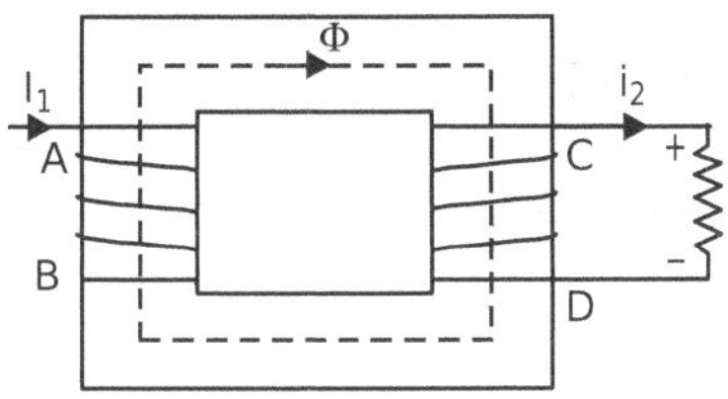

Fig. 5.3

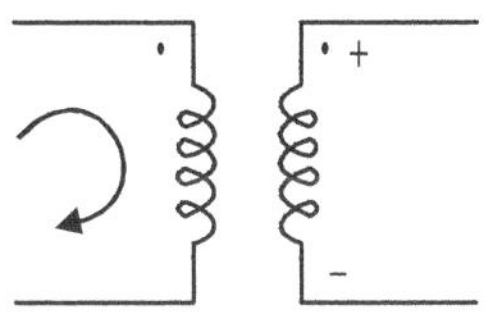

Fig. 5.4

When representing mutual inductances in circuits, the dot notation is frequently employed to depict the sign of the mutual voltage. In Fig. 5.4, the significance of the dot markings is shown. If current enters into the dot and is increasing in coil 1, then the polarity of the voltage is positive at the dot in the second coil. So in the equations 5.8, the sign of the M term will be positive if both currents i_1, i_2 are entering the dots, otherwise it is negative.

5.2 TO PROVE THAT $M_{12} = M_{21} = M$

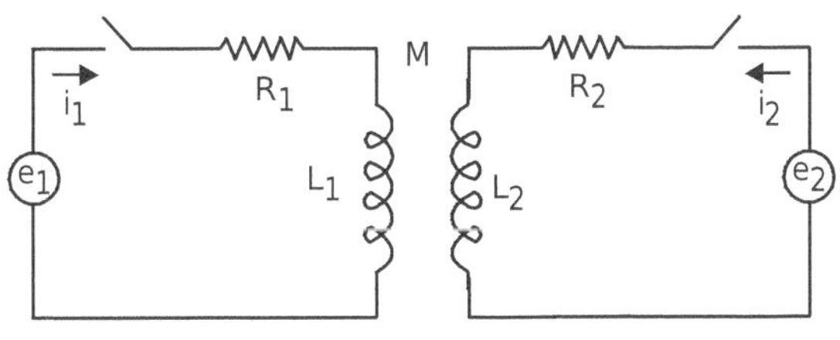

Fig. 5.5

In the circuit shown above, the equations can be written :

$$e_1 = R_1 i_1 + L_1\frac{di_1}{dt} + M_{21}\frac{di_2}{dt}$$

$$e_2 = R_2 i_2 + L_2\frac{di_2}{dt} + M_{12}\frac{di_1}{dt} \quad (5.9)$$

Let us increase the current from 0 to I_1 in coil 1 and 0 to I_2 in coil 2. Then the work done will be $\int e_1 i_1\, dt$ by the source e_1 and $\int e_2 i_2\, dt$ by source 2. Hence the total work done is

$$\int e_1 i_1\, dt + \int e_2 i_2\, dt = \int R_1 i_1^2\, dt + \int R_2 i_2^2\, dt + \int_0^{I_1} L_1 i_1\, di_1 + \int_0^{I_2} L_2 i_2\, di_2$$
$$+ \int_0^{I_2} M_{21} i_1\, di_2 + \int_0^{I_1} M_{12} i_2\, di_1 \tag{5.10}$$

Let us first increase i_1 from 0 to I_1 holding $i_2 = 0$ and then increase i_2 from 0 to I_2, holding $i_1 = I_1$. Then the work done expression above, omitting the resistance terms,

$$W_m = \frac{1}{2}L_1 I_1^2 + \frac{1}{2}L_2 I_2^2 + M_{21} I_1 I_2$$

The above represents the magnetic stored energy in the coils for $i_1 = I_1$ and $i_2 = I_2$.

If, on the other hand, we first increase i_2 from 0 to I_2 keeping $i_1 = 0$ and then increase i_1 from 0 to I_1 keeping i_2 at I_2, we get for the same stored energy.

$$W_m = \frac{1}{2}L_1 I_1^2 + \frac{1}{2}L_2 I_2^2 + M_{12} I_2 I_1$$

Since both W_m values represent the same stored energy, we get

$$M_{12} = M_{21} = M.$$

Hence
$$W_m = \frac{1}{2}L_1 I_1^2 + \frac{1}{2}L_2 I_2^2 + M\, I_1 I_2 \tag{5.11}$$

We can rewrite the equation as

$$\left[\frac{I_1}{I_2}\right]^2 + \frac{2M}{L_1}\left[\frac{I_1}{I_2}\right] + \frac{L_2}{L_1} = \frac{2\,W_m}{L_1 I_2^2}$$

The right hand side contains W_m which is +ve as well as L_1 and I_2^2 which are also positive. So, putting $I_1/I_2 = x$,

$$x^2 + \left(\frac{2M}{L_1}\right)x + \left(\frac{L_2}{L_1}\right) > 0$$

If this quadratic expression is to be positive, i.e., if

$$(x - \alpha)(x - \beta) > 0, \tag{5.12}$$

for all values of x, then α, β will only be imaginary or equal; as otherwise, for some value of x, (between α and β) the above expression will become negative.

Hence the discriminant $(2M/L_1)^2 - 4L_2/L_1 \le 0$

$$M^2 \le L_1 L_2$$

Hence
$$K = M/\sqrt{L_1 L_2} \le 1. \tag{5.13}$$

Hence the coupling co-efficient is always less than 1, at most = 1.

5.3 MUTUAL INDUCTANCE TREATED AS A DEPENDENT SOURCE CIRCUIT

In the following circuit, the currents are not both entering the dots. Hence

$$e_1 = R_1 i_1 + L_1 di_1/dt - M di_2/dt$$

The term $-M di_2/dt$ is represented as a dependent voltage source in circuit 1 since it depends on current i_2. Similarly, since i_1 enters the dot in coil 1, the term $M\,di_1/dt$ appears in coil 2 as shown.

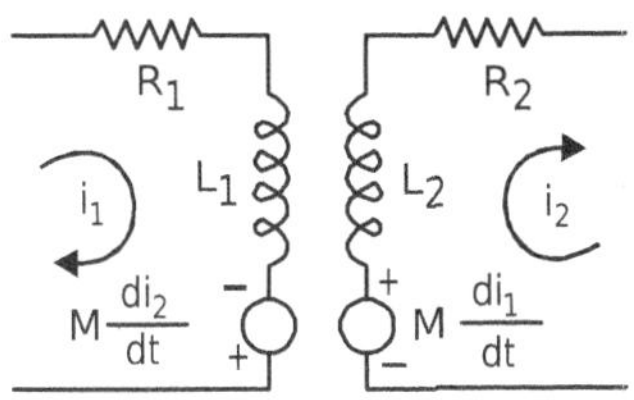

Fig. 5.6

5.4 MUTUAL INDUCTANCE BETWEEN THREE COILS

When three coils are wound so that the magnetic flux due to any one links the other two, we have a set of 3 coupled inductances. The dot markings will have to be different for every pair as shown in Fig. 5.7 (*b*). By knowing the winding sense, the marking can be

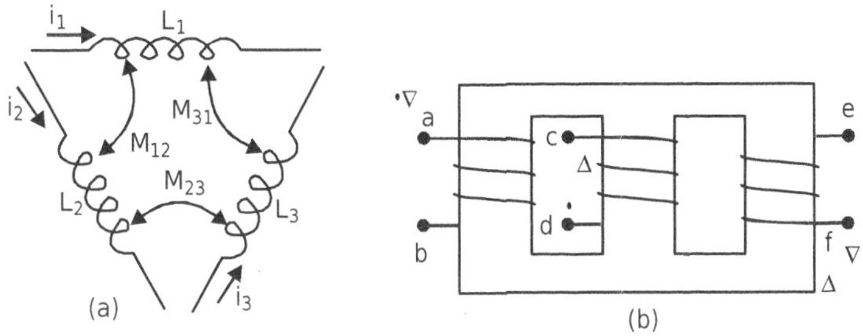

Fig. 5.7 Three coils coupled together.

made as in the case of Fig. 5.3. The equations of the applied voltages v_1, v_2, v_3 to the coils will be written as

$$v_1 = L_1\frac{di_1}{dt} + M_{12}\frac{di_2}{dt} + M_{13}\frac{di_3}{dt}$$

$$v_2 = M_{21}\frac{di_1}{dt} + L_2\frac{di_2}{dt} + M_{23}\frac{di_3}{dt}$$

$$v_3 = M_{31}\frac{di_1}{dt} + M_{32}\frac{di_2}{dt} + L_2\frac{di_3}{dt} \tag{5.14}$$

and $$M_{12} = M_{21};\ M_{23} = M_{32};\ M_{13} = M_{31}.$$

5.5 MUTUAL INDUCTANCES IN SERIES AND PARALLEL

Fig. 5.8 shows the two coils of a mutual inductor connected in series. The voltage across both the coils will be equal to e, the applied voltage. Thus, since both coils carry the same current i,

$$e = L_1 \frac{di}{dt} \pm M \frac{di}{dt} + L_2 \frac{di}{dt} \pm M \frac{di}{dt}$$

$$= (L_1 + L_2 \pm 2\,M)\,(di/dt) \tag{5.15}$$

The ± sign depends on the positions of the dots in the coil. If both currents are entering the dots, then + 2 M applies; otherwise – 2 M. The coils are said to be *series aiding* or *series opposing* respectively.

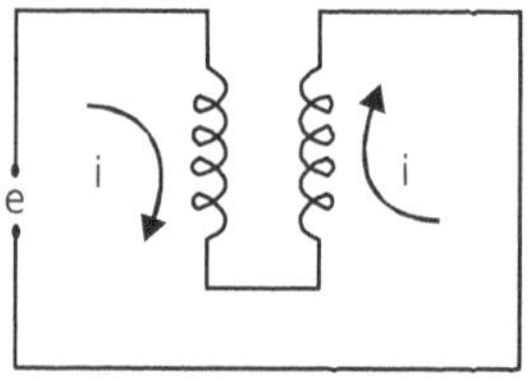

Fig. 5.8 Coils in series.

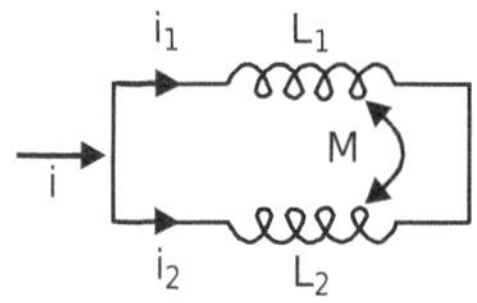

Fig. 5.9 Coils in parallel.

When the same coupled coils are connected in parallel, as in Fig. 5.9, the voltage across each coil is the same.

$$e = L_1(di_1/dt) \pm M(di_2/dt)$$

$$= \pm M(di_1/dt) + L_2(di_2/dt)$$

$$\therefore \qquad \frac{di_1}{dt} = \begin{vmatrix} e & \pm M \\ e & L_2 \end{vmatrix} \div (L_1L_2 - M^2)$$

$$\frac{di_2}{dt} = \begin{vmatrix} L_1 & e \\ \pm M & e \end{vmatrix} \div (L_1L_2 - M^2)$$

$$\frac{di}{dt} = \frac{di_1}{dt} + \frac{di_2}{dt} = \frac{e\left[L_2 - (\pm M) + L_1 - (\pm M)\right]}{L_1L_2 - M^2}$$

Hence the equivalent inductance is given by

$$e \div (di/dt) = \left[L_1L_2 - M^2\right] \div \left[L_1 + L_2 - (\pm 2\,M)\right] \tag{5.16}$$

Example 2. *In the circuit of Fig. 5.5, $L_1 = 0.2$, $L_2 = 0.4$, $M = 0.1$, $R_1 = 2$, $R_2 = 1.6$. If i_1, i_2 are both entering the dots, and their values are such that $i_1 = 2 \sin 4t$, $i_2 = 10 \cos 4t$ (for $t > 0$), find the values of e_1 and e_2 and the stored energy at time t in the coil.*

Solution.

$$e_1 = R_1 i_1 + L_1\, di_1/dt + M\, di_2/dt$$

$$= 4 \sin 4t + 0.2\,(8 \cos 4t) + 0.1\,(-\,40 \sin 4t) = 1.6 \cos 4t$$

$$e_2 = R_2 i_2 + L_2\, di_2/dt + M\, di_1/dt$$

$$= 1.6\,(10 \cos 4t) + 0.4\,(-\,40 \sin 4t) + 0.1\,(8 \cos 4t)$$

$$= 16.8 \cos 4t - 16 \sin 4t$$

Energy stored in the coils at time t is

$$= \frac{1}{2}L_1 i_1^2 + \frac{1}{2}L_2 i_2^2 + M\, i_1 i_2$$

$$= 20 \cos^2 4t + 0.4 \sin^2 4t + \sin 8t.$$

5.6 EQUIVALENT CIRCUITS FOR MUTUAL INDUCTANCE

For two coils, as in Fig. 5.10 (*a*),

$$e_1 = L_1\frac{di_1}{dt} + M\frac{di_2}{dt};$$

$$e_2 = M\frac{di_1}{dt} + L_2\frac{di_2}{dt}$$

These can also be written as

$$e_1 = (L_1 - M)\,di_1/dt + M(di_1/dt + di_2/dt)$$

$$e_2 = M(di_1/dt + di_2/dt) + (L_2 - M)\,di_2/dt \tag{5.17}$$

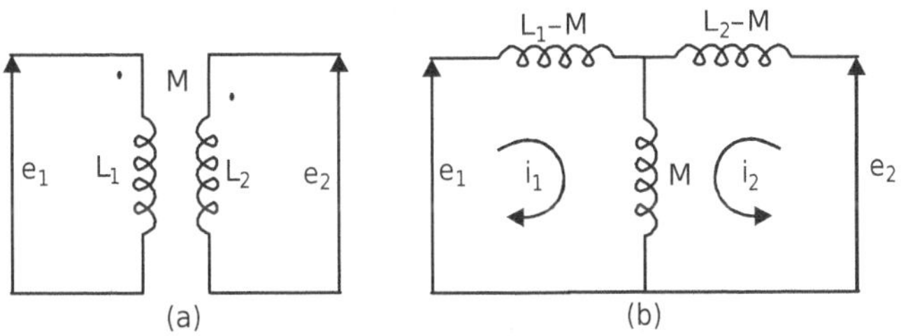

Fig. 5.10 An equivalent circuit for mutual coupled coils.

The circuit of Fig. 5.10 (*b*) also gives the above loop equation (5.17) and the circuit is equivalent to the mutual inductors. Similarly, the following pairs of circuits can be shown to be equivalent.

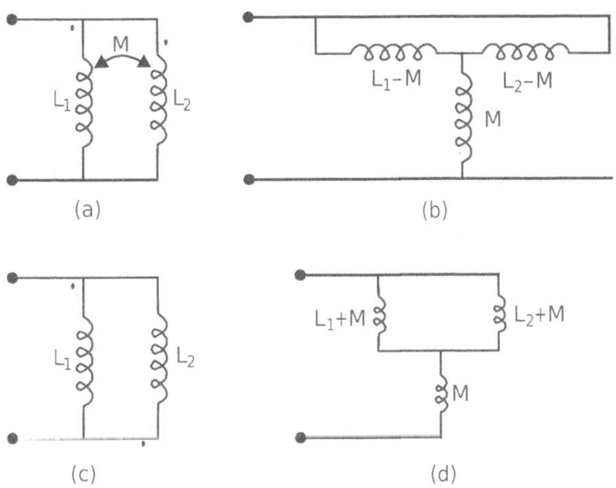

Fig. 5.11 The coupled circuits in (*a*) are equivalent to the circuits (*b*).

5.7 THE IDEAL TRANSFORMER

Two coupled coils in which the coupling is perfect, *i.e.*, K = 1 and in which L_1 and L_2 are very high, is called the ideal transformer.

$$e_1(t) = L_1\, di_1/dt + M\, di_2/dt$$

$$e_2(t) = L_2\, di_2/dt + M\, di_1/dt$$

$$\frac{e_1(t)}{L_1} = \frac{di_1}{dt} + \frac{M}{L_1}\frac{di_2}{dt} \qquad \frac{e_2(t)}{M} = \frac{di_1}{dt} + \frac{L_2}{M}\frac{di_2}{dt}$$

Since $M/L_1 = L_2/M$ when $K = 1$,

$$e_1(t)/L_1 = e_2(t)/M$$

$$\frac{e_1(t)}{e_2(t)} = \frac{L_1}{M} = \frac{N_1\varphi_1/i_1}{N_2\varphi_{1m}/i_1} = \frac{\varphi_1}{\varphi_{1m}}\frac{N_1}{N_2} = \frac{N_1}{N_2} \qquad (\because \varphi/\varphi_{1m} = 1 \text{ if } K = 1)$$

Similarly,

$$\frac{L_1}{M} = \frac{N_1}{N_2} = \frac{M}{L_2} \tag{5.18}$$

Thus *the ratio of the voltage* $e_1(t)/e_2(t)$ *is the turns ratio* N_1/N_2. (5.19)

Also,

$$\frac{di_1}{dt} + \frac{M}{L_1}\frac{di_2}{dt} = \frac{e_1(t)}{L_1}$$

If L_1 is very large, $e_1(t)/L_1 \simeq 0$ so that

$$di_1/dt \quad di_2/dt = -M/L_1 = -N_2/N_1$$

i.e.,

$$i_1/i_2 = -N_2/N_1 \tag{5.20}$$

The *current ratio is the reciprocal of the turns ratio*, if L_1 is very high, as in the ideal transformer. Suppose we have a resistance R connected to the second coil as in Fig. 5.10. Then

$$e_2/i_2 = R \;\therefore\; e_1/i_1 = R(e_1/e_2)(i_2/i_1) = R(N_1/N_2)^2$$

which means that the input resistance is $(N_1/N_2)^2$ times the resistance across the second coil. So the transformer enables the resistance to be transformed at its input terminals.

Example 3. *A loudspeaker of 8 ohms resistance is to be matched to the output of a valve such that it appears as a resistance load of 200 ohms. Find the turns ratio of a suitable transformer to match the load.*

Solution.

$$R_{in} = 200 = R(N_1/N_2)^2 = 8a^2 \qquad (a = \text{turns ratio})$$

$$\therefore \quad a = \sqrt{200/8} = 5.$$

Thus in an ideal transformer the voltage ratio (e_1/e_2) the current ratio (i_1/i_2) and turns ratio (N_1/N_2) are equal.

$$\frac{e_1}{e_2} = \frac{N_1}{N_2} = \frac{i_1}{i_2}$$

The *linear* transformer is also the ideal transformer.

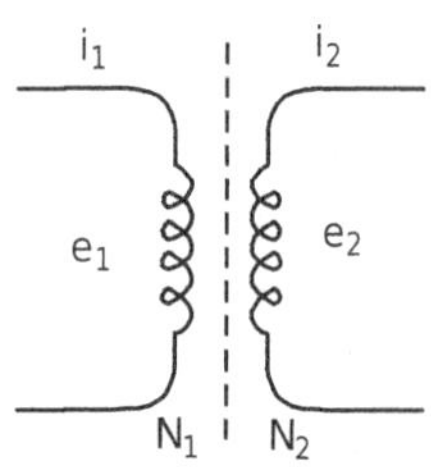

Fig. 5.12 Ideal transformer.

5.8 A.C. CIRCUITS WITH MUTUAL INDUCTANCE—MAGNETICALLY COUPLED COILS

A mutual inductance M introduces an induced voltage M di/dt due to a change in current of di/dt in one of its coils. If i = I sin ωt is the current flowing in one of the coils, the e.m.f. induced in the other will be

$$e = M\, d/dt\, (I \sin \omega t)$$
$$= MI\omega \cos \omega t$$

If we use the complex form of current $i = I\, e^{j\omega t}$,

$$e = M\, d/dt\, (I\, e^{j\omega t})$$
$$= j\omega MI\, e^{j\omega t} = j\omega Mi$$

Hence the mutual reactance which is the ratio of voltage induced in one of the coils due to a change in current in the other is $j\omega$M. The sign $j\omega$M may be positive or negative.

Example 4. *Find the currents I_1, I_2 in the circuit shown below.*

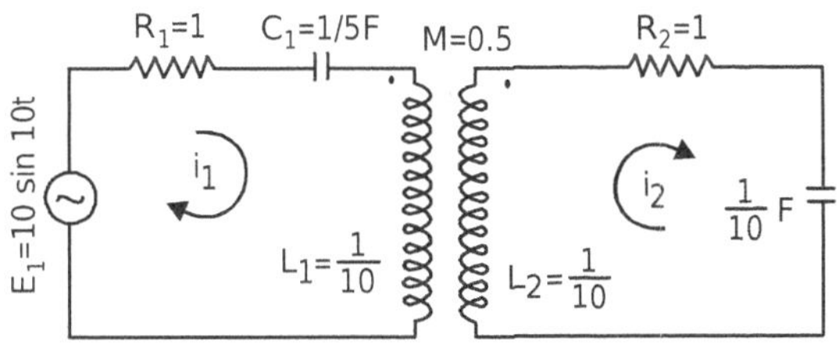

Fig. 5.13

Solution. The e.m.f. applied is E_1 = 10 sin 10t.

The values R_1, R_2, L_1, L_2, C_1, C_2 etc. are marked in the figure. Mark two loop currents as I_1 and I_2. Then writing KVL equations, and taking E_1 = 10 to be the reference phasor,

$$10 + j0 = [R_1 + (1/j\omega C_1) + j\omega L_1]\, I_1 - j\omega M I_2$$

The – sign in $j\omega$M is since I_1 enters the dot and I_2 leaves the dot in the mutual inductance M. Also,

$$j\omega M I_1 = [R_2 + j\omega L_2 + 1/j\omega C_2]\, I_2$$

Substituting ω = 10 and for the given element values,

$$10 = (1 - j0.5 + j1)\, I_1 - j0.5\, I_2 \qquad (a)$$
$$j0.5\, I_1 = (1 + j1 - j)\, I_2 \qquad (b)$$

Substituting for I_2 from (b),

$$10 = (1 + 0.5j)\, I_1 - 0.5j\, (0.5j\, I_1)$$
$$= (1.25 + 0.5j)\, I_1$$
$$I_1 = 10/(1.25 + 0.5j) = 7.4 \angle -21°48'$$
$$I_1 = 7.4 \sin (10t - 21° 48')$$
$$I_2 = j0.5\, I_1 = 3.7 \sin (10t + 90° - 21° 48')$$
$$= 3.7 \sin (10t + 58° 12').$$

Example 5. *X_1 and X_2 are self reactances of the two windings. X_m is the mutual reactance and X_C is the reactance of the capacitor in the coupled circuit shown in Fig. 5.14. Show that this network behaves with respect to the terminals A and B, as open circuit or short circuit according as $X_C = X_2$ or $X_C = (X_1 X_2 - X_m^2)/X_1$.*

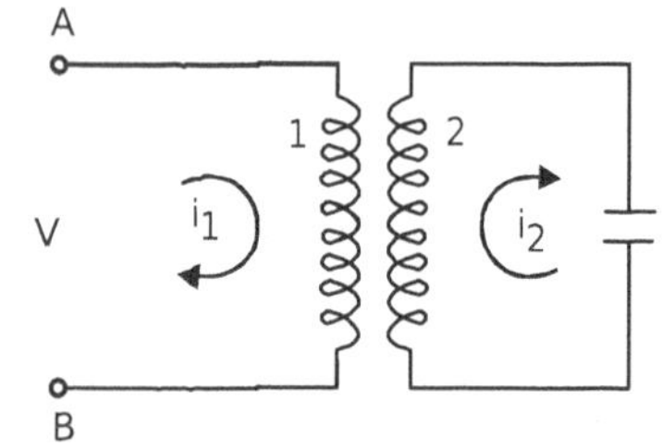

Fig. 5.14

Solution. Writing Kirchhoff's law equation for loops 1 and 2,

$$V = jL_1\omega\, I_1 + Mj\omega\, I_2 \tag{1}$$

$$0 = Mj\omega\, I_1 + L_2 j\omega\, I_2 - I_2\, j/\omega C. \tag{2}$$

Substituting for I_2 from (2) into (1),

$$V = j\omega L_1 I_1 + (Mj\omega)^2 \frac{I_1}{-L_2 j\omega + jX_C}\; j\omega I_1 \left(L_1 - \frac{M^2\omega}{-X_C + L_2\omega} \right)$$

If $X_C = X_2 = L_2\omega$, then the second term within the brackets becomes infinite and so, $I_1 = 0$. Thus $X_C = X_2$ represents open-circuit at the terminals AB.

If $\quad L_1(-X_C + X_2) = M^2\omega.$

We see that I_1 becomes ∞ for a given V. Hence it represents a short-circuit between A and B.

$$-X_C = \left[M^2\omega/L_1\right] - L_2\omega$$

$$= (X_m^2/X_1) - X_2$$

giving $\quad X_C = (X_1X_2 - X_m^2)/X_1.$

Example 6. *L_1 and L_2 represent two coupled coils with a co-efficient of coupling $1/\sqrt{2}$. The polarity is shown by the dots. Calculate for a voltage of 100 sin 1000t, the currents I_1 and I_2.*

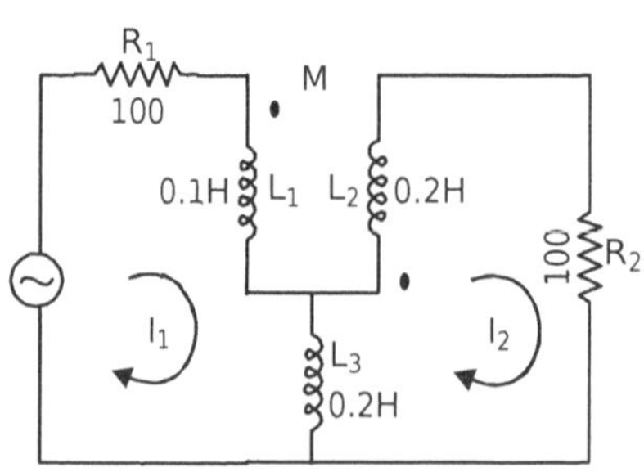

Fig. 5.15

Solution. $$M = \sqrt{L_1L_2}/\sqrt{2} = \sqrt{0.1\times 0.2/2} = 0.1.$$

Let us write the KVL equations for the meshes 1 and 2.

$$\begin{aligned}
100 \sin 1000t &= (R_1 + j\omega L_1 + j\omega L_3)\, I_1 - j\omega L_3 I_2 + j\omega M\, I_2 \\
&= (100 + 100j + 200j)\, I_1 - 200j\, I_2 + 100j\, I_2 \\
&= (100 + 300j)\, I_1 - 100j\, I_2 \qquad (1)\\
0 &= j\omega M I_1 + j\omega L_2 I_2 + RI_2 + j\omega L_3 I_2 - j\omega L_3 I_1 \\
&= 100j\, I_1 + 200j\, I_2 + 100\, I_2 + 200j\, I_2 - 200j\, I_1 \\
&= -100j\, I_1 + (100 + 400j)\, I_2 \qquad (2)
\end{aligned}$$

From (2), $$I_1 = (100 + 400j)\, I_2/100j = j(1 + 4j)\, I_2.$$

$\therefore$ $$\begin{aligned}
100 \sin 1000t &= -(100 + 300j)\, j(1 + 4j)\, I_2 - 100j\, I_2 \\
&= -100j\, I_2\, (7j - 10)
\end{aligned}$$

$$I_2 = \frac{100}{-100j\,(7j-10)} = \frac{7-10j}{7^2+10^2} = 0.0831 \angle -59^\circ\, 45'$$

$\therefore$ $$\begin{aligned}
I_2 &= 0.0831 \sin (1000t - 59^\circ 45') \\
I_1 &= (-4 + j)\, I_2 = 4.12 \angle -14^\circ\, 2' \times .0831 \angle 180 - 59^\circ\, 45' \\
I_1 &= 0.342 \angle 180 - 63^\circ 47'
\end{aligned}$$

i.e., $$I_1 = 0.342 \sin (1000t - 63^\circ\, 47').$$

5.9 TUNED COUPLED CIRCUITS

A tuned coupled circuit is shown below. Such circuits are used for intermediate frequency circuits in radio receivers. The coils L_1L_2 are wound over a ferrite slug and by moving the slug, the co-efficient of coupling K can be varied. The coils 1, 2 are both connected with capacitors C_1 and C_2 so as to be resonant at a particular frequency, the intermediate frequency of the radio receiver, usually 455 Hertz. V_o is the output voltage from the secondary. V_i is the input voltage. The ratio V_o/V_i depends upon the frequency (of V_i) and is a maximum at the particular frequency. So the circuit selects a particular frequency and the tuned coupled circuit is also called a I.F. transformer. Generally C_1 is in parallel with the coil 1, instead of in series as shown in the figure, but the behaviour of the circuit is similar. If there is tuning only in one coil, (*i.e.,* only C_2 is present), it is called a *single tuned* circuit. The circuit shown in a *double tuned* circuit.

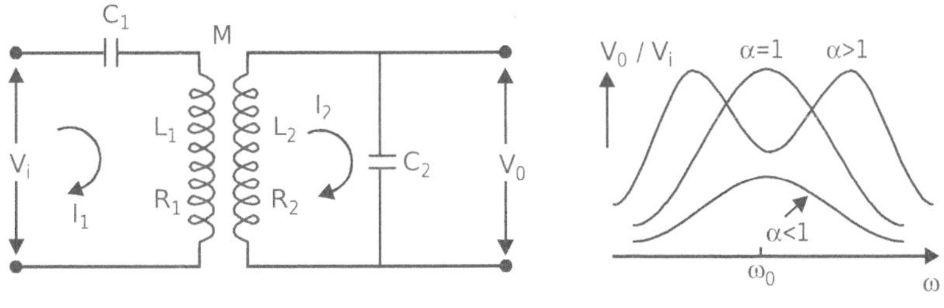

Fig. 5.16 Double tuned circuit.

Assuming currents I_1 and I_2 and if the frequency of V_i is ω,

$$V_i = (R_1 + j\omega L_1 - 1/\omega Cj)\, I_1 + j\omega M\, I_2$$

$$O = j\omega M\, I_1 + (R_2 + j\omega L_2 - 1/j\omega C_2)\, I_2$$

Solving for I_2 and then noting that $V_o = (1/j\omega C_2)\, I_2$,

$$\frac{V_o}{V_i} = \frac{(1/j\omega C_2)\, j\omega M}{\rho_{11}\rho_{22} - (j\omega M)^2} \tag{5.21}$$

where $\rho_{11} = R_1 + j(\omega L_1 - 1/\omega C_1)$ and $\rho_{22} = R_2 + j(\omega L_2 - 1/\omega C_2)$

The circuits of the two coils are tuned using C_1 and C_2 to be resonant at a particular frequency ω_0, and so

$$\omega_0 = \frac{1}{\sqrt{L_1C_1}} = \frac{1}{\sqrt{L_2C_2}} \tag{5.22}$$

The magnification factors Q_1 and Q_2 of the two coils are

$$Q_1 = \omega_0 L_1/R_1 = 1/\omega_0 C_1 R_1$$

$$Q_2 = \omega_0 L_2/R_2 = 1/\omega_0 C_2 R_2 \tag{5.23}$$

Then $\rho_{11} = R_1 + j(\omega L - 1/\omega C_1) = R_1\left[1 + j\left(\frac{\omega L}{R} - \frac{1}{\omega CR}\right)\right]$

$$= R_1\left[1 + jQ\left(\frac{\omega}{\omega_0} - \frac{\omega_0}{\omega}\right)\right]$$

Put $(\omega - \omega_0)/\omega_0 = \delta \qquad \omega/\omega_0 - 1 = \delta \qquad \omega/\omega_0 = 1 + \delta$

$$\rho_{11} = R_1\left[1 + jQ_1\,(1 + \delta - (1+\delta)^{-1})\right] \simeq R_1\left[1 + 2j\delta Q_1\right]$$

Similarly $\rho_{22} = R_2\,(1 + 2j\delta Q_2)$

$$\therefore \quad \frac{V_o}{V_i} = \frac{M/C_2}{R_1R_2(1+2j\delta Q_1)(1+2j\delta Q_2) + \omega^2M^2}$$

$$= \frac{M/C_2R_1R_2}{(1+2j\delta Q_1)(1+2j\delta Q_2) + \omega^2M^2/R_1R_2} \tag{5.24}$$

At resonance, $\omega = \omega_0$, $\delta = 0$; Let $a = \omega_0 M/\sqrt{R_1R_2}$, $a/\omega_0\sqrt{R_1R_2} = M$

$$\therefore \quad \frac{V_o}{V_i} = \frac{M/C_2R_1R_2}{1+a^2} = \frac{(a/\omega_0C_2R_1R_2)\sqrt{R_1R_2}}{1+a^2}$$

$$= \frac{[a\,Q_2/R_1]\sqrt{R_1R_2}}{1+a^2} \tag{5.25}$$

This will be maximum if the derivative of the above expression w.r.t. a is zero, which gives $a = 1$.

Then $\frac{V_o}{V_i} = Q_2\sqrt{R_1R_2}/2R_1$

For frequencies other than resonance,

$$\frac{V_o}{V_i} = \frac{aQ_2\sqrt{R_2/R_1}}{1 + a^2 + 2j\delta(Q_1 + Q_2) - 4\delta^2Q_1Q_2} \tag{5.26}$$

If α is greater than 1, V_o/V_i will not be maximum at ω_0. The plot of this ratio versus frequency for different α values is shown in Fig. 5.16.

Example 7. *A double tuned circuit for a valve amplifier is tuned to a frequency of 450 Hz, the Q factors of the primary and secondary coils being 4 and 6. The primary and secondary inductances are 20 mH and 10 mH respectively. Determine the co-efficient of coupling for maximum secondary e.m.f.*

Solution. The value for M for maximum V_o/V_i is given by

$$\omega_0 M/\sqrt{R_1R_2} = 1$$

$$M = \frac{\sqrt{R_1R_2}}{\omega_0}$$

Also, $$\omega_0 L_1/R_1 = 4;\ \omega_0 L_2/R_2 = 6$$

$$R_1R_2 = \omega_0^2 L_1L_2/(6 \times 4)$$

$\therefore$ $$M = \sqrt{L_1L_2/24}\ .$$

The co-efficient of coupling $K = M/\sqrt{L_1L_2} = 1/\sqrt{24} = 0.21.$

EXERCISES

1. The bandwidth of a series resonant circuit is 0.05 times the resonant frequency. Find the resonant frequency if L/R = 10 milliseconds. [**Ans.** ω = 2000]
2. Determine the angular frequencies for which the following circuit is in resonance. (Fig. 1)

$\left[\textbf{Ans.}\ \text{Rejector at } \omega = 1/\sqrt{2}, \text{ accepter at } \omega = \sqrt{\frac{1}{2}(2+\sqrt{3})}\right]$

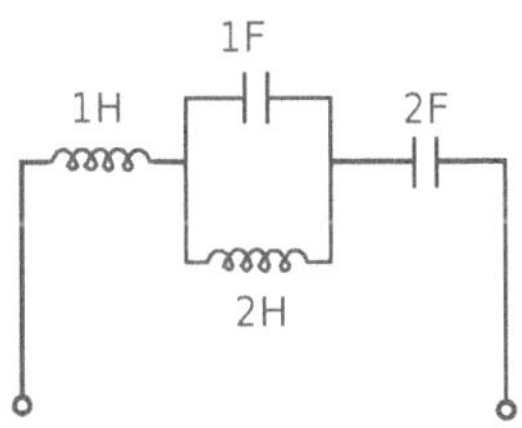

Fig. 1

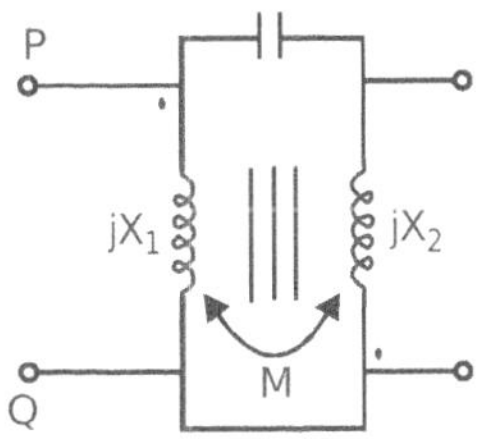

Fig. 2

3. Find the effective reactance at the input terminals of the coupled circuit (Fig. 2).

[**Ans.** $X_m^2 - X_1(X_2 - X_c) \div j(X_1 + X_2 - X_c)$]

4. Find the input impedance at $1/2\pi$ kilocycles/sec for the coupled circuit (Fig. 3).

[**Ans.** Z_{in} = 1.9 $\angle 87°$]

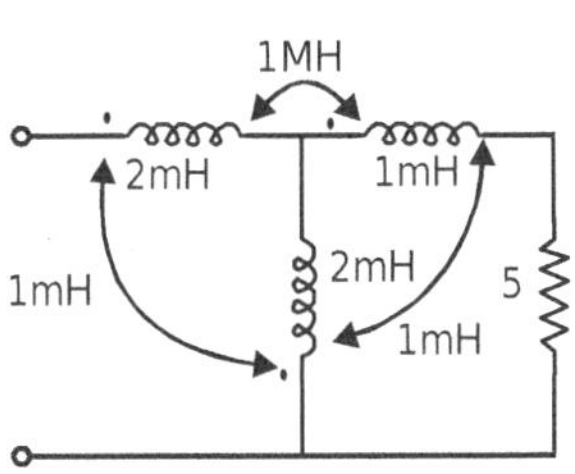

Fig. 3

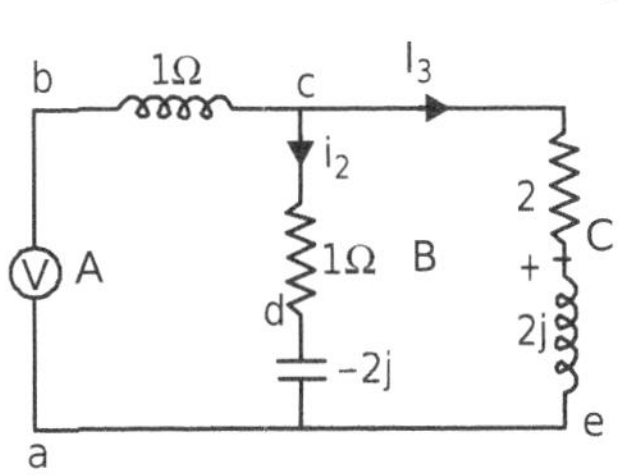

Fig. 4

5. A voltage at frequency $\omega_0 = 1/2\pi$ megacycles is applied to a coupled circuit whose primary $R_1 = 15\ \Omega$, $L_1 = 200\ \mu H$, and $C_1 = 5000$ pF. The coupled secondary has $R_2 = 8$, $L_2 = 100\ \mu H$, $C_2 = 10000$ pF. The Mutual M is 75 μH. Find the effective resistance and reactance seen by the voltage source. Neglecting resistance show that the two frequencies for which the primary curve frequency response exhibits 2 peaks is given by

$$M = \pm[L_1L_2]^{1/2}\left[1-(\omega_0/\omega)^2\right]$$

and hence find their values.

[**Ans.** 720 Ω resistive, $\omega_1 = 128$ KHz, $\omega_2 = 230$ KHz]

6. In Fig. 4, if $i_2 = 1$ A, find V_{ab} Find I_1, I_3. [**Ans.** 3.25 V, 0.8 A, 1.06].

7. In the mutual inductance bridge shown, (Fig. 5), write down the loop equations 1, 2 and 3.

[**Ans.** $[(L_1 + L_3)D + R]\, i_1 + [-R-(M_1 - M_2)D]\, i_2 - (M_2 + L_3)D\, i_3 = 0$

$-[R + (M_1 - M_2)D]\, i_1 + [R + (L_2 + L_4)D]\, i_2 - (L_4 + M_2)D\, i_3 = 0$

$-[(L_3 + M_2)D]\, i_1 - [(L_4 + M_2)D]\, i_2 + (L_3 + L_4 + 2M_2)D\, i_3 = e(t)$

Hence D denotes derivative d/dt.]

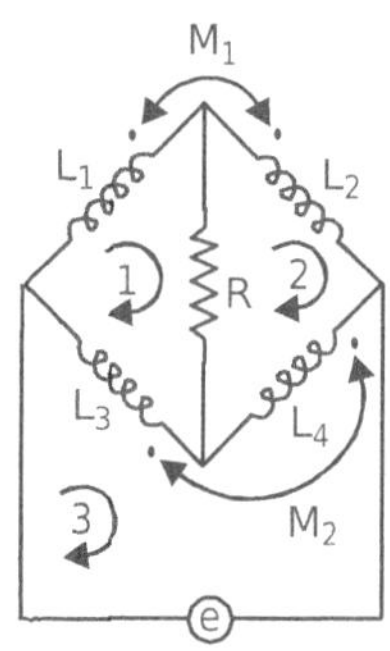

Fig. 5

8. A transformer (Fig. 6) has a primary $L_1 = 4$ H, secondary $L_2 = 1$ H, M = 2 H. Is it an ideal transformer? What is the turns ratio N_1/N_2? If a voltage source S with $e(t) = \sin t$ is applied, find i_1, i_2, the power supplied by S. Instead, if S is a current source, $i = \sin t$, find $e_1(t)$, $e_2(t)$ and the power supplied.

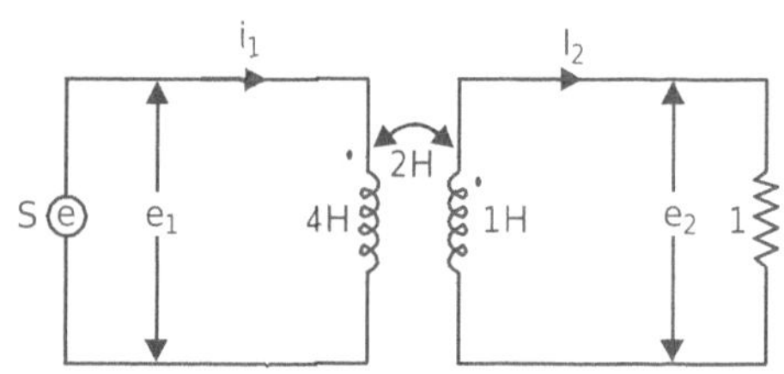

Fig. 6

$$\left[\textbf{Ans. } \text{Yes}; \frac{1}{2}; \frac{1}{4}(\sin t - \cos t); \frac{1}{2}\sin t,\ e_1 = 2(\cos t - \sin t),\ e_2 = \sin t + \cos t\right]$$

9. Establish the following equivalences in the ideal transformers. [Fig. 7 (*a*) to (*c*)].

[**Hint.** (*a*) $LDi_1 - MDi_2 = e_1$ (*i*)

$-MDi_1 + (L_2D + R)i_2 = 0$ (*ii*)

Divide (*i*) by L_1 and (*ii*) by M and use $a = M/L_1 = L_2/M$ = turns ratio.

$$Di_1 - aDi_2 = e_1/L \quad (iii)$$

$$-Di_1 + aDi_2 + (R/M)i_2 = 0 \quad ...(iv)$$

Hence $$i_2 = -(M/L_1)(e_1/R) = -ae_1/R$$

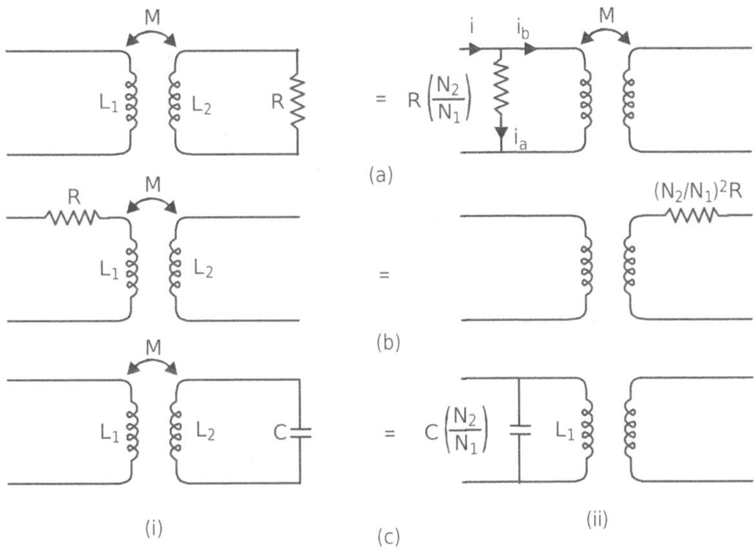

Fig. 7

Put $$i_2 = -ae_1/R \text{ in } (iii),$$

$$i_1 = \frac{e_1}{DL} + \frac{a^2}{R}e_2 = i_b + i_a$$

where i_a, i_b are identified in the equivalent.

(*b*) In the equivalent Fig., we have

$$e_2 = -ae_1 + Ra^2 i_2$$

and so, try to show this relation for Fig. (*ii*) (*b*).

(*c*) is derived similar to (*a*) above.]

10. Find $i(t)$ in circuit in Fig. 8.

[Anna University, May 2007]

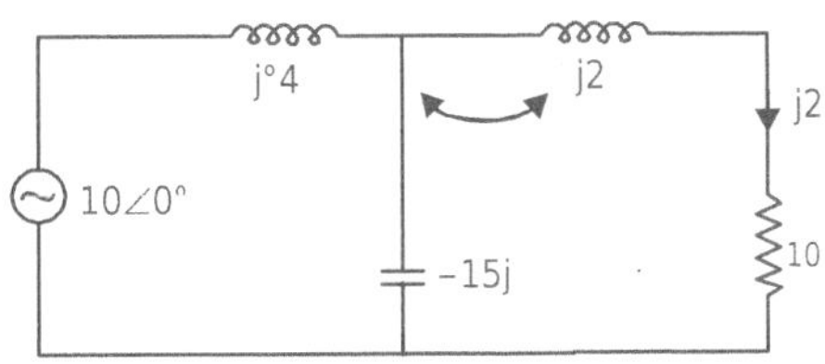

Fig. 8

$\{i_2 = 0.812 \sin(\omega t - 77.2°)\}$

SIMPLE EXERCISES

1. Explain dot notation for mutual inductance.
2. What is the unit of mutual inductance. [Henry].
3. Draw a mutual inductance as a dependant source circuit.
4. What is the formula for a series connection of two inductors which have a mutual coupling ?
5. Define co-efficient of coupling.
6. State the formula for two mutually coupled coils in parallel. From this, if mutual coupling is zero, derive the usual relation for parallel inductors.
7. Write down the transformer equation for voltage and current ratios.
8. Draw the frequency response of a tuned coupled circuit.
9. When will the peak be single and maximum ?

6

NON SINUSOIDAL WAVES AND FOURIER ANALYSIS

6.1 FOURIER SERIES

Fourier Representation for Continuous-Time Signals

Periodic signals: {Fourier Series $\begin{cases}\text{Trigonometric Series}\\ \text{Exponential Series}\end{cases}$

Non-periodic signals: {Fourier Transform

Any periodic wave which is non-sinusoidal can be resolved into a series of sinusoidal waves, by Fourier's series expansion. The function $f(\theta)$ which is periodic with a period 2π will therefore satisfy.

$$f(\theta) = f(\theta + 2n\pi)) \text{ for } n = 0, 1, 2, \ldots \tag{6.1}$$

So, periodic waves repeat themselves in a definite pattern, each of the repeating patterns constituting a cycle. Fourier's series consists in resolving the wave into a sum of sine or cosine functions. The successive terms in the series are having frequencies an integral multiple of the fundamental sine (or cosine) wave. The fundamental wave is a sine (or cosine) wave having the same frequency as the given wave $f(\theta)$. For example, consider the square wave shown in Fig. 6.1(*a*). The fundamental wave for this is shown in Fig. 6.1(*b*), below it. In this case, it is a sine wave. It is shown later that the square wave can be resolved into the following waves, by applying Fourier's series.

$$f(\theta) = \frac{4}{\pi}\left[\sin\theta + \frac{1}{3}\sin 3\theta + \frac{1}{5}\sin 5\theta + \ldots\right] \tag{6.2}$$

The waves (*b*), (*c*), (*d*) ... are the fourier components of the given wave.

In this example of a square wave, we had only sine terms in the expansion. For a general wave, there will be sine terms, cosine terms, and possibly a. *d. c* (steady) terms also. The general formula for expanding a periodic function, having a period 2π, is:

$$f(\theta) = \frac{b_o}{2} + \sum_{n=1}^{n=\infty} [b_n \cos n\theta + a_n \sin n\theta] \tag{6.3}$$

Here $\frac{b_o}{2}$ is the steady term, *i.e.*, a D.C. term. The Σ sign shows that sine and cosine terms are infinite in number. A purely alternating wave—one having equal positive and negative half cycles–will not have the D.C. term. Some waves may have only sine or cosine terms. The co-efficients a_n, b_n are calculated from

$$b_n = \frac{1}{\pi}\int_0^{2\pi} f(\theta)\cos n\theta \, d\theta \tag{6.4}$$

$$n = 0,\ 1,\ 2...$$

$$a_n = \frac{1}{\pi}\int_0^{2\pi} f(\theta)\sin n\theta \, d\theta$$

For the case of the square wave, $f(\theta) = 1$ between 0 and π; and -1 between π and 2π.

(*a*) Given wave $f(\theta)$. One cycle corresponds to 2π. Amplitude is taken to be 1.

(*b*) The fundamental sine wave. Amplitude given by (6.2) as $4/\pi$.

(*c*) The wave $(4/\pi)\,\frac{1}{3}$ sin 3θ, called the third *harmonic*. This is the second term in the expansion. Its frequency is 3 times the fundamental.

(*d*) The fifth harmonic, amplitude is 1/5 of the fundamental.

(*e*) When we add (*b*), (*c*) and (*d*), we get this wave, which nearly resembles the given wave in (*a*). If we added more terms in (6.2)—theoretically upto infinite terms—we get wave (*a*).

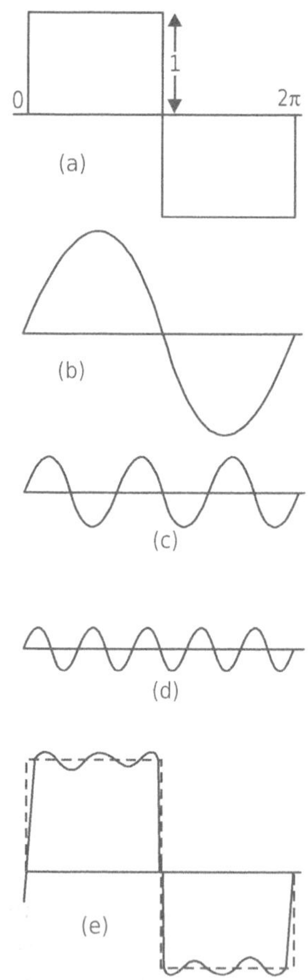

Fig. 6.1

$$b_n = \frac{1}{\pi}\left\{\int_0^{\pi} 1\cos n\theta\, d\theta + \int_{\pi}^{2\pi}(-1)\cos n\theta\, d\theta\right\} = 0$$

$$a_n = \frac{1}{\pi}\left\{\int_0^{\pi} 1\sin n\theta\, d\theta + \int_{\pi}^{2\pi}(-1)\sin n\theta\, d\theta\right\}$$

$$= \frac{1}{\pi}\left\{\left[\frac{\cos n\theta}{n}\right]_0^{\pi} - \left[\frac{\cos n\theta}{n}\right]_{\pi}^{2\pi}\right\}$$

$$= \frac{1}{n\pi}[\cos n\pi - \cos 0 - \cos 2\pi n + \cos n\pi]$$

$$= -\frac{2}{n\pi}[\cos n\pi - 1]$$

Hence cosine terms are absent in the series. For sine terms,

$$a_n = -\frac{2}{n\pi}[\cos n\pi - 1]$$

which is zero if n is even and $\frac{4}{n\pi}$ of n is odd. So the series becomes

$$\frac{4}{\pi}[\sin\theta + 1/3\ \sin 3\theta + 1/5\ \sin 5\theta + ...] \quad (6.5)$$

Example 1. *Find the Harmonics of a triangular wave shown whose period is 1 second and whose maximum value is 100 volts.*

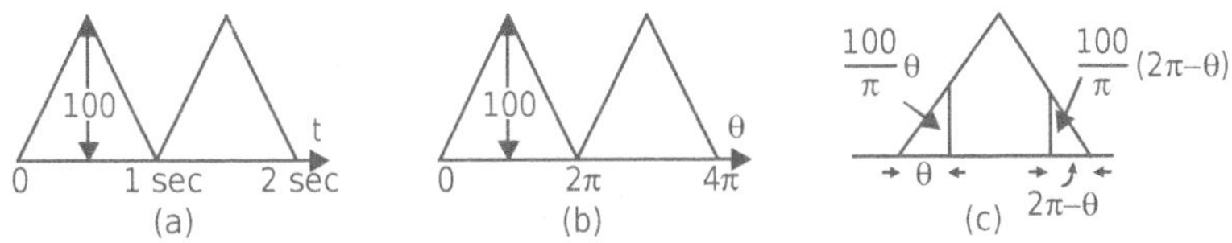

Fig. 6.2

Solution. In forming the Fourier series, let us consider a wave whose period is 2π. Then, by transforming into t, we can rewrite the series. Then let us take the wave shown in Fig. (*b*). Here, one second of t corresponds to 2π radians of θ.

$$t = \theta / 2\pi$$

The equation for the wave must be written. Between θ and π,

$$f(\theta) = (100/\pi)\theta$$

Between π and 2π,

$$f(\theta) = \frac{100}{\pi}(2\pi - \theta)$$

The D.C. term is the average value, which is half the height or 50.

Thus, the values of a_n and b_n are to be found.

$$a_n = \frac{1}{\pi}\int_0^{\pi} f(\theta)\sin n\theta\, d\theta$$

$$= \frac{100}{\pi^2}\left\{\int_0^{\pi}\theta\sin n\theta\, d\theta + \int_{\pi}^{2\pi}(2\pi - \theta)\sin n\theta\, d\theta\right\}$$

The second integral can be rewritten by putting $\alpha = 2\pi - \theta$ as $\sin(2\pi - \alpha) = \sin\alpha$,

$$\int_{\pi}^{0}\alpha\sin n\alpha\, d\alpha = -\int_0^{\pi}\theta\sin n\theta\, d\theta$$

Thus, both first and second integrals are equal but opposite in sign. So, $a_n = 0$. Hence, no sine terms are present.

$$b_n = \frac{100}{\pi^2}\left(\int_0^{\pi} \theta \cos n\theta \, d\theta + \int_{\pi}^{2\pi} (2\pi - \theta)\cos n\theta \, d\theta\right)$$

Integrating by parts,

since $$\int \theta \cos n\theta = \frac{\theta \sin n\theta}{n} + \frac{\cos n\theta}{n^2}$$

The second integral, putting $\alpha = 2\pi - \theta$, becomes

$$-\int_{\pi}^{0} \alpha \cos n\alpha \, d\alpha = \int_0^{\pi} \theta \cos n\theta \, d\theta$$

Thus, both first and second integrals are equal. So,

$$b_n = \frac{200}{\pi^2} \int_0^{\pi} \theta \cos n\theta \, d\theta$$

$$= \frac{200}{\pi^2}\left[\frac{\theta \sin n\theta}{n} + \frac{\cos n\theta}{n^2}\right]_0^{\pi}$$

$$= \frac{200}{\pi^2 n^2}(\cos n\pi - 1)$$

$$= \begin{cases} 0, \text{ if } n \text{ is even} \\ -\dfrac{400}{\pi^2 n^2} \text{ if } n \text{ is odd} \end{cases}$$

The average value, *i.e.*, the $b_0/2$ term is easily seen to be the average height of the triangle, which is 50. Hence

$$f(\theta) = 50 - \frac{400}{\pi^2}\left[\cos \theta + \frac{\cos 3\theta}{3^2} + \frac{\cos 5\theta}{5^2} + \ldots\right]$$

Replacing θ by the variable t,

$$f(\theta) = 50 - \frac{400}{\pi^2}\left[\cos 2\pi t + \frac{1}{3^2}\cos 6\pi t + \ldots\right]$$

6.2 EFFECT OF SYMMETRIES IN THE WAVE

The following simple rules enable us to find out which of the terms may be absent in the series, For example, in the previous example, the sine terms were absent.

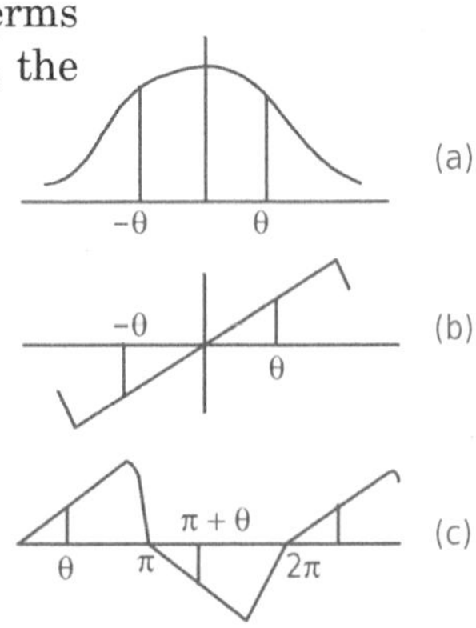

Fig. 6.3

(*a*) The function is the same for both θ and $-\theta$. This is called *zero axis* symmetry. The function is such that.

$$f(\theta) = +f(-\theta)$$

It is called an even function. For *even* functions, it may be shown that Sine terms are absent.

(*b*) This function is one possessing *zero-point* symmetry. Here

$$f(\theta) = -f(-\theta)$$

Hence it also called *odd* function. For odd functions, Cosine functions are absent.

(*c*) The wave is said to have half-wave symmetry *i.e.,* $f(\theta) = -f(\theta + \pi)$. In this case, the wave has no even harmonics.

6.2.1 Numerical Evaluation for Irregular Waves

One period of the given wave is divided into N intervals and ordinates are measured at these intervals as shown.

$$b_0 / 2 \ = \ \text{average value} = \text{Area under the curve/base.}$$

The area can be found using trapezoidal rule so that

$$b_0 / 2 \ = \ [(f_0 + f_N)/2 + f_1 + f_2 + f_3 + \ldots f_{N-1}] \div N$$

Next consider b_n.

$$b_n \ = \ (1/\pi)\int_0^{2\pi} f(\theta)\cos n\theta\, d\theta$$

$$= \ (1/\pi) \text{ Area under } f(\theta)\cos n\theta \text{ over one period.}$$

(*a*) Centre point method

To evaluate this, find (*a*) the average ordinate in any interval which is $(f_i + f_{i+1})/2$ and $d\theta$ is $2\pi/N$.

(*b*) Find the cosine of the angle $n\theta$ at the middle of this interval; So $\cos n\theta_i$ is found where

$$\theta_i \ = \ n\left(i + \frac{1}{2}\right)2\pi/N.$$

(*c*) Multiply (*a*) and (*b*), sum up and then divide by π.

$$b_n \ = \ \frac{1}{N}\sum_{i=0}^{N-1}(f_i + f_{i+1})\ \cos\left[\left(i + \frac{1}{2}\right)2\pi n/N\right]$$

Similarly $$a_n \ = \ \frac{1}{N}\overset{N-1}{\underset{0}{}}(f_i + f_{i+1})\sin\ \ i + \frac{1}{2}\ \ 2\pi n/N \qquad (6.6)$$

The numerical method is to be employed when the irregular wave can not be expressed as a mathematical function.

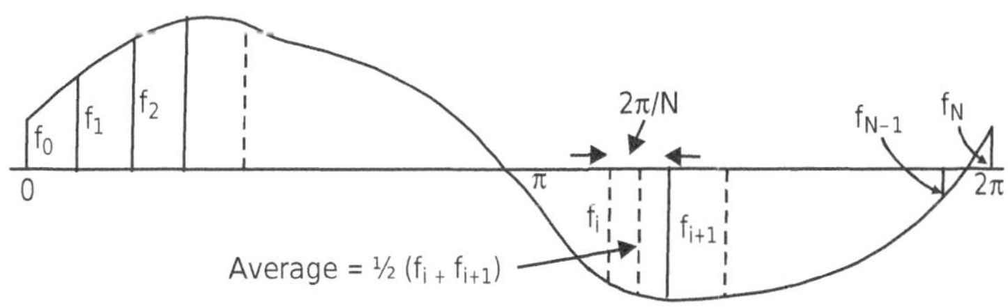

Fig. 6.4 Showing ordinates for irregular wave.

Example 2. *Find the fundamental, 3rd harmonic and fifth harmonics of the following complex wave, with half wave symmetry, whose ordinates at 30° intervals starting from 0° are given in order as :*

0, 1, 3, 4, 2, 1, 0, –1, –2, –4, –3, –1, 0 ....

Solution. The function can be roughly sketched. It will be found that it possesses half wave symmetry and is an *odd* function because $f(x) = -f(-x)$. Hence the series will have only sine terms of odd order. Equation (6.6) is applied. Here N = 12, as $2\pi/N = 30°$.

For the fundamental, $n = 1$. The summation for the first 6 ordinates multiplied by 2 gives the total sum.

$$a_1 = 2\left[\sum_0^5 (f_i + f_{i+1}) \sin\left(i+\frac{1}{2}\right)30°\right] \div 12$$

$$= (1/6)[(0+1)\sin 15 + (1+3)\sin 45 + (3+4)\sin 75 + (4+2)\sin 105 + (2+1)\sin 135 + (1+0)\sin 165]$$

$$= 1/6[0.26 + 4 \times 0.707 + 7 \times 0.97 + 6 \times 0.97 + 3 \times 0.707 + 1 \times 0.26] = 3.026$$

$$a_3 = 1/6\left[\sum_0^5 (f_i + f_{i+1})\sin\left(i+\frac{1}{2}\right)3\times 30°\right]$$

$$= 1/6\,[1 \sin 45 + 4 \sin 135 + 7 \sin 225 + 6 \sin 315 + 3 \sin 405 + 1 \sin 495]$$

$$= 1/6[0.707 + 4 \times 0.707 - 7 \times 0.707 - 0.707 \times 6 + 3 \times 0.707 + 0.707 \times 1]$$

$$= -0.471$$

$$a_5 = 1/6\Sigma(f_i + f_{i+1})\sin\left(i+\frac{1}{2}\right)5\times 30°$$

$$= 1/6[1 \sin 75 + 4 \sin 225 + 7 \sin 375 + 6 \sin 165 + 3 \sin 315 + 1 \sin 105]$$

$$= 1/6\,[0.97 - 4 \times 0.707 + 7 \times 0.26 + 6 \times 0.26 - 3 \times 0.707 + 0.97]$$

$$= -0.37/6 = -0.06.$$

Hence,
$$f(\theta) = 3.026 \sin\theta - 0.471 \sin 3\theta - 0.06 \sin 5\theta + \ldots$$

Direct Point Method

Instead of using the mid-interval, some books use the directly given intervals. For *e.g.*,

$$a_1 = \frac{2}{6}[0 \sin 0° + 1 \sin 30° + 3 \sin 60° + 4 \sin 90° + 2 \sin 120° + 10 \sin 150°]$$

$$= 3.1.$$

6.3 THE R.M.S. VALUE OF A COMPLEX WAVE

If a complex wave $i(\theta)$ has harmonics of both sine and cosine terms,

$$i(\theta) = \Sigma b_n \cos n\theta + a_n \sin n\theta$$

$$= \Sigma\sqrt{a_n^2 + b_n^2}\sin(n\theta + \varphi_n) \quad (\tan\varphi = b_n/a_n)$$

$$= I_1 \sin(\theta + \varphi_1) + I_2 \sin(2\theta + \varphi_2) + \ldots$$

where I_1, I_2, I_3... are the maximum values of the harmonics and $I_n = [a_n^2 + b_n^2]^{1/2}$. The power produced in a resistor R by each harmonic current is $I_n^2R/2$; so, the total power will be $\Sigma I_n^2 R/2$.

$$\therefore \qquad I_{RMS} \text{ of a complex wave } = \sqrt{\Sigma I_n^2/2} \qquad (6.7)$$

If there is a D.C. component also, it becomes $\sqrt{I_{dc}^2 + \Sigma I_n^2/2}$.

Example 3. *A voltage given by*

$$v = 100 \sin \omega t + 50 \sin (3\omega t + 30°) + 20 \sin (\omega t - 90°)$$

is applied to an R-L circuit of impedance $(3 + j4)$ *to the fundamental* ω. *Find the current and the power absorbed. Find also the overall p.f.*

Solution. The current due to each of the harmonics is found in the usual way.

The current due to the fundamental component

$$= 100/(3 + j4) = 20\angle -53°8'$$

The impedance for 3rd harmonic is $3 + j(3 \times 4) = 3 + j12$ and so, the third harmonic current is $50/(3+j12) = 4.042 \angle -76°$.

The impedance due to 5th harmonic $= 3 + j(5 \times 4) = 3 + j20$ so that 5th harmonic current is $20/(3 + j20) = 1 \angle -81°30'$.

Hence
$$i = 20 \sin(\omega t - 53°8') + 4.042 \sin(3\omega t + 30 - 76°)$$
$$+1 \sin (5\omega t - 90 - 81°30')$$

The R.M.S. current can be found as

$$\sqrt{(20/\sqrt{2})^2 + (4.04/\sqrt{2})^2 + (1/\sqrt{2})^2} = 14.44 \text{ A}.$$

The power $= R\, I_{rms}^2 = 3 \times 14.44^2 = 627$ W.

$$\text{The power factor } = 627/(\text{RMS Volts} \times \text{RMS Current})$$
$$= 627 \div [\sqrt{(100^2 + 50^2 + 20^2)/2} \times 14.44]$$
$$= 0.53.$$

Example 4. *A 3-phase 50 c/s alternator has a phase voltage* $e = 100 \sin \omega t + 10 \sin 3\omega t + 5 \sin 5\, \omega t$. *Determine the line voltages if the alternator is*

(a) star connected *(b) Delta connected.*

Solution. (*a*) The line voltage in the star connection are got from the phase voltage e_1, e_2, e_3 as

$$V_{L1} = e_1 - e_2, V_{L2} = e_2 - e_3\ V_{L3} = e_3 - e_1.$$

Thus the line voltage 1 – 2 is

$$V_L = 100[\sin \omega t - \sin (\omega t - 120°)]$$
$$+ 10 [\sin 3\omega t - \sin 3 (\omega t - 120)]$$
$$+ 5 [\sin 5\omega t - \sin 5 (\omega t - 120)]$$

$$= 100\sqrt{3} \sin (\omega t + 30°) + 10(0) + 5 [\sin 5\omega t - \sin (5\omega t - 240)]$$
$$= 100\sqrt{3} \sin (\omega t + 30) + 5\sqrt{3} \sin (5\omega t - 30°)$$

Note that there is no third harmonic in the line voltages in star connection.

The RMS line voltage = $\sqrt{(100\sqrt{3})^2/2+(5\sqrt{3})^2/2} = 122.6$ Volt.

(*b*) In this case, the line voltage is the same as the phase voltage,

So $$V_{L1} = 100 \sin \omega t + 10 \sin 3\omega t + 5 \sin 5\omega t$$

The RMS line voltage = $\sqrt{\frac{1}{2}[100^2+10^2+(5)^2]} = 70.8$ Volt.

Example 5. *Given the set of twelve values of the periodic function, find the harmonics upto 5th.*

[1 3 5 7 3 1 –1 –3 –5 –7 –3 –1]

Solution. Use directly the points instead of centres of points.

Sum of all points = 1 + 3 + 5 + ... – 1 = 0

Hence the D.C. term $b_0 = 0$.

We can evaluate the cos $n\theta$ and sin $n\theta$ terms using a computer program (BASIC or C) as below:

```
Read X                 ; X (1) To X (12) the points
For K = 1 to 6         ; 12 points
For J = 1 to 12        ; 6 term of Fourier Series
                 b(k) = b(k) + X(J) * cos ((j – 1) * (k – 1) *  2π/12)
                 a(k) = a(k) + X(J) * sin ((j – 1) * (k – 1) * 2π/12)
NEXT J
NEXT K
                 a(k) = a(k)/12 * 2
                 b(k) = b(k)/12 * 2
PRINT A(k), B(k)
```

The values of the Fourier co-efficients for are:

$b_1 = 7.46/6 \qquad a_1 = 31.856/6$

$b_3 = -2/6 \qquad a_3 = -6/6$

$b_5 = 0.536/6 \qquad a_5 = 1 + 4.14/6.$

and $b_0 = 0.$

6.4 THE COMPLEX FORM OF FOURIER SERIES

The function $f(\theta)$ can be rewritten, after substituting the exponential forms for *sine* and *cosine* functions as

$$f(\theta) = \frac{1}{2}b_0 + \Sigma\left\{b_n.\frac{1}{2}(e^{jn\theta}+e^{-jn\theta}) - ja_n.\frac{1}{2}(e^{jn\theta}-e^{-jn\theta})\right\}$$

$$= \frac{1}{2}b_0 + \Sigma\left[\frac{1}{2}(b_n - ja_n)e^{jn\theta} + (b_n + ja_n)e^{-jn\theta}\right]$$

Let us find the value of the co-efficient $C_n = \frac{1}{2}(b_n - ja_n)$.

Then, from equations (6.4) giving b_n and a_n, we get

$$C_n = \frac{1}{2\pi}\int_0^{2\pi} f(\theta)(\cos n\theta - j\sin n\theta)d\theta$$

$$= \frac{1}{2\pi}\int_0^{2\pi} f(\theta)e^{-jn\theta}$$

Similarly, $$C_{-n} = \frac{1}{2}(b_n + ja_n) = \frac{1}{2\pi}\int_0^{2\pi} f(\theta)\, e^{jn\theta}\, d\theta \qquad \text{(for } n = 0\text{)}$$

The factor $$\frac{1}{2}b_0 = \frac{1}{2\pi}\int_0^{2\pi} f(\theta)d(\theta) = \frac{1}{2\pi}\int f(\theta)e^{j0\theta}\,d\theta = C_0$$

Hence $$f(\theta) = C_0 + \Sigma C_n e^{jn\theta} + C_{-n}e^{-jn\theta}$$

$$= \sum_{-\infty}^{\infty} C_n e^{jn\theta} \tag{6.8}$$

where $$C_n = \left(\frac{1}{2\pi}\right)\int_0^{2\pi} f(\theta)e^{-jn\theta}d\theta \tag{6.9}$$

Note that n takes all values right from $-\infty$ to $+\infty$, inclusive of zero. The above is the complex form of Fourier's series.

Example 6. *For a saw tooth wave of Fig. 6.5, find the complex form of the series.*

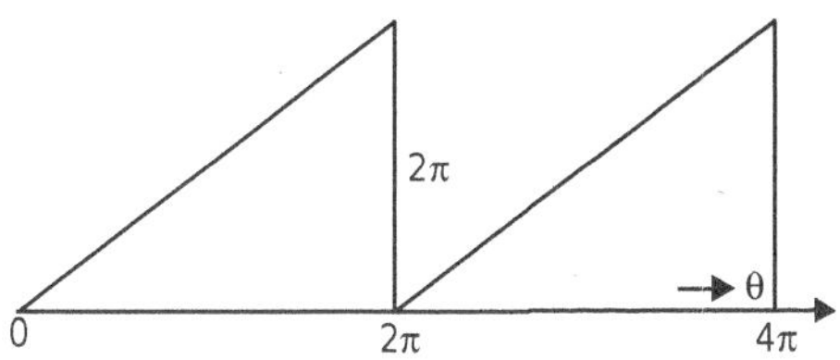

Fig. 6.5

Solution. $$f(\theta) = (\theta) \qquad (0 < \theta < 2\pi)$$

$$C_n = \left(\frac{1}{2\pi}\right)\int_0^{2\pi} \theta e^{-jn\theta}d\theta$$

$$= j/n^* \qquad \text{(for all } n \text{ except 0)}$$

$$- \pi \qquad \text{(for } n = 0\text{)}$$

---

* $$\frac{1}{2\pi}\int_0^{2\pi} \theta d\left(\frac{e^{-jn\theta}}{-jn}\right) = \frac{1}{2\pi}\left[\theta.\frac{e^{-jn\theta}}{-jn} - \int\frac{e^{-jn\theta}}{-jn}d\theta\right]_0^{2\pi}$$

$$= \frac{1}{2\pi}\left[\frac{\theta.e^{-jn\theta}}{-jn} + \frac{e^{-jn\theta}}{(-jn)^2}\right]_0^{2\pi}$$

$$= \frac{1}{2\pi}\left[-\frac{2\pi - 0}{jn} + \frac{(1-j0)-(1-j0)}{(-jn)^2}\right]$$

$$= j/n.$$

Hence the series is, from (6.7),

$$f(\theta) = \pi + \sum_{-\infty}^{\infty}\left(\frac{j}{n}\right)e^{jn\theta}$$

$$= -j\frac{e^{-j3\theta}}{3} - j\frac{e^{-j2\theta}}{2} - je^{-j\theta} + \pi + j\,e^{j\theta}$$

$$+ \frac{je^{j2\theta}}{2} + \frac{je^{j3\theta}}{3} + \dots$$

Clubbing the corresponding negative and positive terms,

$$f(\theta) = \pi - 2\left(\sin\theta + \frac{1}{2}\sin 2\theta + \frac{1}{3}\sin 3\theta + \dots\right)$$

A Table of Fourier series for several useful waveforms is given below. Among these, the *pulse waveforms* (5) find use in digital communication and hence the harmonic content of a pulse series is related to the transmission line bandwidth. Another waveform is the *rectifier signal* (9) waveform. Its harmonics are required for designing the ripple filters.

Table : Ready Reckoner for Fourier Series

Nos.	*Periodic waveform*	*Fourier series*	*RMS value A*
1.	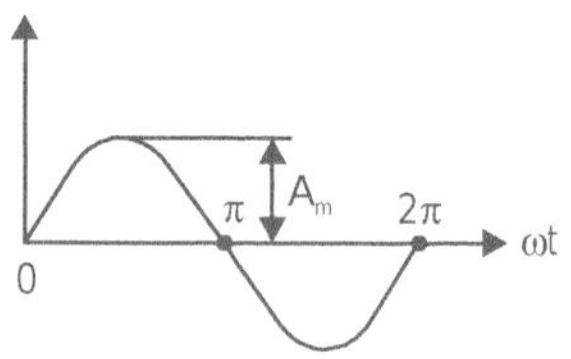	$f(\omega t) = A_m \sin \omega t$	$\frac{A_m}{\sqrt{2}}$
2.	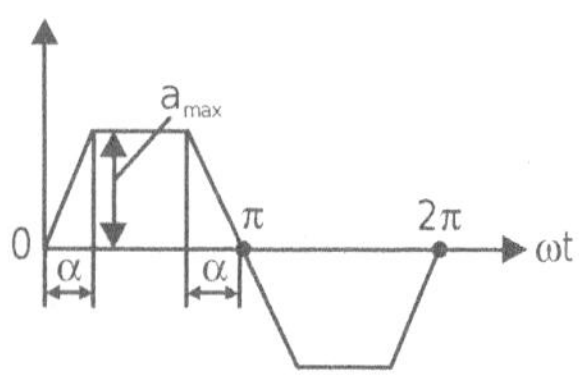	$f(\omega t) = \frac{4a_{max}}{\alpha\pi}\left(\sin\alpha \times \sin\omega t + \frac{1}{9}\sin 3\alpha \times \sin 3\omega t + \frac{1}{25}\sin 5\alpha \times \sin 5\omega t + \dots + \frac{1}{k^2} \times \sin k\alpha \sin k\omega t + \dots\right)$	$a_{max}\sqrt{1 - \frac{4\alpha}{3\pi}}$
3.	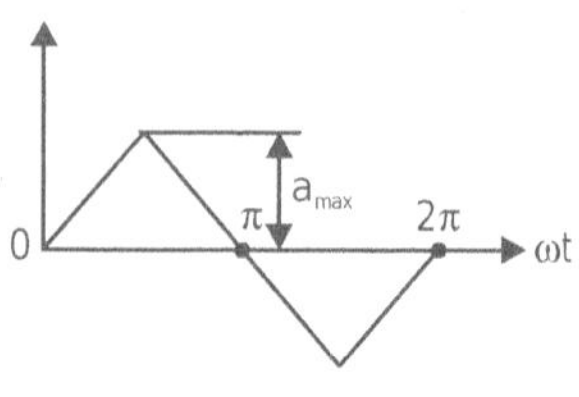	$f(\omega t) = \frac{8a_{max}}{\pi^2}\left(\sin\omega t - \frac{1}{9}\sin 3\omega t + \frac{1}{25} \times \sin 5\omega t - \dots + \frac{(-1)^{\frac{k-1}{2}}}{k^2} - \sin k\omega t + \dots\right)$	$\frac{a_{max}}{\sqrt{3}}$

4.	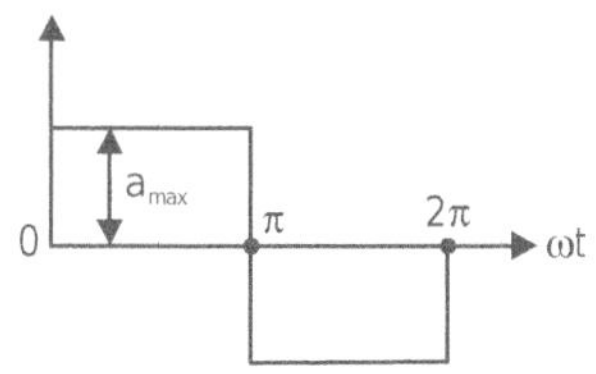	$f(\omega t)=\frac{4a_{max}}{\pi}\left(\sin\omega t+\frac{1}{3}\sin 3\omega t+\frac{1}{5}\times\sin 5\omega t+\frac{1}{k}\sin k\omega t+\ldots\right)$	a_{max}
5.	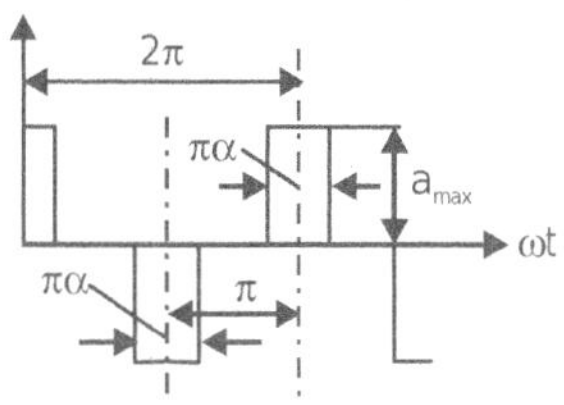	$f(\omega t)=\frac{4a_{max}}{\pi}\left[\sin\frac{\alpha\pi}{2}\times\cos\omega t+\frac{1}{3}\sin\frac{3\alpha\pi}{2}\times\cos 3\omega t+\frac{1}{5}\times\sin\frac{5\alpha\pi}{2}\cos 5\omega t+\ldots\right]$	$a_{max}\sqrt{\alpha}$
6.	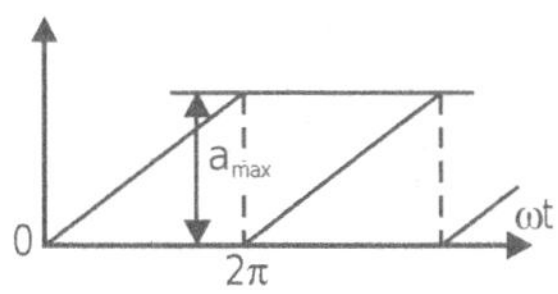	$f(\omega t)=a_{max}\left[\frac{1}{2}-\frac{1}{\pi}\times\left(\sin\omega t+\frac{1}{2}\sin 2\omega t+\frac{1}{3}\sin 3\omega t+\ldots\right)\right]$	$\frac{a_{max}}{\sqrt{3}}$
7.	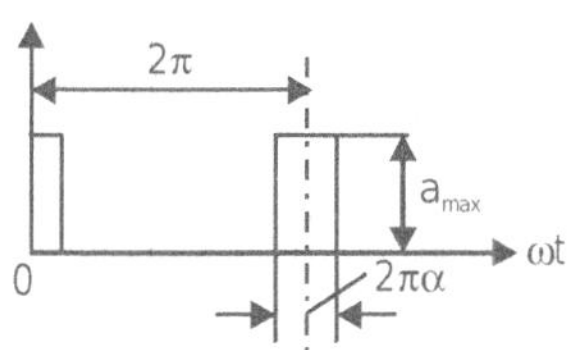	$f(\omega t)=a_{max}\left[\alpha+\frac{2}{\pi}\times\left(\sin\alpha\pi\cos\omega t+\frac{1}{2}\sin 2\alpha\pi\cos 2\omega t+\frac{1}{3}\sin 3\alpha\pi\times\cos 3\omega t\right)\right]$	$a_{max}\sqrt{\alpha}$
8.	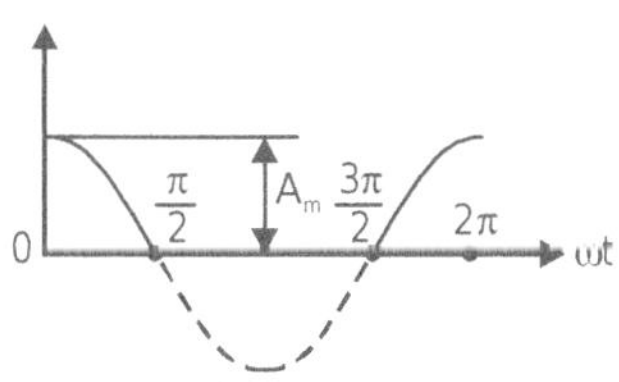	$f(\omega t)=\frac{2A_m}{\pi}\left(\frac{1}{2}+\frac{\pi}{4}\cos\omega t+\frac{1}{1.3}\times\cos 2\omega t-\frac{1}{3.5}\cos 4\omega t+\frac{1}{5.7}\cos 6\omega t+\ldots\right)$	$\frac{A_m}{2}$
9.	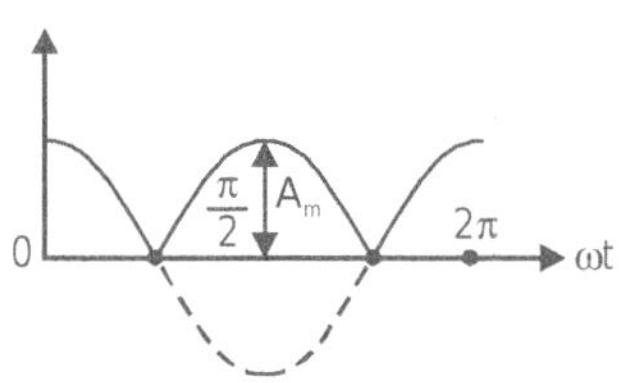	$f(\omega t)=\frac{4A_m}{\pi}\left(\frac{1}{2}+\frac{1}{1.3}\cos 2\omega t-\frac{1}{3.5}\cos 4\omega t+\frac{1}{5.7}\cos 6\omega t-\ldots\right)$	$\frac{A_m}{\sqrt{2}}$

10.	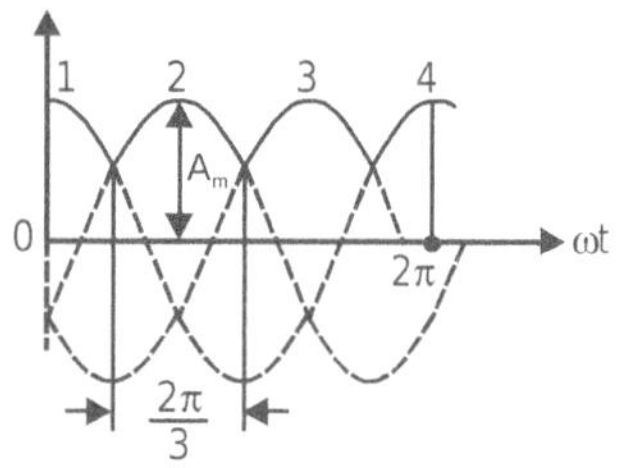	$f(\omega t) = \dfrac{3\sqrt{3A_m}}{\pi} \times \left(\dfrac{1}{2} + \dfrac{1}{2.4}\cos 3\omega t - \dfrac{1}{5.7}\cos 6\omega t + \dfrac{1}{8.10}\cos 9\omega t - \dots\right)$	$A_m\sqrt{\dfrac{1}{2} + \dfrac{3\sqrt{3}}{4\pi}}$
11.	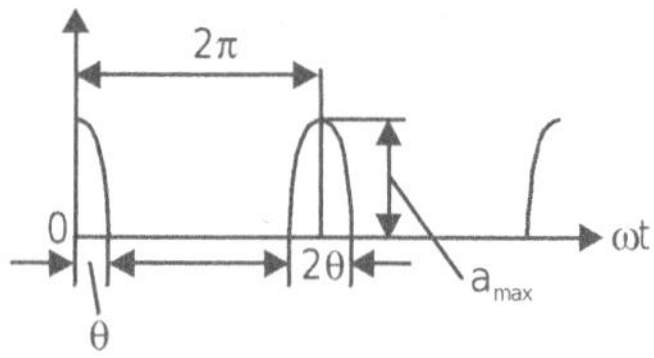 $f(\omega t) = a_{max} \times \dfrac{\cos \omega t - \cos \theta}{1 - \cos \theta}$ at $f > 0$.	$f(\omega t) = a_{max} \times \left(\alpha_0 + \sum_{k=1}^{\infty} a_k \cos k\omega t\right)$ where $\alpha_0 = \dfrac{1}{\pi(1 - \cos \theta)} \times (\sin \theta - \theta \cos \theta)$; $a_k = \dfrac{1}{n\pi(1 - \cos \theta)} \times \left[\dfrac{\sin (n-1)\theta}{n-1} - \dfrac{\sin(n+1)\theta}{n+1}\right]$	$a_{max}\sqrt{\dfrac{\theta\left(\dfrac{1}{2} + \cos^2\theta\right) - \dfrac{3}{4}\sin 2\theta}{\pi(1 - \cos \theta)^2}}$

6.5 TRANSFORM METHODS–FOURIER TRANSFORM

The response of a system is obtained from actual physical measurements. The input for obtaining such response can be a *step input* (such as the sudden plunging of a thermometer into hot water) or a *ramp* (uniformly increasing) input (such as heating a container with liquid) or a *sinusoidal input* (such as an oscillatory swing). The nature of the response will be a time response.

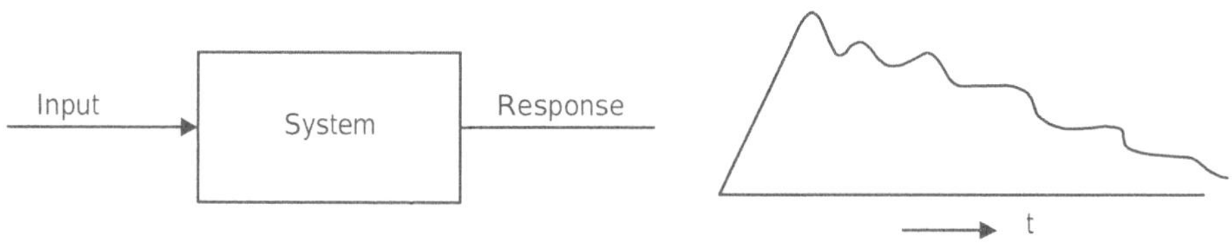

Fig. 6.6

The study of the response after finding out the several frequency components in it is called the spectral response. This is done by the well-known Fourier transformation of $f(t)$. The time varying signal is said to be in the 'time domain', while the frequency values after transformation are said to be in the 'frequency domain.'

It is not necessary that the signal should be a time-varying signal. Signals that vary in space, either in one dimension or two dimensions, are also usually analysed in terms of the frequency components. For example, the variation of gas concentration in the y-direction in an absorption (packed) column is a signal which is a length or height oriented signal $f(x)$.

Studies of patterns in space such as particle distribution are involving two dimensional signals. In highly viscuous flow and flow of heterogeneous particles in a liquid are analysed in one or more cross sections of the flow as space oriented signals, $f(x, y)$. The study of these are now-a-days done by ultrasonic imaging techniques. We know that even the flow of blood is analysed in the ultrasonographic scanners. They do a Fourier transform of the reflected echo signals and obtain the values of the several frequency components.

A graphical plot of the frequencies as frequency versus amplitude is a frequency spectrum. The amplitude is usually comprising the amplitudes of the 'Cosine' and 'sine' parts, which is mathematically called as the 'Complex' amplitude (since $\cos\theta + j\sin\theta = \exp(j\theta)$. The magnitude spectrum is got by (real part^2 + imagining part^2)$^{1/2}$.

When we find the frequency amplitudes in several time slots and plot a time-varying spectrum, we need a three dimensional figure. Such a plot is called a **spectrogram plot.**

6.5.1 Fourier Series—A Train of Rectangular Pulses

Pulse train is very much used in communication of PCM data. Its Fourier series we shall study. Let the pulse duration be τ and its period be T. Let a_{max} be the height of the pulses.

$$\alpha = \tau/T, \text{ a ratio.}$$

The Fourier Series for this pulse train is

$$f(\omega t) = a_{max}\left[\alpha + \frac{2}{\pi}\sum_{k=1}^{\infty}\frac{\sin k\alpha\pi\ \cos k\omega t}{k}\right]$$

Representing each harmonic as a sum of two sinusoids, one corresponding to a positive and another to a negative frequency, we may re-unite this as

$$f(\omega t) = \frac{a_{max}}{\pi}\sum_{k=-\infty}^{\infty}\frac{\sin k\alpha\pi}{k}\cos k\omega t$$

The constant component is $\alpha\pi$ from

$$\underset{k\to 0}{\text{Lt}}\ \frac{\sin k\alpha\pi}{k} \to \alpha\pi,$$

If the period T is varied (or the frequency is varied) the effect on the series in noted in figure below. Note that the lower frequency pulse train has the finest frequency spectrum.

Putting $\omega_0 - 2\pi/\tau - \omega/\alpha$, we may write for $f(\omega t)$:

$$f(\omega t) = \frac{\alpha a_{max}}{\pi}\sum_{k=-\infty}^{\infty}\frac{\sin k\alpha\pi\ \cos k\alpha\omega_0 t}{k\alpha}$$

The respective amplitude spectra plotted for different values of α are shown in Fig. 6.7. The ratio of the harmonic amplitude to the coefficient a is plotted as ordinate, and the ratio of the harmonic frequency $k\omega$ to the fundamental frequency ω_0 as absissa.

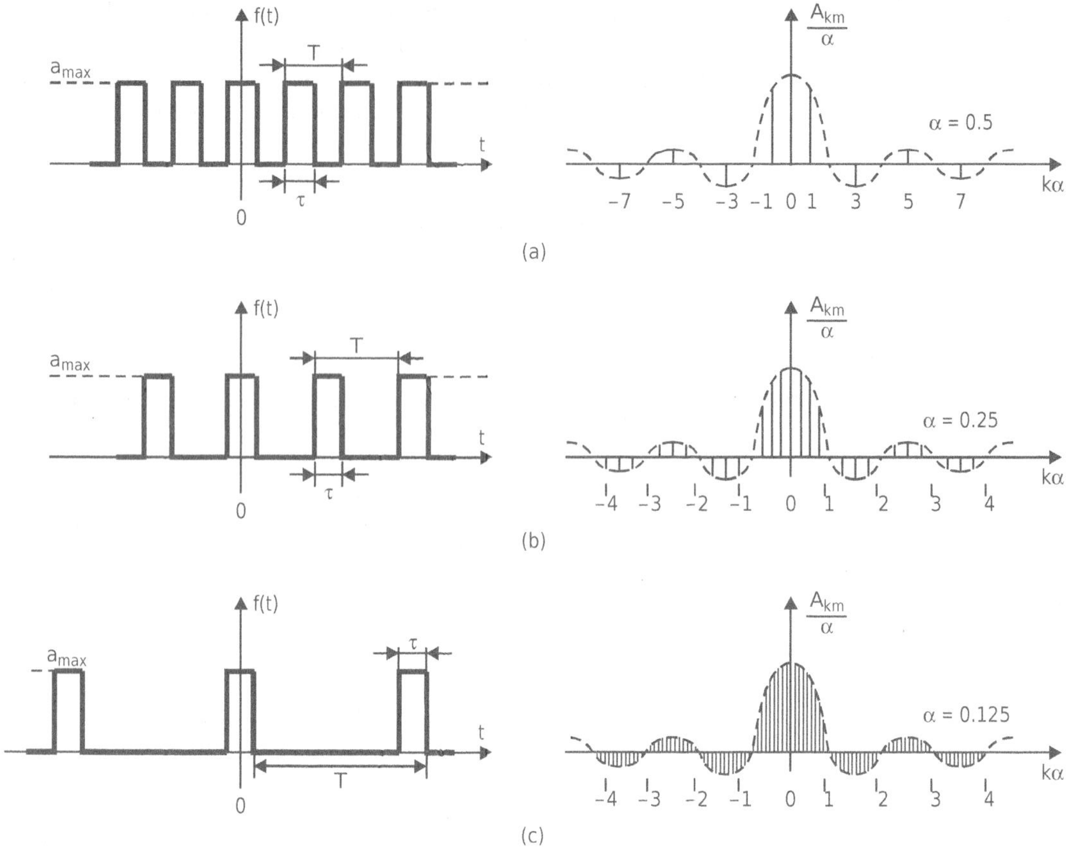

Fig. 6.7

6.6 INTRODUCTION TO THE FOURIER TRANSFORM

The Fourier transform is defined as the integral of the product of the function $f(x)$ or $f(t)$ (if t is the time variable) with the exponential function $\exp(-j2\pi fx)$ or $\exp(-2\pi jft)$ and over the limits $-\infty$ to $+\infty$ for x or t.

Where x is the independent variable in the function $f(x)$:

$$F(f) = \int_{-\infty}^{\infty} f(x)\exp\{-j2\pi fx\}\, dx \tag{6.10}$$

Where t is the independent variable in the function $f(t)$:

$$F(f) = \int_{-\infty}^{\infty} f(t)\exp\{-j2\pi ft\}\, dt \tag{6.11}$$

Here $j = \sqrt{-1}.$

What we get on the left hand side is the function F with its independent variable as f, the so-called 'frequency'. As f varies, $F(f)$ takes different values. If $f(x)$ is a continuous function of x, then $F(f)$ is also a continuous function of F.

More often we get functions $f(t)$ as periodically varying in t, with a period T. In that case, $F(f)$ takes discrete values, rather than a continuous form. For example, if $f(t) = 5 \sin 2t$, then it is a periodic function of frequency $2/2\pi$, and this is the only component in F(*f*) at $f = 2/2\pi$. The value is 5.

Note that the F(*f*) plot is discrete in this case, having values only at $\pm 1/\pi$ and zero elsewhere. That is because of the periodicity of the *f*(*t*) function.

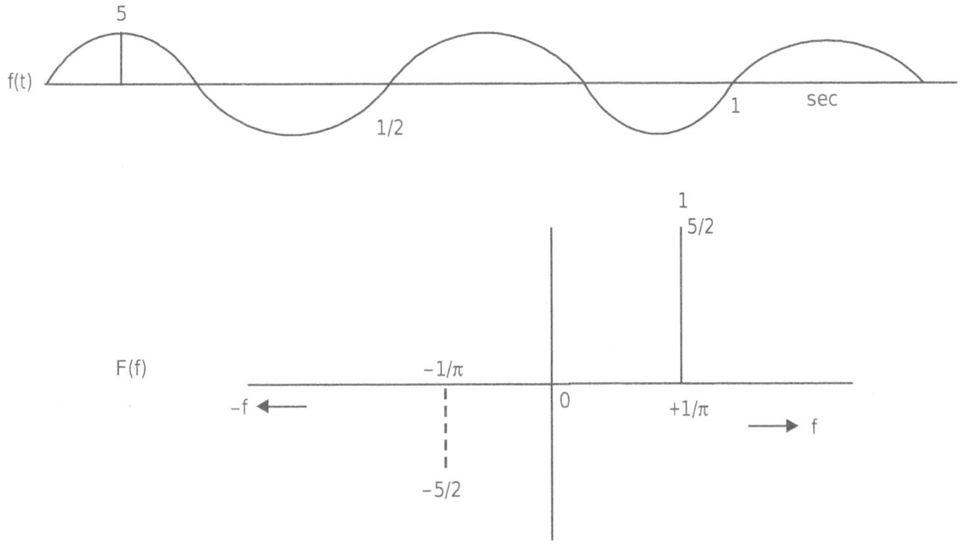

Fig. 6.8 Function in time and its frequency spectrum plot.

Note that the F(*f*) plot is discrete in this case, having values only $\pm 1/\pi$ and zero elsewhere. That is because of the periodicity of the *f*(*t*) function.

Note that the complex exponential function is simply

$$\exp[-j2\pi ft] = \cos 2\pi ft - j \sin 2\pi ft$$

indicating that F(*f*) tells about the various sines and cosines present in *f*(*t*).

Given F(*f*), the frequency spectrum, the function *f*(*x*) or *f*(*t*) as the case may be, would be obtained by the *Inverse Fourier Transform.*

$$f(x) = \int_{-\infty}^{\infty} F(f)\exp\{j2\pi fx\}\, dx \tag{6.12}$$

or

$$f(t) = \int_{-\infty}^{\infty} F(f)\exp\{2\pi ft\}\, dt \tag{6.13}$$

depending on what you name the independent variable in *f*, as *t* or as *x*.

Consider a function of *x* as shown in Fig. 6.8, which has a value A for $x = 0$ to X, and value 0 elsewhere.

Using the formula F(*f*),

$$F(f) = \int_{-\infty}^{\infty} f(x)\exp\{-j2\pi fx\}\, dx$$

$$= \int_{0}^{x} A \exp[-2\pi jfx]\, dx$$

$$= \frac{-A}{j2\pi f}\left[e^{-j2\pi fx}\right]_0^X \qquad (6.14)$$

$$= \frac{-A}{j2\pi f}\left[e^{-j2\pi fX} - 1\right]$$

$$= \frac{A}{j2\pi f}\left[e^{j\pi fX} - e^{-j\pi fX}\right]e^{-j\pi fX}$$

$$= \frac{A}{\pi f}\sin(\pi f X)e^{-j\pi fX}$$

This is a complex function, since it contains $e^{-j\pi fX}$. The complex form means that it has a cosine and a sine parts,

$$\exp[-j\pi fX] = \cos \pi fX - j\sin \pi fX$$

Both the spectra, the cos and sine, are usually drawn one below another and they are referred to as the real and 'imaginary' spectra.

Since a complex number $(a + jb)$ has a magnitude $\sqrt{a^2 + b^2}$ and a phase $\tan^{-1}$ (b/a), we can find the magnitude and draw the magnitude spectrum, as $|F(f)|$. The values in the negative f region are just the same as in the positive part.

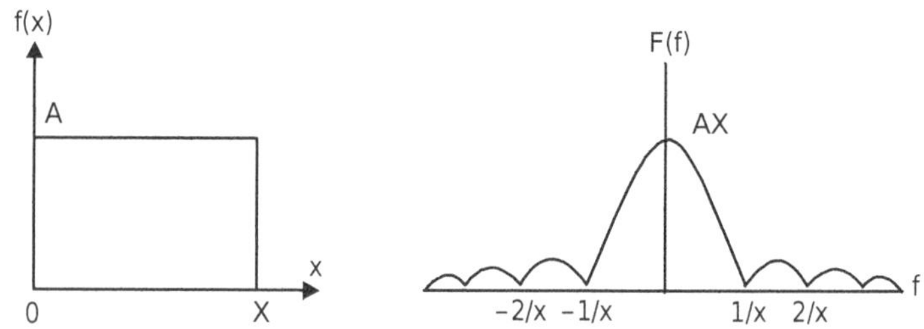

Fig. 6.9 Fourier spectrum of a rectangular function is a sinc function.

Thus the Fourier spectrum of the rectangle is

$$|F(f)| = AX\left|\left(\frac{\sin \pi fX}{\pi fX}\right)\right| \qquad (6.15)$$

1-D FOURIER TRANSFORM IN NMR AN EXAMPLE

In the analysis of chemicals, at least two important analytical instruments directly make use of Fourier transforms. These are the nuclear magnetic resonance spectrometer and the F.T. Infra-red spectrometer.

The NMR spectrum from the instrument is obtained after taking a Fourier transform of the free-induction-decay (F.I.D.) signals from the instrument coils. The coils pick up the signals after a burst of radio frequency pulses have been applied to the chemical sample kept within a highly powerful magnetic cavity. Two signals are obtained actually from the electronic phase sensitive detectors of the instrument, which are called the in-phase and 'quadrature' components. These two F.I.D. signals happen to be the real and imaginary parts of the complex signal $(u + jv)$ that has to be Fourier transformed. Upon such transformation, the frequency components in the signal relate to the chemical groups present in the sample chemical.

Fig. 6.10 shows the F.I.D. signal from a typical spectrometer output for an organic hydrocarbon. The two spectra-real and imaginary parts – are combined in the magnitude spectrum, which is drawn only for the positive values of frequency; that too, with a frequency axis in chemical terms as ppm.

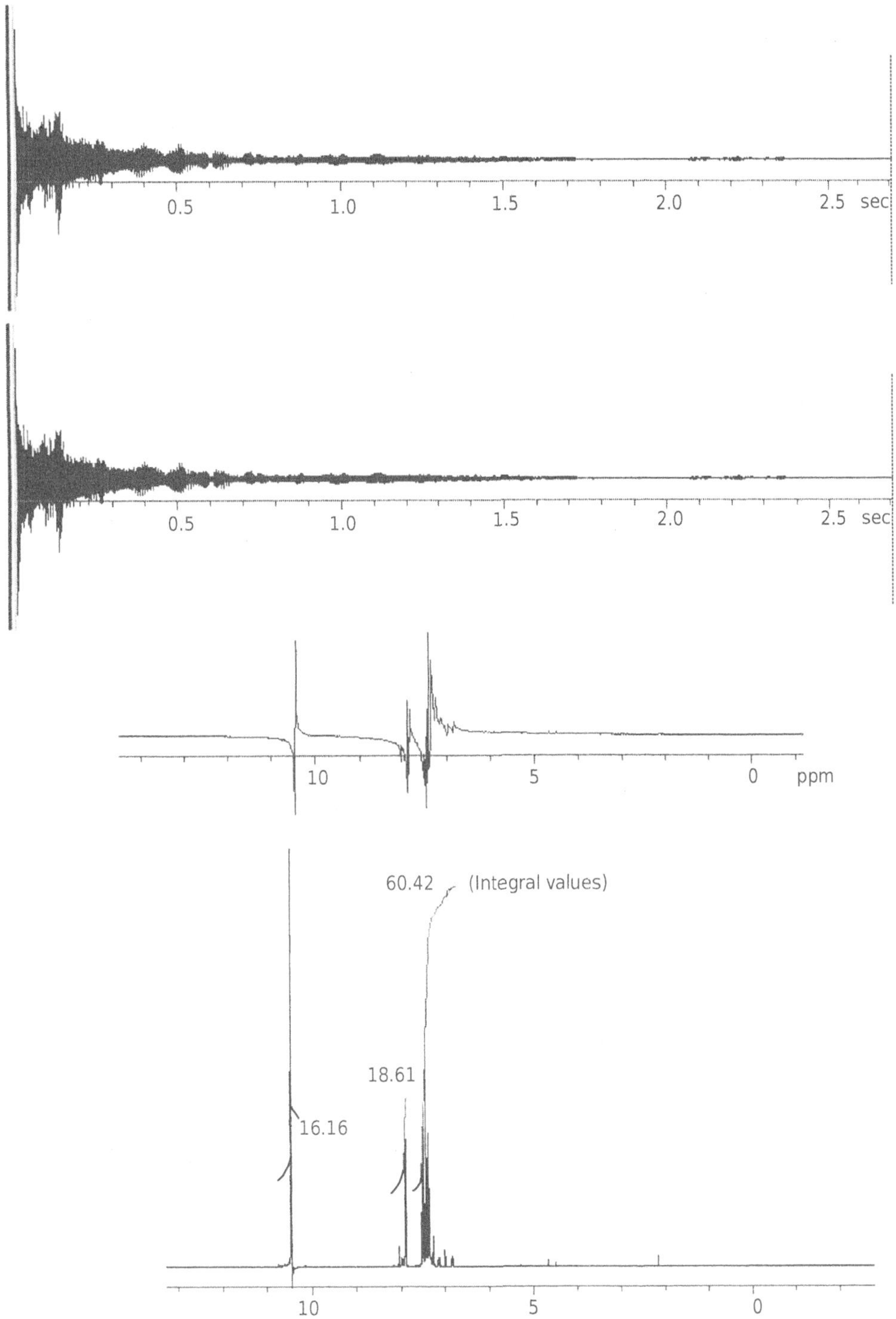

Fig. 6.10

(*a*) Real part of the N.M.R. free induction decay signal.

(*b*) Quadrature part of the free induction decay signal.

(*c*) Spectrum of real part in p.p.m.

(*d*) Spectrum after phasing, or phase shifting so that the negative parts are rendered positive.

Additional Problem

Example 1. *Express $f(x) = x$ $(-\pi < x < \pi)$ as a Fourier series with period 2π.*

Solution. $f(x) = x$ is an odd function.

Hence in the expansion, the cosine terms are absent.

$$\therefore \quad x = \Sigma b_n \sin nx$$

where
$$b_n = \frac{2}{\pi}\int_0^{\pi} x \sin nx\, dx$$

$$= \frac{2}{\pi}\left[\frac{-x \cos nx}{n}\right]_0^{\pi} + \int_0^{\pi} \frac{\cos nx}{n} dx$$

$$= -\frac{2}{\pi}\cos n\pi = -\frac{2}{\pi}(-1)^n = \frac{(-1)^{n+1}2}{n}$$

$$\therefore \quad x = 2\left(\sin x - \frac{1}{2}\sin 2x + \frac{1}{3}\sin 3x - \ldots\right).$$

Example 2. *If*
$$f(x) = -x \text{ in } -\pi < x < 0.$$
$$= x \text{ in } 0 \le x < \pi$$

expand $f(x)$ as a Fourier series in the interval $-\pi$ to π.

Deduce that
$$\frac{\pi^2}{8} = 1 + \frac{1}{3^2} + \frac{1}{5^2} + \frac{1}{7^2} + \ldots$$

Solution.

We easily see that $f(x)$ is an even function by drawing the graph of the function and noting that it is symmetrical with respect to the y axis.

$$\therefore \quad f(x) = \frac{a_o}{2} + \sum_{n=1}^{\infty} a_n \cos nx$$

where $a_n = \dfrac{2}{\pi}\displaystyle\int_0^{\pi} f(x) \cos nx\, dx (n = 0, 1, 2\ldots)$

$$a_0 = \frac{2}{\pi}\int_0^{\pi} f(x)\, dx = \frac{2}{\pi}\int_0^{\pi} x\, dx = \frac{2}{\pi}\left[\frac{x^2}{2}\right]_0^{\pi} = \pi$$

$$a_n = \frac{2}{\pi}\int_0^{\pi} x \cos nx\, dx.$$

$$= \frac{2}{\pi}\left\{\left[\frac{x \sin nx}{n}\right]_0^\pi - \int_0^\pi \frac{\sin nx}{n} dx\right\}$$

$$= \frac{2}{n^2\pi}\left[\cos nx\right]_0^\pi = \frac{2}{n^2\pi}(\cos n\pi - 1)$$

$$= \frac{2}{n^2\pi}\left\{(-1)^n - 1\right\}$$

When n is odd, $a_n = \frac{4}{n^2\pi}$

When n is even, $a_n = 0$

$$\therefore \quad f(x) = \frac{\pi}{2} - \frac{4}{\pi}\left[\frac{\cos x}{1^2} + \frac{\cos 3x}{3^2} + \frac{\cos 5x}{5^2} + \ldots\right]$$

When $x = 0,\ f(x) = 0.$

$$\therefore \quad 0 = \frac{\pi}{2} - \frac{4}{\pi}\left(\frac{1}{1^2} + \frac{1}{3^2} + \frac{1}{5^2} + \ldots\right).$$

$$\therefore \quad \frac{1}{1^2} + \frac{1}{3^2} + \frac{1}{5^2} + \ldots = \frac{\pi^2}{8}.$$

Spectrum of Periodic Signals

We begin with the plots of Fourier spectra using some examples.

Example 1. *Determine the fundamental frequency of the following signals, and plot their sinusoidal spectra (both magnitude and phase) and their exponential spectra.*

(a) $x(t) = 2 + 3\cos(0.2t) + \cos(0.25t + \pi/2) + 4\cos(0.3t - \pi)$

(b) $x(t) = 1 + 10\cos(2\pi(60)t + \pi/8) + 2\cos(2\pi(300)t - \pi/4)$.

Solution. (*a*) $x(t) = 2 + 3\cos(0.2t) + \cos(0.25t + \pi/2) + 4\cos(0.3t - \pi)$

Frequencies are 0.2 rad/sec, 0.25 rad/sec, 0.3 rad/sec

The greatest common divisor is 0.05 rad/sec (all the frequencies are divided by 0.05 in an integer number of ways)

$$\Rightarrow \quad \omega_0 = 0.05 \text{ rad/sec}$$

Sinusoidal spectrum

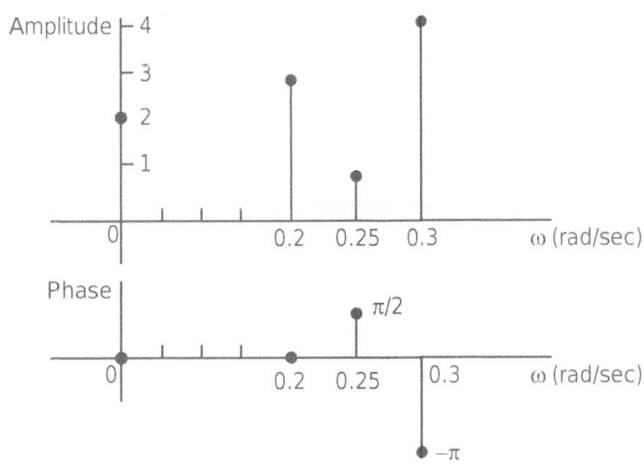

Exponential spectrum

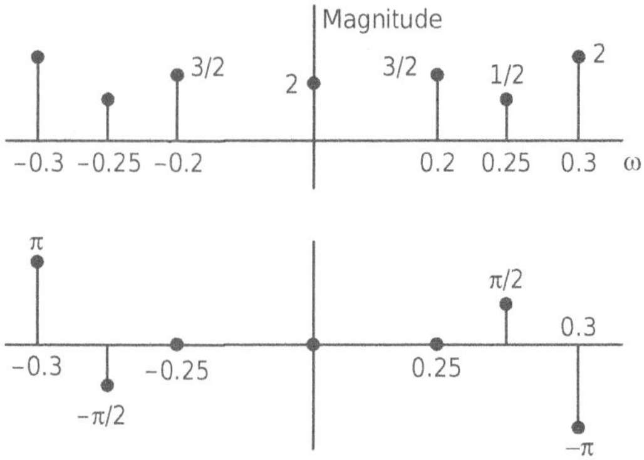

(b) $x(t) = 1 + 10 \cos(0.2\pi(60)t + \pi/8) + 2 \cos(2\pi(300)t - \pi/4)$

Greater common divisor is $2\pi(60)$ rad/sec

or 60Hz $\Rightarrow$ $f_0 = 60$ Hz

Sinusoidal spectrum

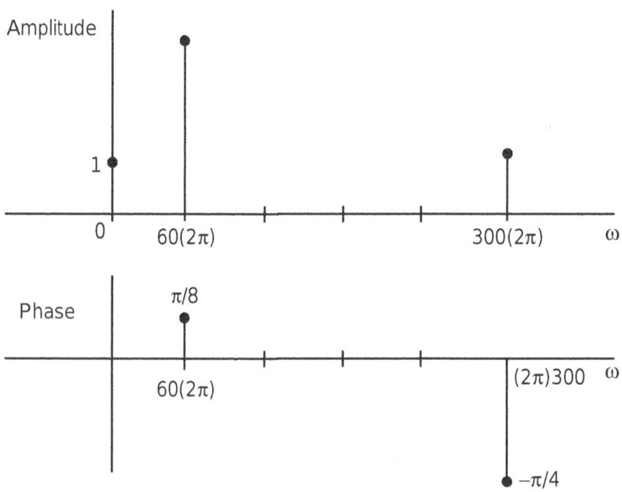

Exponential spectrum

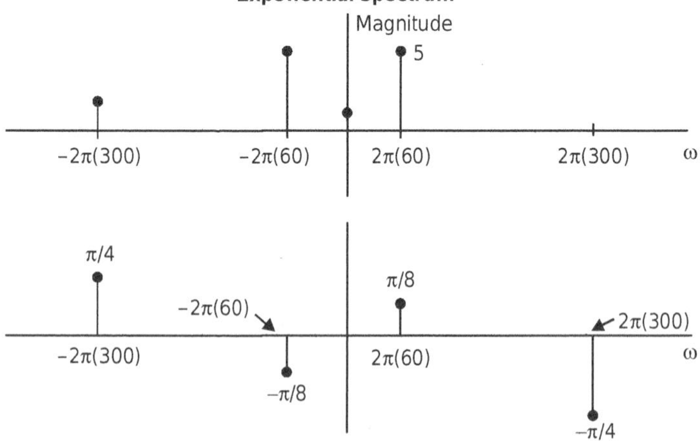

Example 2. *Find an expression for x(t) and plot the spectrum for each graph.*

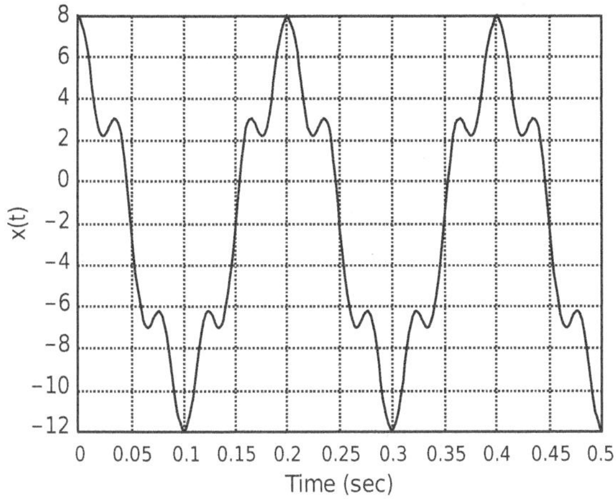

(a)

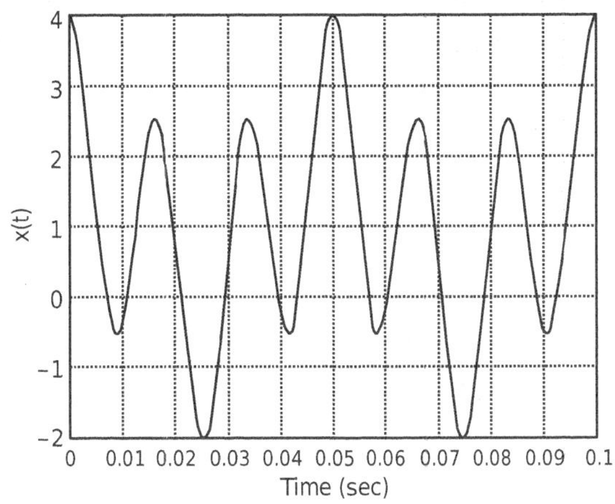

(b)

Solution.

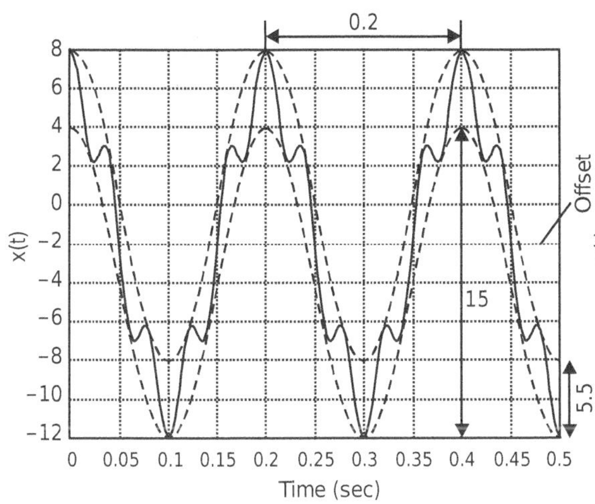

(a)

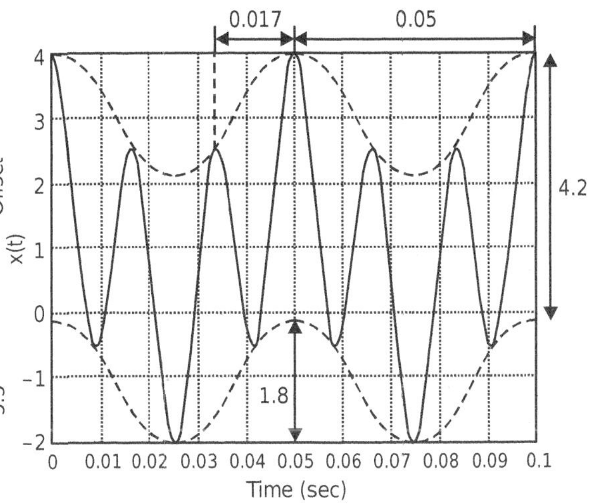

(b)

(a) Low frequency signal: A = 15/2 = 7.5, T = 0.2, So

$$\omega = \frac{2\pi}{0.2} = 10\pi, \phi = 0 \text{ (peak at } t = 0)$$

High frequency: A = 5.5/2 = 2.75, T = 0.04 $\rightarrow \omega = \dfrac{2\pi}{0.04} = 50\pi$ (5 cycles of high frequency in 0.2 sec), $\phi = 0$

$$\text{Offset} = -2$$

$$x(t) = -2 + 7.5 \cos(10\pi t) + 2.75 \cos(50\pi t)$$

(b) Low frequency : A = 1.8/2 = 0.9, T = 0.05 $\Rightarrow \omega = 40\pi$, $\phi = 0$

High frequency: A = 4.2/2 = 2.1, T = 0.017 (or 3 cycles in 0.05 sec), $\phi = 0$, $\omega = 118\pi$

$$\text{Offset} = 1$$

$$x(t) = 1 + 0.9 \cos(40\pi t) + 2.1 \cos(118\pi t)$$

These are estimates based on the plot, exact values may differ.

Example 3. *Draw the spectrum for the following signal, and make a rough sketch of x(t).*

$$x(t) = 10 \cos(200\pi t) \cos(2000\pi t)$$

Solution.

$$x(t) = 10\left[\frac{e^{j200\pi t} + e^{-j200\pi t}}{2}\right]\left[\frac{e^{j2000\pi t} + e^{-j2000\pi t}}{2}\right]$$

$$= \frac{10}{4}\left[e^{j2200\pi t} + e^{-j2200\pi t} + e^{j1800\pi t} + e^{-j1800\pi t}\right]$$

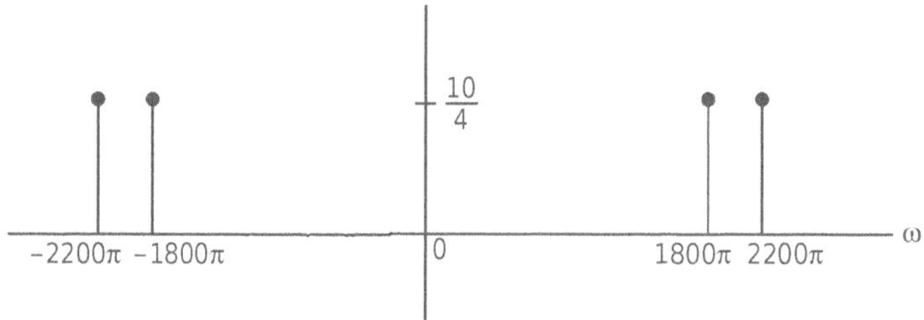

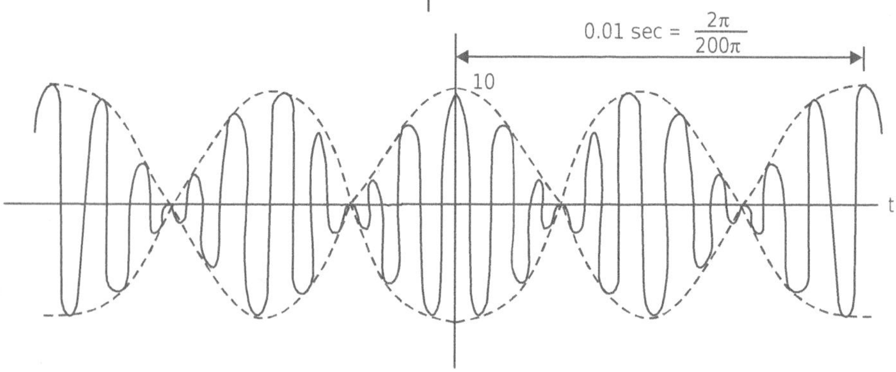

Waveform-based Problems

Example 1. *Represent the following using the* ***Trigonometric Fourier Series*** *representation.*

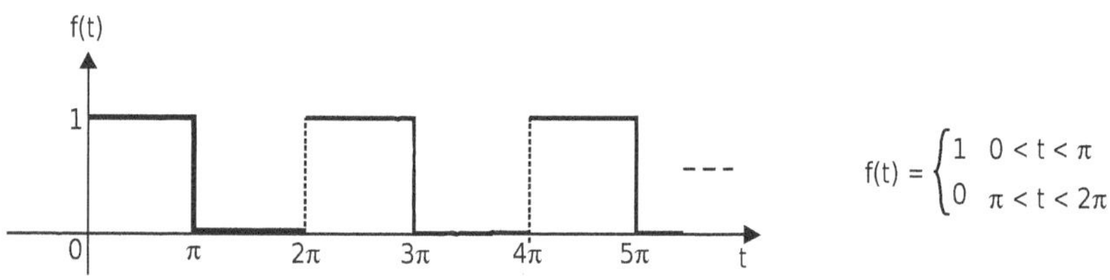

Solution.

For the signal the period is:

$$T_0 = 2\pi, \text{ and } \omega_0 = \frac{2\pi}{T_0} = 1$$

The average or **DC** component for *f*(*t*) is given by:

$$a_0 = \frac{1}{2\pi}\int_0^{2\pi} f(t)\,dt = \frac{1}{2\pi}\int_0^{\pi} dt = \frac{1}{2\pi}[t]_0^{\pi} = \frac{1}{2}$$

Note that $f(t) = 0$ in the time interval π to 2π.

The a_n Fourier co-efficients are:

$$a_n = \frac{1}{\pi}\int_0^{\pi} \cos n\,\omega_0 t\, dt = \frac{1}{n\pi}\left[\sin nt\right]_0^{\pi} = \frac{\sin n\pi}{n\pi} = 0$$

Remember that $\omega_0 = 1$ in the example.

The b_n Fourier co-efficients are:

$$b_n = \frac{1}{\pi}\int_0^{\pi} \sin n\,\omega_0 t\, dt = -\frac{1}{\pi}\left[\frac{1}{n}\cos nt\right]_0^{\pi} = -\frac{[\cos n\pi - 1]}{n\pi}$$

For $n = 0, 2, 4, 6, \ldots$ we have $b_n = 0$, and for $n = 1, 3, 5, 7, \ldots, b_n = 2/n\pi$, then

$$b_n = \begin{cases} 0 & \text{for } n \text{ even} \\ \dfrac{2}{n\pi} & \text{for } n \text{ odd} \end{cases}$$

and we can represent $f(t)$ as the trigonometric series (infinite series):

$$f(t) = \frac{1}{2} + \frac{2}{\pi}\left[\sin t + \frac{1}{3}\sin 3t + \frac{1}{5}\sin 5t + \ldots\right]$$

Taking into account the DC component and the first three terms of the series, this is, $n = 1, 3$, and 5; we obtain the, **DC component, First, Third and Fifth Harmonic of $f(t)$.**

Example 2. *For the following signal:*

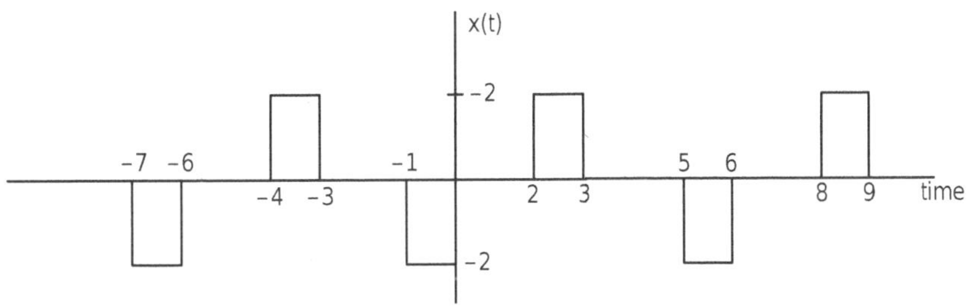

(a) Find the Fourier series

(b) Plot the spectra versus frequency, $\omega = n\omega_0$.

Solution.

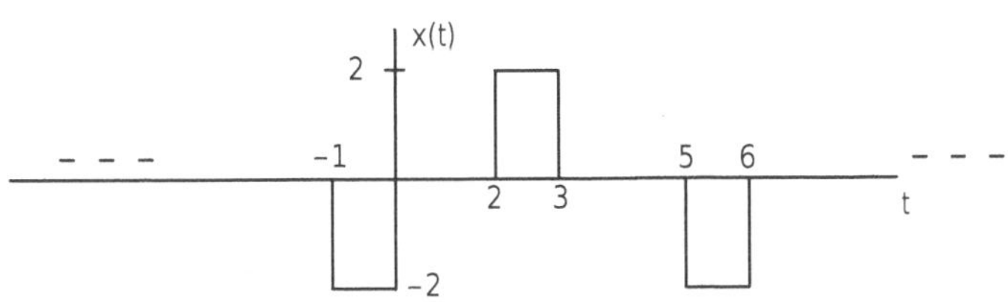

T = 6, (from –1 to +5), $\omega_0 = 2\pi/6 = \pi/3$, we use 'complex form' of Fourier series.

$$x(t) = \sum_{n=-\infty}^{\infty} C_n e^{jn\omega_0 t}$$

$$C_n = \frac{1}{6}\int_{-1}^{5} x(t)e^{-jn\omega_0 t}dt$$

$$= \frac{1}{6}\int_{-1}^{0} -2e^{-jn\omega_0 t}dt + \frac{1}{6}\int_{2}^{3} 2e^{-jn\omega_0 t}dt$$

$$= \frac{-1}{3}\frac{-1}{jn\omega_0}e^{-jn\omega_0 t}\Big|_{-1}^{0} + \frac{1}{3}\frac{-1}{jn\omega_0}e^{-jn\omega_0 t}\Big|_{2}^{3}, \; n \neq 0$$

$$= \frac{1}{jn\pi}(1-e^{jn\pi/3}) - \frac{1}{jn\pi}\left(e^{-jn\pi} - e^{-jn2\pi/3}\right), \; n \neq 0$$

$$= \frac{1}{jn\pi}\left[e^{jn\pi/6}\left(e^{-jn\pi/6} - e^{jn\pi/6}\right) - e^{-jn5\pi/6}\left(e^{-jn\pi/6} - e^{jn\pi/6}\right)\right]$$

$$= \frac{2}{n\pi}\left[-e^{jn\pi/6}\sin\left(\frac{n\pi}{6}\right) + e^{-jn\frac{5\pi}{6}}\sin\left(\frac{\pi n}{6}\right)\right]$$

$$= \frac{2}{n\pi}\sin\left(\frac{\pi n}{6}\right)\left[-e^{jn3\pi/6} + e^{-jn3\pi/6}\right]e^{-2\pi/6nj}$$

$$\boxed{C_n = \frac{-4j}{n\pi}\sin\left(\frac{\pi n}{6}\right)\sin\left(\frac{n\pi}{2}\right)e^{-jn\pi/3}}$$

if $n = 0$

$$C_0 = \frac{1}{6}\int_{-1}^{0} -2dt + \frac{1}{6}\int_{2}^{3} 2dt = 0$$

for $n = 1$ $\quad C_1 = -0.5513 - 0.318j = 0.64e^{-j2.6}, \; C_{-1} = \overline{C_1}$

$n = 2$ $\quad C_2 = 0$

$n = 3$ $\quad C_3 = 0 - j\,0.42 = 0.42e^{-j1.6}, \; C_{-3} = \overline{C_3}$

$n = 4$ $\quad C_4 = 0$

$n = 5$ $\quad C_5 = 0.1103 - 0.0637j = 0.127e^{-j0.52}, \; C_{-5} = \overline{C_5}$

Since the values are complex, the spectrum is shown for magnitude and phase separately as below: (phase in Radious).

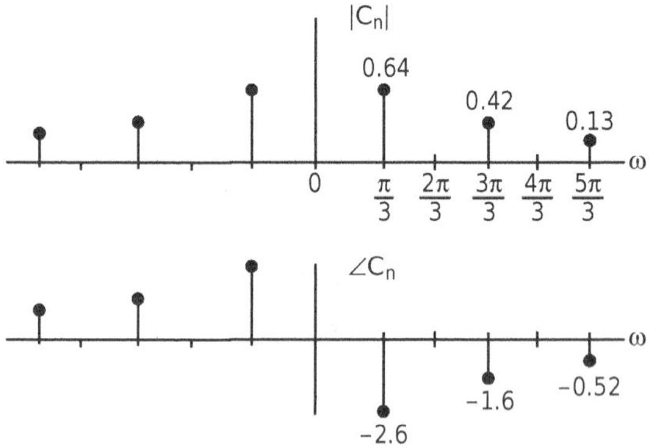

Spectrum

Some of the work can be done using MATLAB:

```
» n = 1:15;
» cn = –4 * j./n / pi.*sin(pi * n / 6).*sin(n * pi / 2).*exp(–j * n * pi / 3);
» n = –15: —1;
» c_n = –4 * j./n / pi.*sin(pi * n / 6).*sin(n * pi / 2).*exp(–j * n * pi / 3);
» cn = [c_n 0 cn];
» n = –15:15;
» subplot (221), stem (n, abs (cn))
» title ('|c_n|')
» subplot (222), stem (n, angle (cn))
» title ('angle (c_n) in rad')
```

To plot the Fourier series to check your answers:

```
T = 6;
ω0 = 2*pi/T;
t = –1.5*T:T/1000:1.5*T; % in steps of T/1000.
N = input ('Number of harmonics');
c0 = 0;
x = c0*ones (1, length (t)); % dc component
for n = 1 : N,
    cn = –4*j/n/pi*sin(pi*n/6)*sin(n*pi/2)*exp(–j*n*pi/3);
    c_n = conj(cn);
    x = x + cn*exp(j*n*ω0*t) + c_n*exp(–j*n*ω0*t);
end
plot (t, x)
title (['N = ', num2str(N)])
```

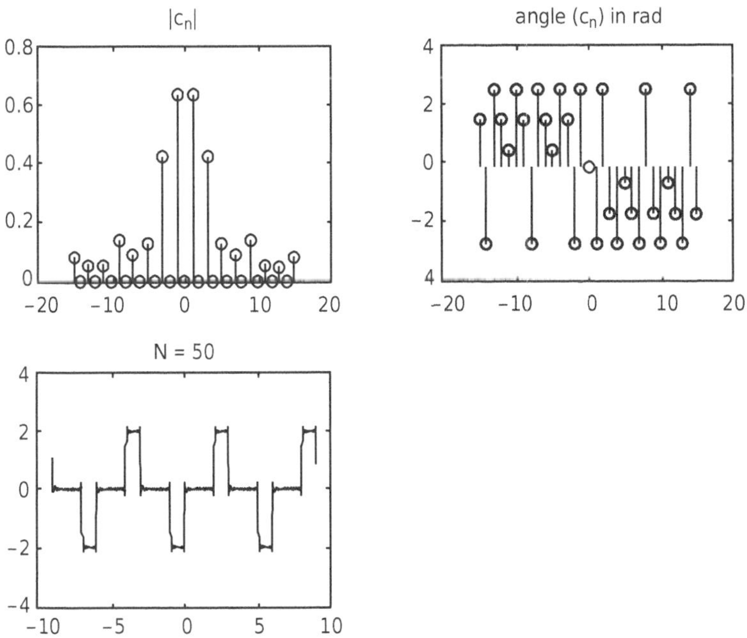

Example 3. *Repeat problem 1 for the following signal:*

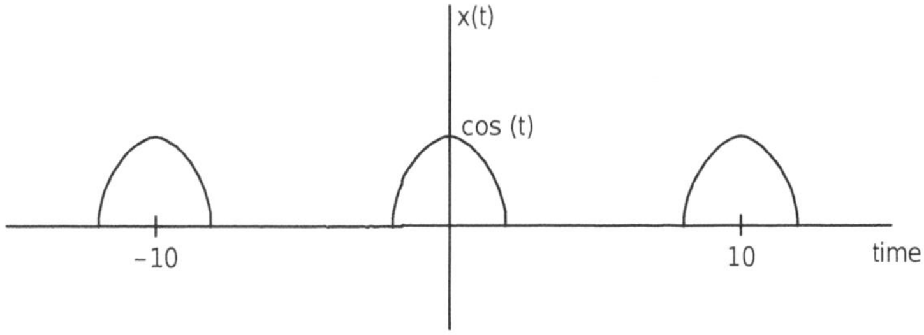

Solution.

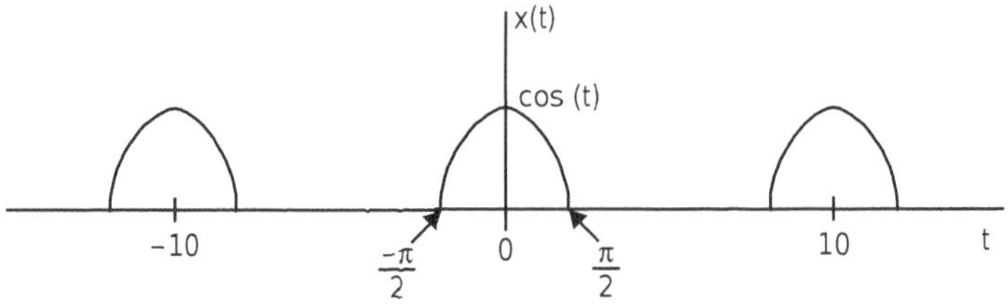

$T = 10 \Rightarrow \omega_0 = 2\pi/10 = \pi/5$

$$x(t) = \sum_{n=-\infty}^{\infty} C_n \, e^{jn\omega_0 t}$$

$$C_n = \frac{1}{T}\int_T x(t)\, e^{-jn\omega_0 t} dt$$

$$= \frac{1}{10}\int_{-\pi/2}^{\pi/2} \cos(t)\, e^{-jn\omega_0 t} dt$$

$$= \frac{1}{20}\int_{-\pi/2}^{\pi/2} \left(e^{jt} + e^{-jt}\right) e^{-jn\omega_0 t} dt$$

$$= \frac{1}{20}\int_{-\pi/2}^{\pi/2} e^{jt(1 - n\omega_0)} + e^{-jt(1 + n\omega_0)} dt$$

$$= \frac{1}{20}\,\frac{1}{j(1-n\omega_0)}\left[e^{j\pi/2(1-n\omega_0)} - e^{-j\pi/2(1-n\omega_0)}\right]$$

$$+ \frac{-1}{20j(1+n\omega_0)}\left[e^{-\pi/2j(1+n\omega_0)} - e^{j\pi/2(1+n\omega_0)}\right]$$

$$= \frac{1}{20j}\,\frac{1}{(1-n\omega_0)}\left[je^{-j\pi/2n\omega_0} + je^{+j\pi/2n\omega_0}\right]$$

$$- \frac{1}{20j(1+n\omega_0)}\left[je^{-j\pi/2n\omega_0} - je^{j\pi/2n\omega_0}\right]$$

$$= \frac{1}{10(1-n\omega_0)}\cos\left(\frac{\pi}{2}n\omega_0\right) + \frac{1}{10(1+n\omega_0)}\cos\left(\frac{\pi}{2}n\omega_0\right)$$

$$= \cos\left(\frac{\pi}{2}n\omega_0\right)\frac{1}{10}\left[\frac{1}{1-n\omega_0} + \frac{1}{1+n\omega_0}\right]$$

$$= \cos\left(\frac{\pi}{2}n\omega_0\right)\frac{1}{10}\frac{2}{1-(n\omega_0)^2}$$

$$C_n = \cos\left(\frac{\pi}{2}n\omega_0\right)\frac{1}{5(1-(n\omega_0)^2)}$$

for $$\omega_0 = \frac{\pi}{5};\ C_n = \cos\left(\frac{\pi^2 n}{10}\right)\frac{1}{5\left(1-\left(\frac{n\pi}{5}\right)^2\right)}$$

$$C_0 = 1/5$$

$$C_1 = \cos\left(\frac{\pi^2}{10}\right)\frac{1}{5\left(1-\left(\frac{\pi}{5}\right)^2\right)} = 0.1822$$

$$C_2 = 0.1355$$

$$C_3 = 0.0771; \qquad C_{-n} = \overline{C}_n = C_n$$

$$C_4 = 0.026$$

$$C_5 = -0.005$$

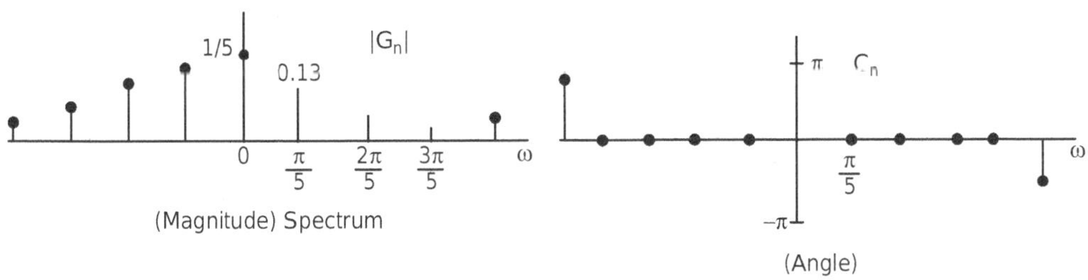

(Magnitude) Spectrum

(Angle)

Some of the work can be done using MATLAB:

```
» n = –10 : 10;
» cn = cos (pi/2*n*ω0)/5./(1 – (n*ω0).^2) ;
» subplot (221), stem (n, abs (cn))
» title ('|c_n|')
» subplot (222), stem (n, angle (cn))
» title ('angle (c_n) in rad')
```

will give first plots shown below. To check your answer, you can plot the truncated series and see if it converges correctly.

```
T = 10;
ω0 = 2*pi/T;
t = –1.5 : T/1000 : 1.5*T;
N = input ('Number of harmonics');
c0 = 1/5;
x = c0*ones (1, length (t)); % dc component
for n = 1 : N,
    cn = cos (pi/2*n*ω0)/5/(1 – (n*ω0)^2);
    c_n = cn;
    x = x + cn*exp (j*n*ω0*t) + c_n*exp (–j*n*ω0*t);
end
plot (t, x)
title (['N = ', num2str (N)])
```

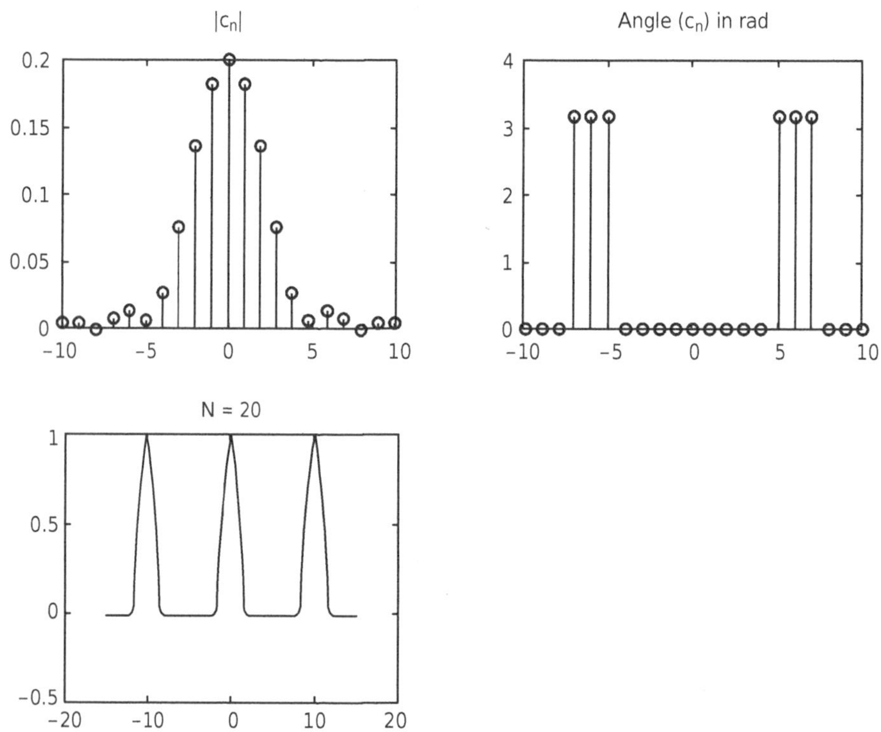

Example 4. *Compute the Fourier series for the following signals:*

(a) $x(t) = 2 + 4 \cos(50t + \pi/2) + 12 \cos(100t - \pi/3)$

(b) $x(t) = 4 \cos(2\pi(1000)t) \cos(2\pi 750000t)$

(c)

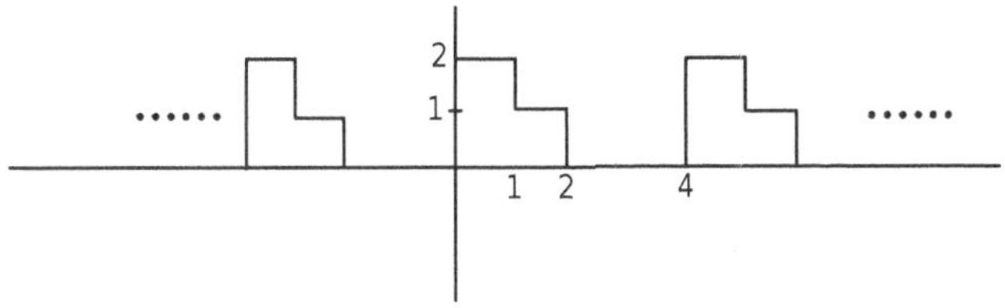

(d)

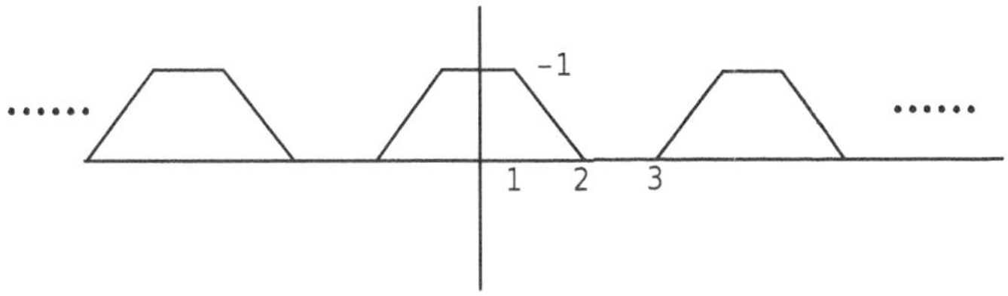

Solution. (*a*) $$x(t) = 2 + 4\cos(50t + \pi/2) + 12\cos(100t - \pi/3)$$

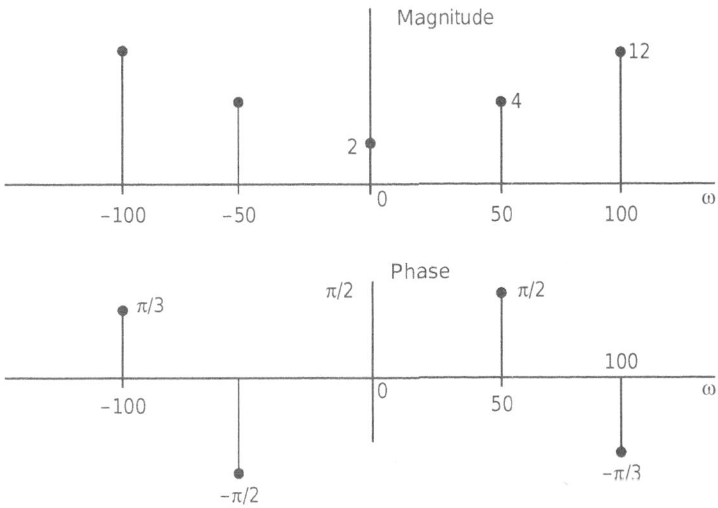

(*b*) $$x(t) = 4\cos(2\pi(1000)t)\cos(2\pi(750000)t)$$

$$= 4\left[\frac{e^{j2\pi(1000t)} + e^{-j2\pi(1000t)}}{2}\right]\left[\frac{e^{j2\pi(750000t)} + e^{-j2\pi(75000t)}}{2}\right]$$

$$= e^{-j2\pi(749000)t} + e^{-j2\pi(749000)t} + e^{j2\pi(751000t)} + e^{-j2\pi(751000t)}$$

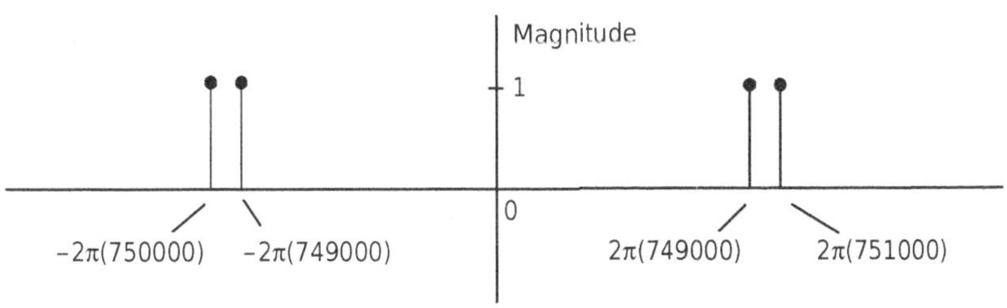

Phase is zero

ω_0 is greatest common factor,

$$\omega_0 = 2\pi(1000)$$

(c)

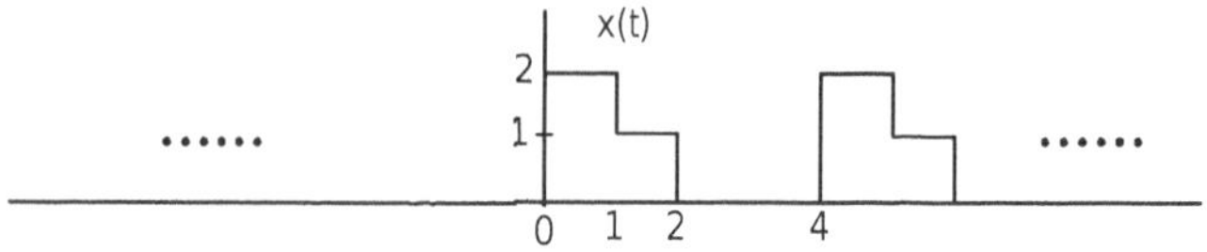

T = 4, so

$$\omega_0 = 2\pi/4 = \pi/2$$

$$C_0 = \frac{1}{4}\int_0^2 x(t)dt = \frac{3}{4}$$

$$C_n = \frac{1}{4}\int_0^2 x(t)\, e^{-jn\pi/2t} dt$$

$$= \frac{1}{4}\int_0^1 2e^{-jn\pi/2t} dt + \frac{1}{4}\int_1^2 e^{-jn\pi/2t} dt$$

$$= \frac{1}{2}\frac{1}{-jn\pi/2} e^{-jn\pi/2t}\Big|_0^1 + \frac{1}{4}\frac{1}{-jn\pi/2} e^{-jn\pi/2t}\Big|_1^2$$

$$= \frac{1}{-jn\pi}\left[e^{-jn\pi/2} - 1\right] + \frac{1}{-j2n\pi}\left[e^{-jn\pi} - e^{-jn\pi/2}\right]$$

This can be simplified

$$e^{-jn\pi} = (-1)^n$$

$$C_n = \frac{1}{jn\pi}\left[1 - \frac{1}{2}(-1)^n - \frac{1}{2}e^{-jn\pi/2}\right]$$

$$C_0 = 3/4$$

$$x(t) = \sum_{n=-\infty}^{\infty} C_n e^{jn\pi t/2}$$

(d)

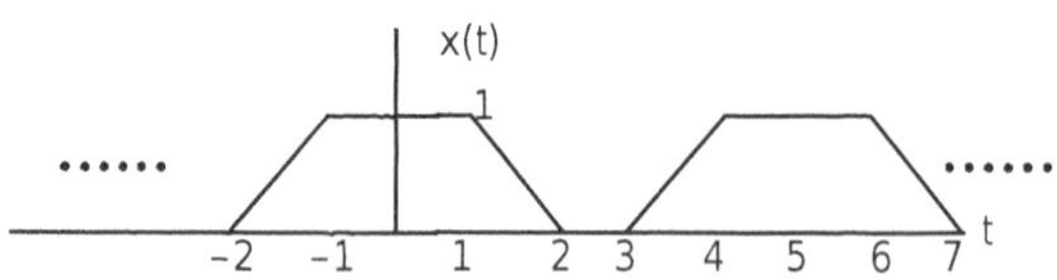

T = 5

$$\omega_0 = \frac{2\pi}{5}$$

$$C_0 = \frac{1}{5}\int_{-2}^{2} x(t)dt = \frac{3}{5}$$

$$C_n = \frac{1}{5}\int_{-2}^{-1}(t+2)\,e^{-j\omega_0 nt}dt + \frac{1}{5}\int_{-1}^{1}e^{-j\omega_0 nt}dt + \frac{1}{5}\int_{1}^{2}(-t+2)\,e^{-jn\omega_0 t}dt$$

Recall $\int te^{at}dt = \frac{e^{at}}{a^2}(at-1)$

$$C_n = \frac{1}{5}\left[\frac{e^{-jn\omega_0 t}}{(-jn\omega_0)^2}(-jn\omega_0 t - 1) - \frac{2}{jn\omega_0}\,e^{-jn\omega_0 t}\Bigg|_{-2}^{-1}\right.$$

$$\left.+\frac{-1}{j\omega}e^{-jn\omega_0 t}\Bigg|_{-1}^{1} + \frac{-e^{-jn\omega_0 t}}{(-jn\omega_0)^2}(-jn\omega_0 t - 1)\,\frac{-2}{jn\omega_0}e^{-jn\omega_0 t}\Bigg|_{1}^{2}\right]$$

$$C_n = \frac{1}{5}\left[\frac{e^{jn\omega_0}}{-n^2\omega_0}(jn\omega_0 - 1) - \frac{2}{jn\omega_0}\,e^{jn\omega_0} + \frac{e^{jn\omega_0^2}}{\omega_0^2 n^2}(jn\omega_0 2 - 1) + \frac{2}{jn\omega_0}e^{j\omega_0 n^2} - \frac{1}{j\omega_0 n}e^{-jn\omega_0}\right.$$

$$+\frac{1}{j\omega_0 n}e^{jn\omega_0} + \frac{e^{jn\omega_0^2}}{n^2\omega_0^2}(-jn\omega_0 2 - 1) - \frac{2}{jn\omega_0}e^{-jn\omega_0^2}$$

$$\left.-\frac{e^{-jn\omega_0}}{n^2\omega_0^2}(-jn\omega_0 - 1) + \frac{2}{jn\omega_0}\,e^{-jn\omega_0}\right]$$

$$= \frac{1}{5}e^{jn\omega_0}\left(\frac{-j}{n\omega_0} + \frac{1}{n^2\omega_0^2} + \frac{2j}{n\omega_0} - \frac{j}{\omega_0 n}\right) + \frac{1}{5}e^{-jn\omega_0}\left(\frac{j}{n\omega_0} + \frac{j}{n\omega_0} + \frac{1}{n^2\omega_0^2} - \frac{2j}{n\omega_0}\right)$$

$$+\frac{1}{5}e^{j2n\omega_0}\left(\frac{j2}{\omega_0^2 n^2} - \frac{1}{\omega_0^2 n^2} - \frac{2j}{n\omega_0}\right) + \frac{1}{5}e^{-j2n\omega_0}\left(\frac{-j2}{n\omega_0} - \frac{1}{n^2\omega_0^2} + \frac{2j}{n\omega_0}\right)$$

$$= \frac{1}{5n^2\omega_0^2}\left(e^{jn\omega_0} + e^{-jn\omega_0 t}\right) - \frac{1}{5n^2\omega_0^2}\left(e^{j2n\omega_0} + e^{-j2n\omega_0}\right)$$

$$C_n = \frac{2}{5n^2\omega_0^2}\left[\cos(n\omega_0) - \cos(2n\omega_0)\right],\ C_0 = \frac{3}{5}$$

$$x(t) = \sum_{n=-\infty}^{\infty} C_n e^{jn2\pi/5t}$$

EXERCISES

1. Find the Fourier series expansion of the waveform.

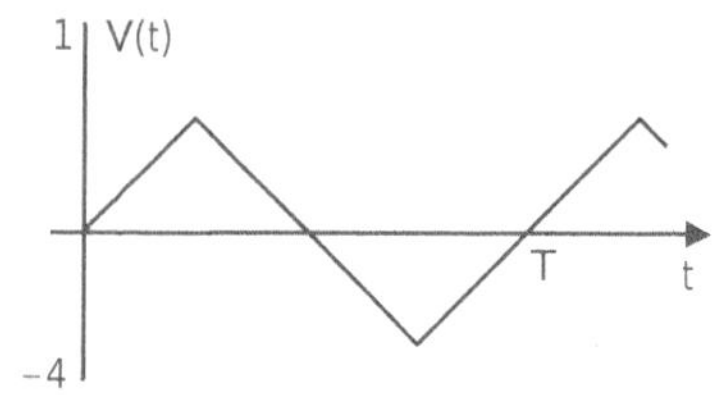

The voltage is applied to an RC series circuit of R = 1 kilo-ohm and C = 1 μF. If the amplitudes of the fundamental and 3rd harmonic components of current in the circuit in the steady state are in the ratio 5 : 1, determine the period T of the waveform.

[2π millisecs.]

2. Find the harmonics of the unit trapezoidal wave. So, deduce (*a*) the harmonics of a square wave putting $\alpha = 0$ and (*b*) for triangular wave, putting $a = \pi/2$.

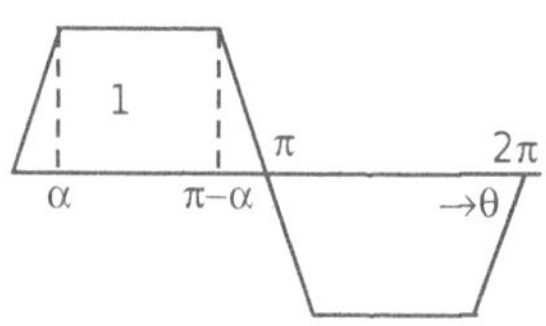

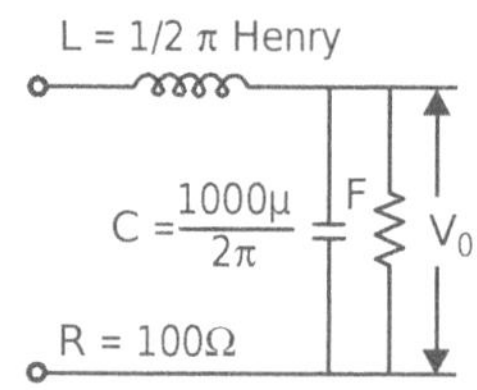

$$\left\{\frac{4}{\pi\alpha}\left(\sin\alpha\sin\theta + \frac{1}{3^2}\sin 3\alpha\sin 3\theta + \ldots\right)\right\}$$

3. If a sinusoidal a.c. wave is passed via a rectifier, the negative half-cycles do not send any current via the rectifier, and a half sinusoid wave is obtained. Find its harmonics.

[**Hint:** Vide Table (Wave 8)] $\left[\textbf{Ans. } \frac{1}{2}, \frac{2}{3\pi}, \frac{-2}{15\pi}, \frac{2}{35\pi}\right]$

4. Find the harmonics of the full wave sinusoid.

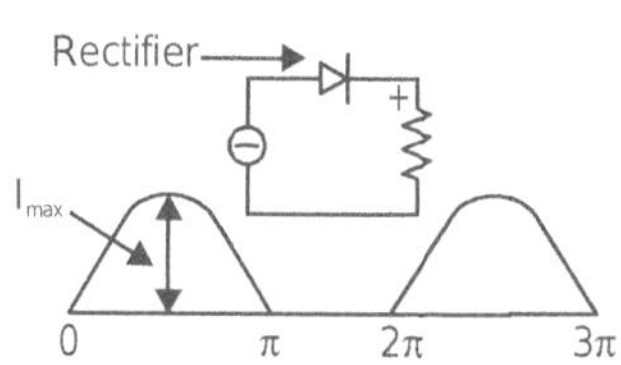

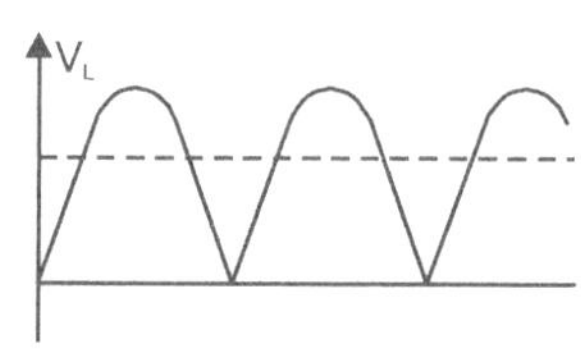

$$\left[\textbf{Ans. } \frac{1}{2} \text{ wave : } V\left\{\frac{1}{\pi} + \frac{1}{2}\sin\theta - \frac{2}{\pi}\Sigma\frac{\cos n\theta}{n^2 - 1}\right\} n = 2, 4, 6 \ldots\right.$$

$$\left.\text{Full wave : } V\left\{\frac{2}{\pi} - \frac{4}{\pi}\Sigma\frac{\cos n\theta}{n^2 - 1}\right\} n = 2, 4, 6 \ldots \infty\right]$$

5. If the half wave rectified wave of 100 volts max., is applied to the filter circuit, find the output V_0 upto the 4th harmonic.

$$\left[\textbf{Ans.} 100\left[\frac{1}{\pi} - \frac{1}{3}\sin\theta + \frac{2}{27\pi}\cos 2\theta + \frac{2}{600\,\pi}\cos 4\theta + \ldots\right]\right]$$

6. A pulse train of unit amplitude and at 100 Hz has a width of 0.5 ms. Find its 1, 2, 3 harmonics. (vide Table–7)

$$\left[\textbf{Ans. } \alpha = \frac{0.5}{10}, \frac{2}{\pi}\sin\alpha\pi, \frac{1}{\pi}\sin 2\alpha\pi, \frac{2}{3\pi}\sin 3\alpha\pi\right]$$

7. Show that in the range $(0, \pi)$, the sine series for $\pi x - x^2$ is $\frac{8}{\pi}\left(\sin x + \frac{1}{3^3}\sin 3x + \frac{1}{5^3}\sin 5x + \ldots\right)$.

8. Find a Fourier cosine series corresponding to the function $f(x) = x$, defined in the interval $(0, \pi)$.

9. Find the Fourier sine series and the Fourier cosine series corresponding to the function,

$$f(x) = \pi - x \text{ when } 0 < x < \pi$$

defined in the interval 0 to π.

10. Show that, when $0 < x < \pi$.

$$f(x) = \sin 2x + \frac{1}{2}\sin 4x + \frac{1}{4}\sin 8x + \frac{1}{5}\sin 10x \ldots$$

$$= \frac{2}{\sqrt{3}}\left(\cos x - \frac{1}{5}\cos 4x + \frac{1}{7}\cos 7x - \frac{1}{11}\cos 11x \ldots\right).$$

where $$f(x) = \frac{\pi}{3}\left(0 < x < \frac{\pi}{3}\right)$$

11. Determine the first three harmonics of the Fourier series for the values.

x	30°	60°	90°	120°	150°	180°	210°	240°	270°	300°	330°	360°
y	2.34	3.01	3.68	4.15	3.69	2.20	0.83	0.51	0.88	1.09	1.19	1.64

[**Ans.** 2.1; –0.28, –.018 1.61, –0.5, + 0.2]

12. The values of y, a periodic function of x, are given below for twelve equidistant values of x covering the whole period. Express y in a Fourier series as far as the second harmonic if the first value is for $x = 30°$.

3.5, 6.09, 7.82, 5.58, 8.43, 7.73,
6.98, 6.19, 6.04, 5.55, 5.01, 3.35,

[**Ans.** 6.27 –1.99 cos x – 0.67 cos $2x$ + 0.75 sin x – 0.69 sin $2x$]

13. The following 12 values of $f(\theta)$ correspond to equidistant values of the angle $\theta°$ in the range (0° to 360°) :

θ	0°	30°	60°	90°	120°	150°	180°	210°	240°	270°	300°	330°
$f(\theta)$	10.5	20.2	26.4	29.3	27.0	21.5	12.8	1.6	–11.2	–18.0	–15.8	–3.5

Determine approximately, the Fourier expansion for $f(\theta)$ in terms of 0, up to the terms involving sin 3θ and cos 3θ.

[**Ans** 8.4 + 23.1 sin θ –1.7 cos θ + 1.1 sin θ = 20 + 3.1 cos 2θ –0.6 sin 3θ + 0.45 cos 3θ]

14. Find the Fourier series for the following signal. Also, sketch the approximation if a large number of terms are kept in the series (say N = 30).

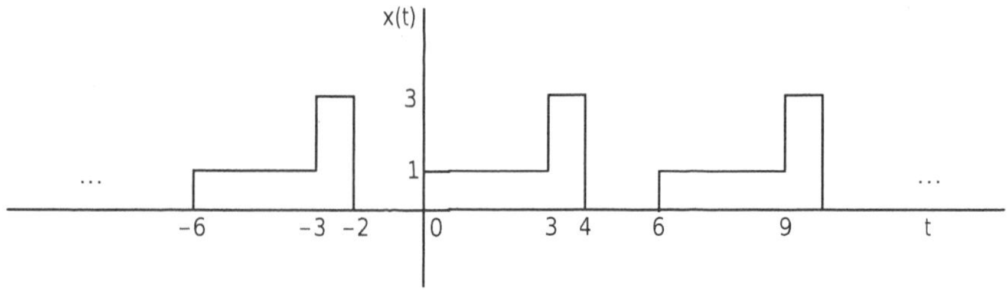

[**Ans.**

$$x(t) = \sum_{m=-\infty}^{\infty} C_n \, e^{jn\omega_0 t}; \qquad T = 6, \; \omega_0 = \frac{2\pi}{6} = \frac{\pi}{3}$$

$$C_n = \frac{1}{6}\int_0^6 x(t)\, e^{-jn\omega_0 t}\, dt$$

$$= \frac{1}{6}\int_0^3 e^{-jn\frac{\pi}{3}t}\, dt + \frac{1}{6}\, 3\int_3^4 e^{-jn\frac{\pi}{3}t}\, dt$$

$$= \frac{1}{6}\,\frac{-1}{j\frac{\pi}{3}n}\, e^{-jn\frac{\pi}{3}t}\Bigg|_0^3 + \frac{1}{2}\,\frac{-1}{jn\frac{\pi}{3}}\, e^{-jn\frac{\pi}{3}t}\Bigg|_3^4, \; n \neq 0$$

$$C_n = \frac{-1}{j2\pi n}\left(e^{-jn\pi} - 1\right) - \frac{3}{j2\pi n}\left(e^{-jn4\frac{\pi}{3}} - e^{-jn\pi}\right), \; n \neq 0$$

if $n = 0$,

$$C_0 = \frac{1}{6}\int_0^6 x(t)\, dt = 1$$

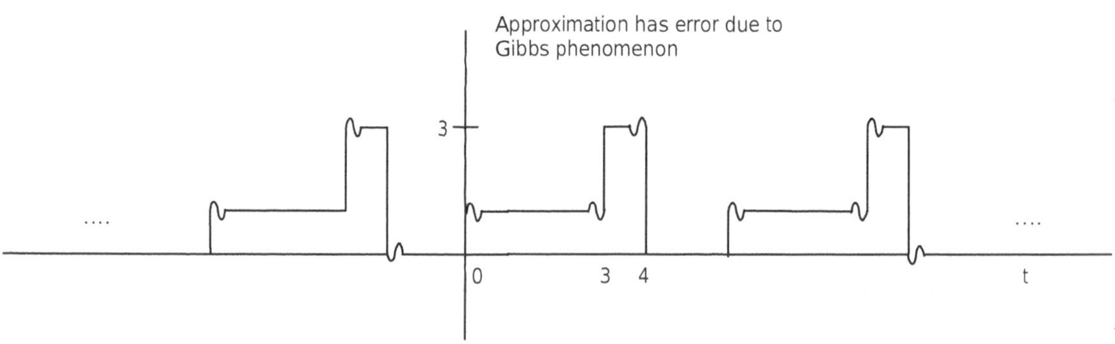

]

15. For the following plot, find the Fourier series and plot its spectra.

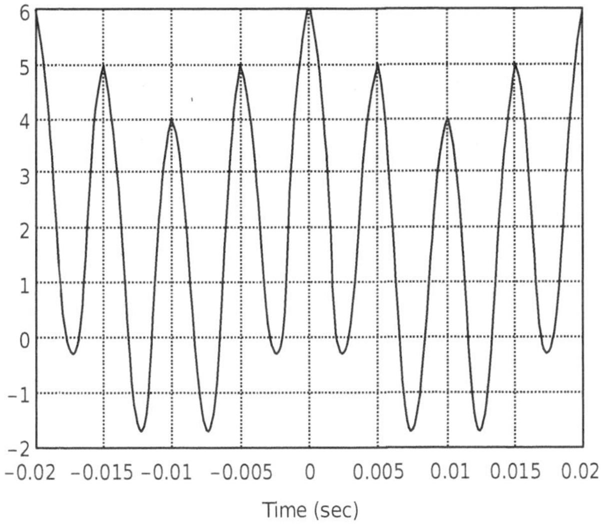

[Ans.

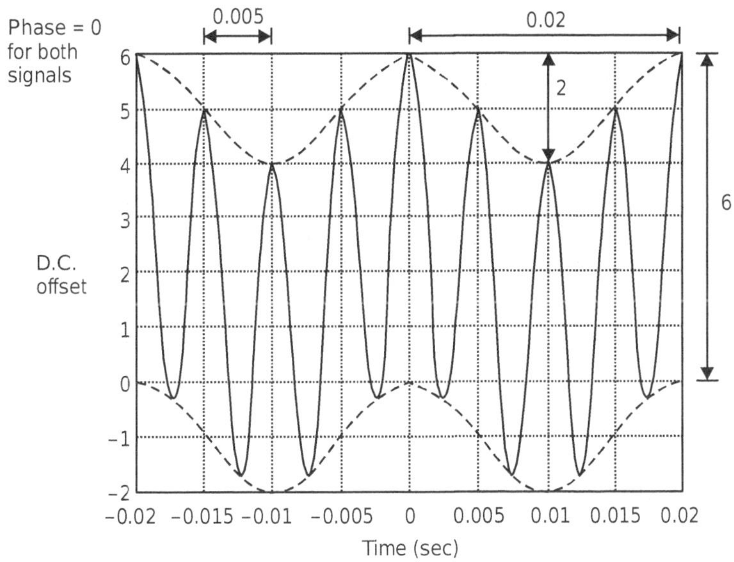

$$x(t) = 2 + \cos\left(\frac{2\pi}{0.02}t\right) + 3\cos\left(\frac{2\pi}{0.005}t\right)$$

$$= 2 + \cos(100\pi t) + 3\cos(400\pi t)]$$

16. Time signals and their corresponding spectra are shown below. However, they are in random order. Match them up.

1. .... 2. .... 3. .... 4. .... 5. ....

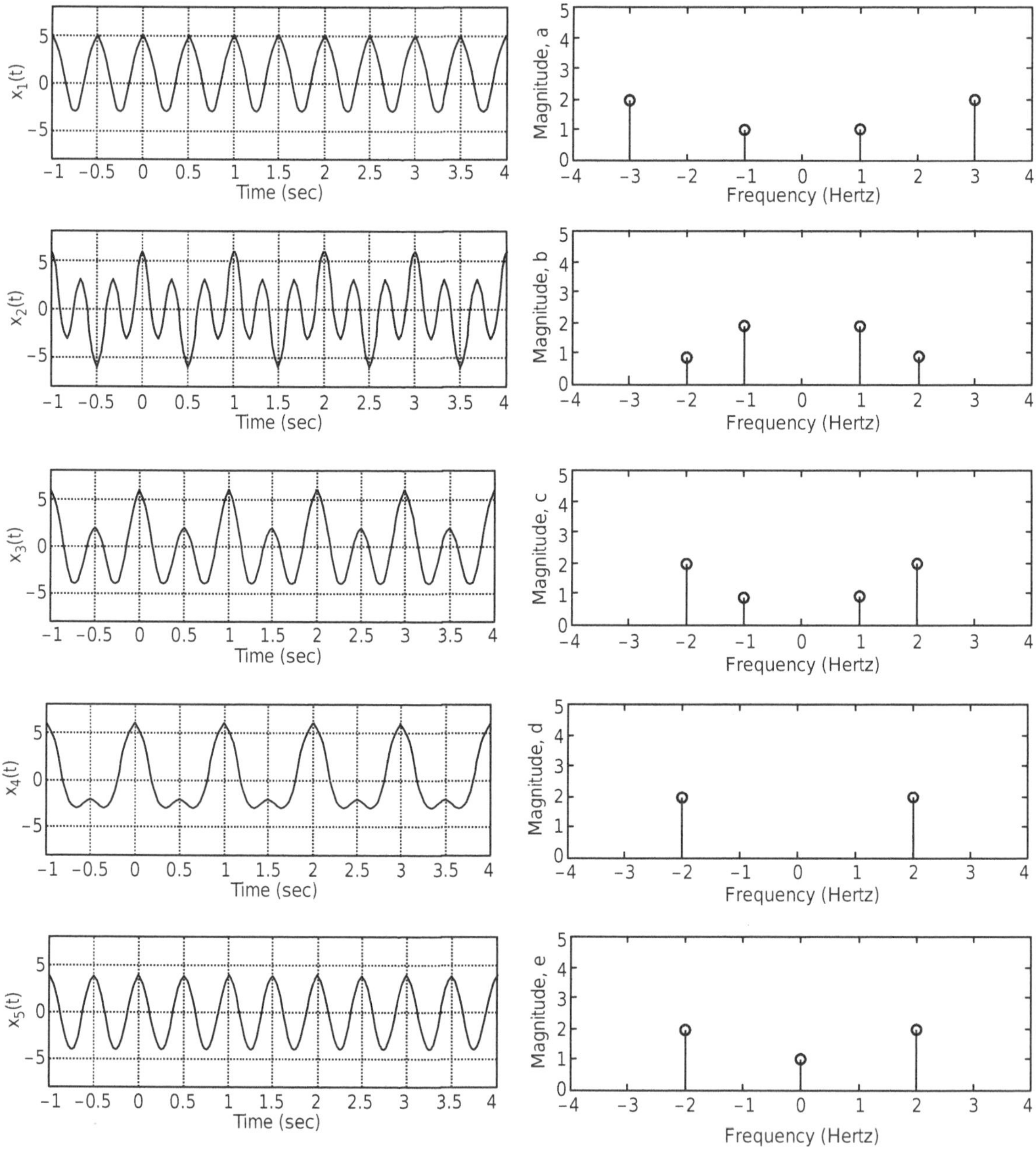

[**Ans.** Time signals and their corresponding spectra are shown below. However, they are in random order. Match them up.

1. (*e*) 2. (*a*) 3. (*c*) 4. (*b*) 5. (*d*)

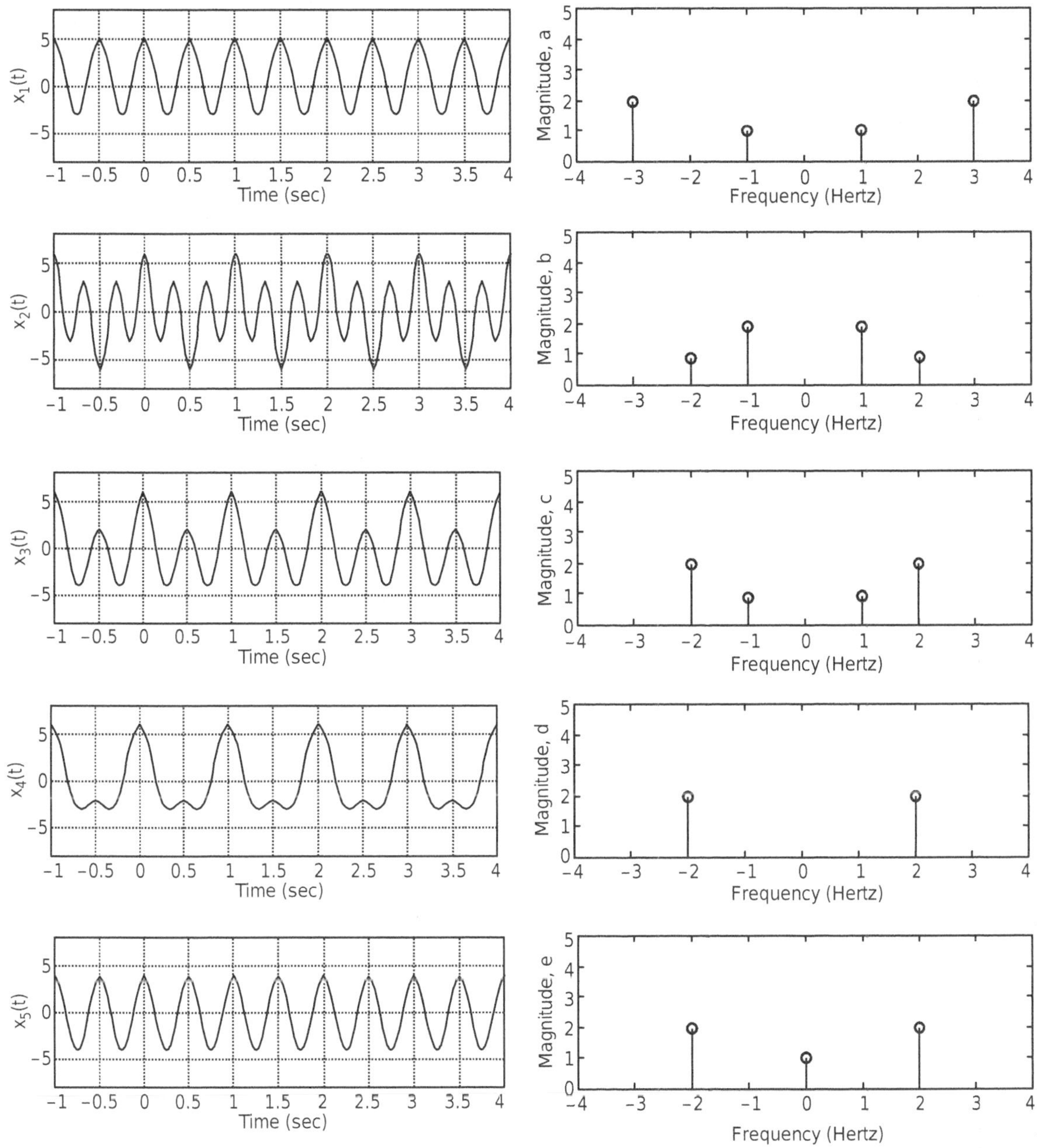

]

SIMPLE EXERCISES

1. A square wave of 100V peak to peak of frequency 1 KHz is given to a 10 ohms resistor. What are the third, fifth and seventh harmonic peak to peak currents in it ?

$$\left[\textbf{Ans. } \frac{40}{3\pi}, \frac{40}{5\pi}, \frac{40}{7\pi}\right]$$

2. For a periodic function $f(t)$ derive the general Fourier series.

$$\left[\textbf{Ans. } f(t) = \frac{b_0}{2} + \sum_{n=1}^{\infty} b_n \cos n\theta + a_n \sin n\theta\right.$$

where

$$b_n = \frac{1}{\pi}\int_0^{2\pi} f(t) \sin nt\, dt$$

$$\left. a_n = \frac{1}{\pi}\int_0^{2\pi} f(t) \cos nt\, dt \right]$$

3. Determine the integrals $\int x \sin nx\, dx$ and $\int x \cos nx\, dx$ and thereby find the Fourier Series expansion of a sawtooth wave of peak value unity. [See Table No. 6]

4. Explain zero point symmetry, zero axis symmetry and half wave symmetry of periodic waves. What are the Fourier series terms for each of these ?

[**Ans.** sine terms nil, cosine terms nil and even harmonics nil.]

5. Write a Basic or 'C' language program to evaluate the discrete Fourier Series for a set of N equidistant data values, assumed to be periodic after these points.

[N points make one waveform.]

Apply the program to a set of points falling on a triangle and compare the harmonics got with the values as per analytical solution for a triangular wave [See Table No. 3].

6. A single rectifier and a bridge rectifier are used commonly for AC to DC conversion. (*i*) Find the Fourier series of a half-wave rectified AC of V volts peak. (*ii*) Find the Fourier series of a full wave rectified signal of same peak value.

Compare the harmonics of the two.

[See Table Nos. 8 and 9]

	DC	ω	2ω	3ω	4ω	5ω	6ω
1/2 wave	V/π	V/2	$\frac{2}{3}\frac{V}{\pi}$	0	–2/15π	0	$\frac{2}{35}\frac{V}{\pi}$
Full wave	2V/π	0	$\frac{4}{3}\frac{V}{\pi}$	0	–4/15π	0	$\frac{4}{35}\frac{V}{\pi}$

7. Define the RMS value of complex waveform in terms of its harmonics. [Eqn. 6.7]

8. Derive the Fourier Series of a pulse train and plot the Fourier co-efficients as a spectrum. What does the figure indicate with respect to the bandwidth occupied by a pulse ?

[**Ans.** sin $k\alpha$ p/k. A wide bandwidth is occupied.]

9. Explain the difference between Fourier series and Fourier transform. Write down the relation for the Fourier transform of a signal $f(t)$ and the Inverse Fourier transform of a frequency spectrum F(ω). [Eqn. 6.11, 6.13

For a periodic function, Fourier Series is same as Fourier Transform.]

10. What are the important waveforms in signal communication ? Draw (sketch) their Fourier spectrum. [(1) Square wave (2) Pulse, Fig. 6.7]

7

FOURIER TRANSFORMS

7.1 FOURIER SERIES FOR NON-PERIODIC SIGNALS

For periodic signals, such as a square wave, we saw that it can be represented as a series of sine and cosine waves, from the fundamental and all the harmonics.

$$f(t) = \sum a_n \sin nt + \sum b_n \cos nt \tag{7.1}$$

For a square wave, which is continuous and also periodic, with a frequency, say 50 Hz, we will have n values from 50 Hz, and in multiples of 50 Hz, as

50, 100, 150, 200, 250, 300, ...

We have found the series for a square wave and noted that the harmonics are lesser and lesser in amplitude, being

$$1,\ 0,\ \frac{1}{3},\ 0,\ \frac{1}{5},\ 0,\ ....$$

for the above frequencies. Note that the even harmonics are zero.

Consider a non-periodic signal. It is also a signal which exists for a certain span of time. For instance, Fig. 6.10 gives the NMR instrument signal. This is a signal which lasts for 2.5 seconds and comprises a lot of frequency components in it. This was got by the instrument from the chemical in a combined RF and magnetic field, after exciting the chemical with a powerful RF pulse. It is an one-time signal, and is not a periodic signal.

For such a signal, we can consider the period to be infinite.

For the periodic signal, of period T, the basic or fundamental frequency is $\frac{1}{T}$, and $f_0 = \frac{1}{T}$.

All Fourier series components one in *'multiples'* of this f_0. Thus, we have seen 50, 100, 150, 200, 250, 300, ... in the above example. Note that intermediate frequencies do not exist, (such as 75 or 80 or whatever).

In the non-periodic signal, since $f_0 \to 0$, or infinitesimally small, we will have *all* frequencies in the Fourier components. The difference between the several components diminishes and so we will have a continuous 'spectrum' of frequency components possible for a non-periodic signal.

In that case, the summation in (7.1) becomes an integration.

$$f(t) = \int_0^{\infty} \left\{ X_r(f) \cos 2\pi ft + X_i(f) \sin 2\pi ft \right\} dt \tag{7.2}$$

Note that the a_n and b_n have been written as $X_r(f)$, $X_i(f)$, where f varies from 0 to ∞ continuously instead of in steps as for Fourier series. 'r' and 'i' are *real* and *imaginary* parts of Fourier components, meaning cosine and sine component amplitudes.

Note that the summation is for all frequencies and hence df is in the integration.

The above equation is usually written in terms of the complex notation for sinusoids,

$$e^{j\omega t} = e^{j2\pi ft} = (\cos \omega t + j \sin \omega t) \quad (7.3)$$

Also, the integral should be actually from $-\infty$ to $+\infty$, because $e^{j\omega t}$ refers to positive values of frequencies, while $e^{-j\omega t}$ refers to negative values. And we know that

$$\frac{(e^{j\omega t} + e^{-j\omega t})}{2} = \cos \omega t$$

$$\frac{(e^{+j\omega t} - e^{-j\omega t})}{2j} = \sin \omega t \quad (7.4)$$

Therefore (7.2) is re-written as

$$\boxed{x(t) = \frac{1}{2\pi}\int_{-\infty}^{\infty} X(\omega)\, e^{j\omega t}\, d\omega} \quad (7.5)$$

We have changed from f to ω, the angular frequency in the above, putting $\omega = 2\pi f$ and $d\omega = 2\pi df$ in (7.2) and changing from sine-cosine form into complex exponential form using exp ($j\omega t$). The division by 2π is also noteworthy.

If the function is $f(t)$ instead of $x(t)$,

$$f(t) = \frac{1}{2\pi}\int_{-\infty}^{\infty} F(\omega)\, e^{j\omega t}\, d\omega \quad (7.6)$$

What we saw in (7.5) is really the inverse Fourier transform. The relation that finds the frequencies of a given signal is the Fourier transform. From

$$f(t) = \frac{1}{2\pi}\int_{-\infty}^{\infty} F(\omega)\, e^{j\omega t}\, d\omega, \quad (7.7)$$

let us find the value of F(ω) now. Form the integral below

$$F(\omega) = \int_{-\infty}^{\infty} f(t)\, e^{-j\omega t}\, dt \quad (7.8)$$

Take the right hand side of (7.6). Substitute (7.7) in it

$$\frac{1}{2\pi}\int_{-\infty}^{\infty} F(\omega)\, e^{j\omega t}\, d\omega = \frac{1}{2\pi}\int_{-\infty}^{\infty} e^{j\omega t}\, d\omega \int_{-\infty}^{\infty} f(x)\, e^{-j\omega x}\, dx$$

Note we have changed from t to x in making this substitution, because we already have t in $e^{j\omega t}$.

Interchanging the order of integration,

$$I = \frac{1}{2\pi}\int_{-\infty}^{\infty} F(\omega)\, e^{j\omega t}\, d\omega = \frac{1}{2\pi}\int_{-\infty}^{\infty} f(x)\, dx \int_{-\infty}^{\infty} e^{j\omega(t-x)}\, d\omega$$

The second integral is the $\delta(t - x)$ function.

$$\int_{-\infty}^{\infty} e^{-j\omega(t-x)}\, d\omega = \delta(t - x).$$

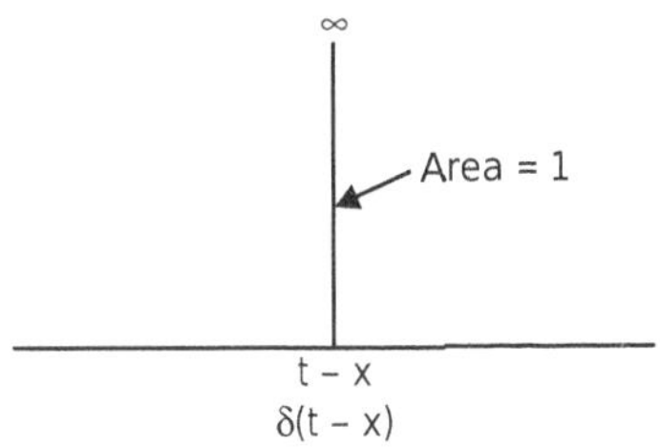

Fig. 7.1

So,
$$I = \int_{-\infty}^{\infty} f(x)\,\delta(t-x)\,dx$$
$$= f(t) \qquad \text{(provided } f(t) \text{ is 'continuous')}$$

because the product $f(x)\ \delta(t - x)$ becomes zero for all x except when $x = t$ and then product is just unity.

Thus the inter-relationship between (7.6) and (7.7) is proved.

7.2 FOURIER TRANSFORM PLOTS

Any signal can be written as a sum of scaled sinusoids, as was expressed in Eq. (7.1). That equation is usually written using complex exponential notation :

$$x(t) = \int_{-\infty}^{\infty} \hat{x}(f)\, e^{j\omega t}\, d\omega \tag{7.9}$$

The complex exponential notation, remember, is just a shorthand for sinusoids and co-sinusoids, but it is mathematically more convenient. The $\hat{x}(\omega)$ are the Fourier transform coefficients for each frequency component ω. These coefficients are complex numbers and can be expressed either in terms of their real (cosine) and imaginary (sine) parts or in terms of their amplitude and phase.

A second equation tells you how to compute the Fourier transform coefficients, $\hat{x}(\omega)$, from the input signal :

$$\hat{x}(\omega) = \mathsf{F}\,\{x(t)\} = \frac{1}{2\pi}\int_{-\infty}^{\infty} x(t)\, e^{-j\omega t}\, dt \tag{7.10}$$

These two equations are inverses of one another. Eq. (7.9) is used to compute the Fourier transform coefficients from the input signal, and then Eq. (7.8) is used to reconstruct the input signal from the Fourier coefficients.

The equations for the Fourier transform are rather complex (no pun intended). The best way to get an intuition for the frequency domain is to look at a few examples. Figure 7.2 plots sinusoidal signals of two different frequencies, along with their Fourier transform amplitudes. A sinusoidal signal contains only one frequency component, hence the frequency plots contain impulses. Both sinusoids are modulated between plus and minus one, so the impulses in the frequency plots have unit amplitude. The only difference between the two sinusoids is that one has 4 cycles per second and the other has 8 cycles per second. Hence the impulses in the frequency plots are located at 4 Hz and 8 Hz, respectively.

Figure 7.3 shows the Fourier transforms of a sinusoid and a cosinusoid. We can express the Fourier transform coefficients either in terms of their real and imaginary parts, or in terms of their amplitude and phase. Both representations are plotted in the figure. Sines and cosines of the same frequency have identical amplitude plots, but the phases are different.

Note sine function has a negative imaginary part for its negative frequency component as $\sin\theta = (e^{j\theta} - e^{-j\theta})/2j$, the second term is negative.

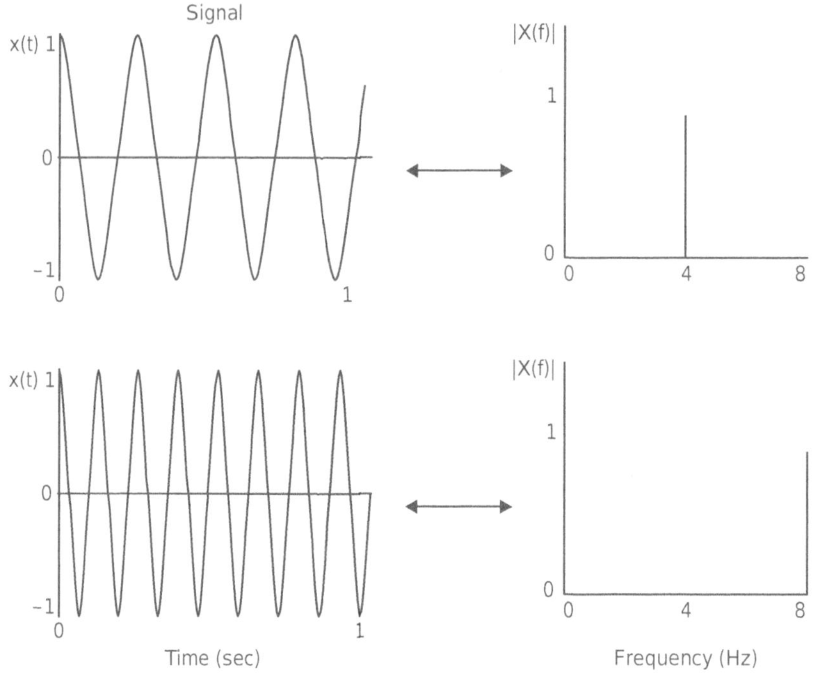

Fig. 7.2 Fourier transforms of sinusoidal signals of two different frequencies.

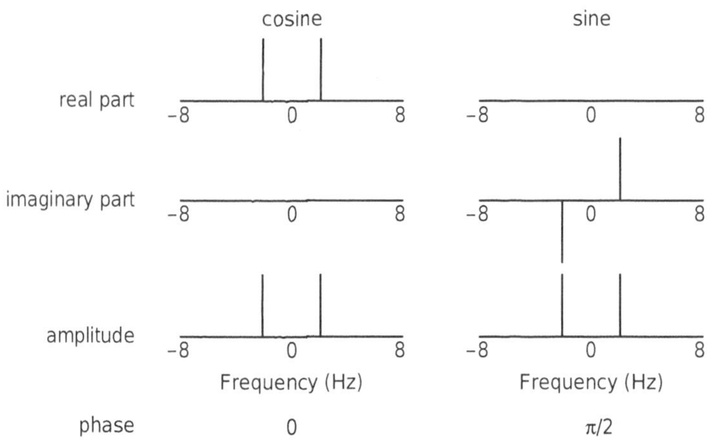

Fig. 7.3 Fourier transforms of sine and cosine signals. The amplitudes and the same, but the phases are different.

Do not be put off by the negative frequencies in the plots. The equations for the Fourier transform and its inverse include both positive and negative frequencies. This is really just a mathematical convenience. The information in the negative frequencies is redundant with that in the positive frequencies. Since $\cos(-f) = \cos(f)$, the negative frequency components in the real part of the frequency domain will always be the same as the corresponding positive frequency components. Since $\sin(-f) = -\sin(f)$ the negative frequency components in the imaginary part of the frequency domain will always be minus one times the corresponding positive frequency components. Often, people plot only the positive frequency components, as was done in Fig. 7.2, since the negative frequency components provide no additional information. Sometimes, people plot only the amplitude. In this case, however, there is information missing.

There are a few facts about the Fourier transform that often come in handy. The first of the properties is that the Fourier transform is itself a linear system, which one can check by making sure that Eq. (7.9) obeys both homogeneity and additivity. This is important because it makes it easy for us to write the Fourier transforms of lots of things. For example, the Fourier transform of the sum of two signals is the sum of the two Fourier transforms:

$$\mathrm{F}\{x(t)+y(t)\} = \mathrm{F}\{x(t)\}+\mathrm{F}\{y(t)\} = \hat{x}(\omega)+\hat{y}(\omega), \tag{7.11}$$

where we have used F{.} as a shorthand notation for "the Fourier transform of". The linearity of the Fourier transform was one of the tricks that made it easy to write the transforms of both sides of Eq. 7.6.

A second fact, known as the convolution property of the Fourier transform, is that the Fourier transform of a convolution equals the product of the two Fourier transforms :

$$\mathrm{F}\{h(t)*x(t)\} = \mathrm{F}\{h(t)\}\,\mathrm{F}\{x(t)\} = \hat{h}(\omega)\hat{x}(\omega) \tag{7.12}$$

This property was also used to write the Fourier transform of Eq. (7.6). Indeed this property is central to much of the discussion. Above it is emphasized that for a shift-invariant linear system (*i.e.*, convolution), the system's responses are always given by shifting and scaling the frequency components of the input signal. This fact is expressed mathematically by the convolution property above, where $\hat{x}(\omega)$ are the frequency components of the input and $\hat{h}(\omega)$ is the frequency response, the (complex-valued) scale factors that shift and scale each frequency component.

7.3 THE FOURIER TRANSFORM OF AN IMPULSE FUNCTION

The Fourier transform of a unit impulse function $\delta(t)$ is given by

$$\mathrm{F}[\delta(t)] = \int_{-\infty}^{\infty} \delta(t)\, e^{-j\omega t}\, dt \tag{7.13}$$

From the sampling property of an impulse function, it is evident that the integral on the right-hand side of Eq. (7.13) is unity. Hence

$$\mathrm{F}[\delta(t)] = 1 \tag{7.14}$$

Thus the Fourier transform of a unit impulse function is unity.

It is therefore evident that an impulse function has a uniform spectral density over the entire frequency interval. In other words, an impulse function contains all frequency components with the same relative amplitudes.

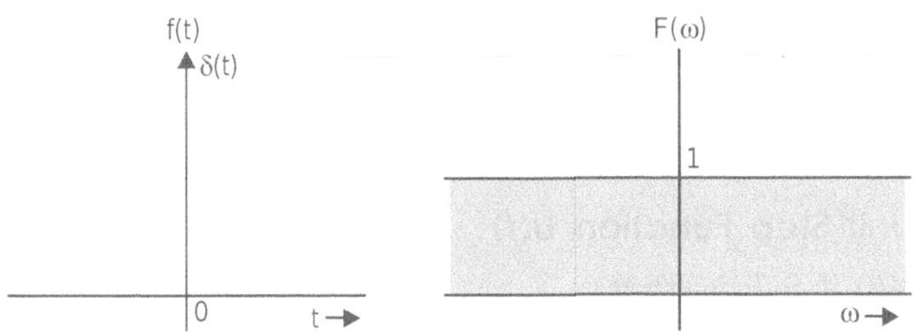

Fig. 7.4 An impulse function and its transform.

1. A Constant—What is its F.T.?

Let $f(t) = k$, a constant.

A constant is nothing but a zero frequency component. So, we only have a zero frequency in the F.T. spectrum. There is only one line at $f = 0$. Its amplitude is $2\pi k$ or $2\pi k\delta(\omega)$, since $\delta(\omega)$ has a value unity only at $\omega = 0$.

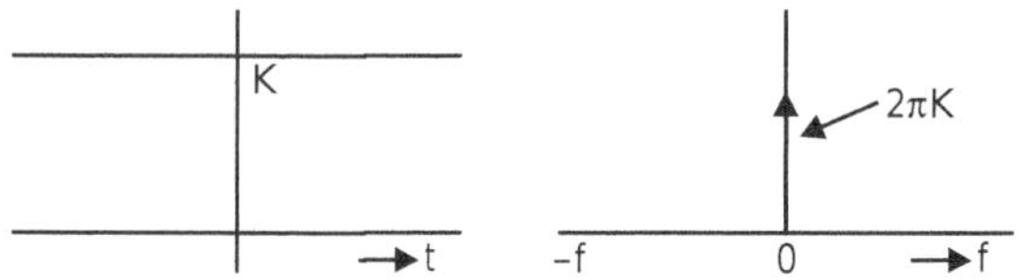

Fig. 7.5 Constant term and its Fourier transform.

Note that 2π comes in, that is because of ω to t relation.

2. Transform of sgn (*t*)

The signum function abbreviated as sgn (t) is defined by

$$\text{sgn}(t) = \begin{matrix} 1 & t > 0 \\ -1 & t < 0 \end{matrix} \qquad (7.14a)$$

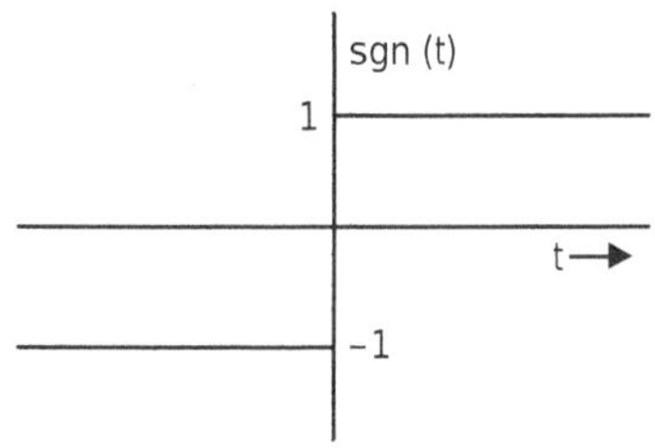

Fig. 7.6

It can be seen easily that

$$\text{sgn}(t) = 2u(t) - 1 \qquad (7.14b)$$

($u(t)$ is step function),

The Fourier transform of sgn (t) can be easily obtained when we observe that

$$\text{sgn}(t) = \lim_{a \to 0}\left[e^{-at}u(t) - e^{at}u(-t)\right] \qquad (7.15)$$

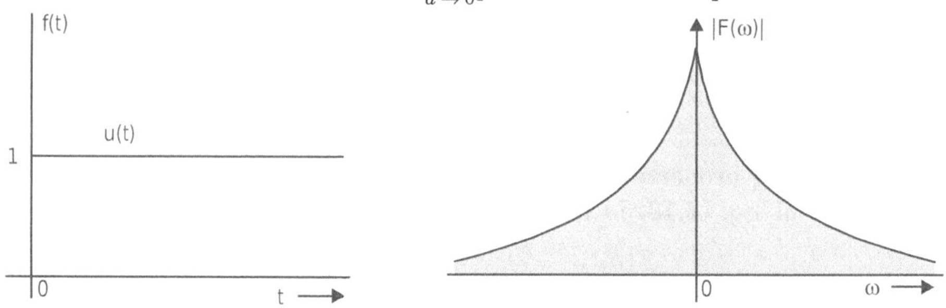

Fig. 7.7. The unit step function and spectral density function for signum function.

$$F[\text{sgn}(t)] = \lim_{a \to 0}\left[\int_0^{\infty} e^{-at}e^{-j\omega t}dt - \int_{-\infty}^{0} e^{at}e^{-j\omega t}dt\right]$$

$$= \lim_{a \to 0}\left[\frac{-2j\omega}{a^2 + \omega^2}\right] = \frac{2}{j\omega} \qquad (7.16)$$

The transform is a reciprocal function shown.

3. Transform of Unit Step Function *u*(*t*)

From Eq. (7.14*b*) it follows that

$$u(t) = \frac{1}{2}\left[1 + \text{sgn}(t)\right]$$

Hence $$\mathrm{F}[u(t)] = \frac{1}{2}\{\mathrm{F}[1] + \mathrm{F}[\mathrm{sgn}\,(t)]\}$$

Using Eqs. (7.13*b*) and (7.16), we obtain

$$\mathrm{F}[u(t)] = \pi\delta(\omega) + \frac{1}{j\omega} \tag{7.17}$$

The spectral density function contains an impulse at $\omega = 0$ (Fig. 7). Thus the function $u(t)$ contains a large d-c component as expected. In addition, it has other frequency components. The function $u(t)$ appears to be a pure d-c signal, and hence frequency components other than $\omega = 0$ may appear rather strange. However, the function $u(t)$ is not a true d-c signal. It is zero for $t < 0$, and there is an abrupt discontinuity at $t = 0$, giving rise to other frequency components. For a true d-c signal, $f(t)$ is constant for the entire interval $(-\infty, \infty)$. We have already seen from Fig. 5 that such a signal indeed has no frequency components other than d-c ($\omega = 0$).

4. Eternal Sinusoidal Signals cos $\omega_0 t$ and sin $\omega_0 t$

We shall now consider the sinusoidal signals cos $\omega_0 t$ and sin $\omega_0 t$ in the entire interval $(-\infty, \infty)$. These signals do not satisfy the condition of absolute integrability; yet their Fourier transforms exist and can be found by a limiting process analogous to that used for a constant function $f(t) = A$. We shall first assume these functions to exist in the interval $-\tau/2$ to $\tau/2$ and zero outside this interval. In the limit, τ will be made infinity. The procedure is now demonstrated.

$$\mathrm{F}(\cos\omega_0 t) = \lim_{\tau\to\infty}\int_{-\tau/2}^{\tau/2}\cos\omega_0 t\, e^{-j\omega t}\,dt$$

$$= \lim_{\tau\to\infty}\frac{\tau}{2}\left\{\frac{\sin[(\omega-\omega_0)\tau/2]}{(\omega-\omega_0)\tau/2} + \frac{\sin[(\omega+\omega_0)\tau/2]}{(\omega+\omega_0)\tau/2}\right\}$$

and using (6.15) of a sampling pulse.

$$= \lim_{\tau\to\infty}\left\{\frac{\tau}{2}\,\mathrm{Sa}\left[\frac{\tau(\omega-\omega_0)}{2}\right] + \frac{\tau}{2}\,\mathrm{Sa}\left[\frac{\tau(\omega+\omega_0)}{2}\right]\right\} \tag{7.18}$$

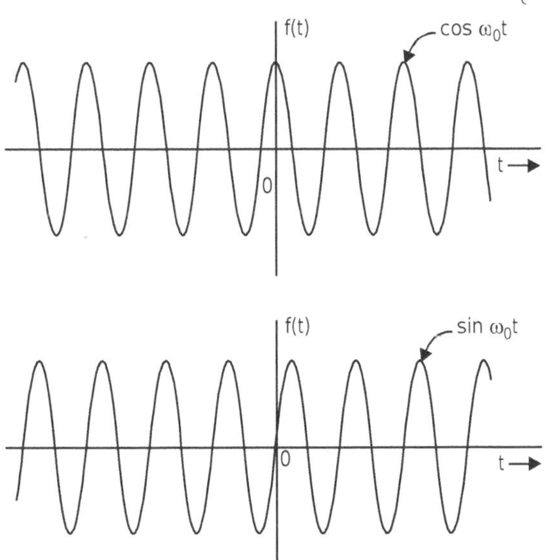

Fig. 7.8

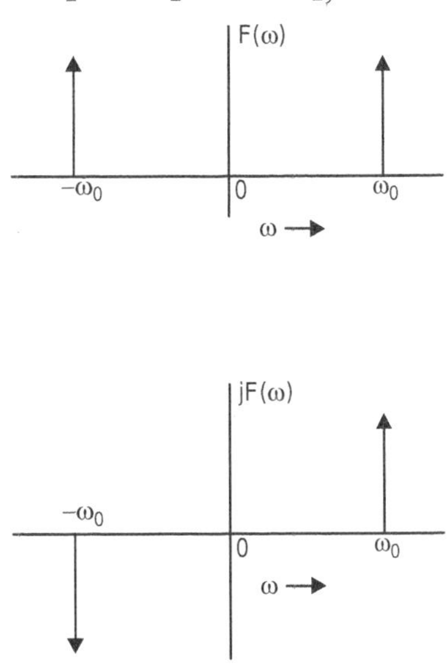

Fig. 7.9

In the limit the sampling function *Sa* ( ), becomes an impulse function and we have

$$F(\cos \omega_0 t) = \pi[\delta(\omega - \omega_0) + \delta(\omega + \omega_0)] \quad (7.19)$$

Similarly, it can be shown that

$$F(\sin \omega_0 t) = j\pi[\delta(\omega + \omega_0) - \delta(\omega - \omega_0)] \quad (7.20)$$

Therefore the Fourier spectrum for these functions consists of two impulses at ω_0 and $-\omega_0$, respectively.

5. Transform of a Sinusoidal Burst Signal

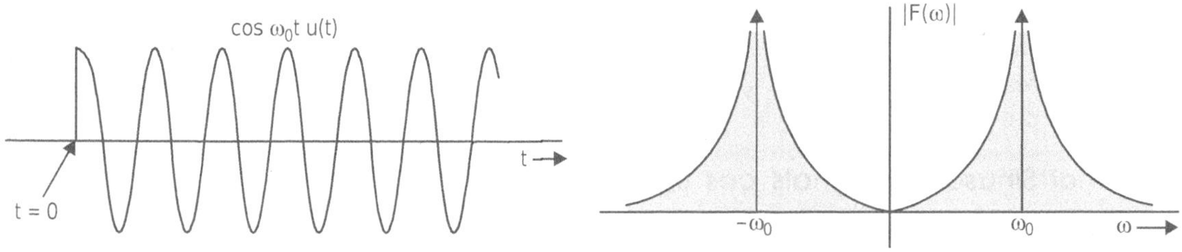

Fig. 7.10. F-T-spectrum of a signal cos $\omega_0 t.u(t)$.

A sudden onset of a cosine signal or sine signal is usual when we gate a signal through a switch for output.

For a cosine signal, when we give a burst at $t = 0$, since there is a sudden change at $t = 0$ by the step function $u(t)$, the transform shows smear on either side of ω_0, as shown.

Like this, whenever sine/cosine signals do not exist for all time, as in bursts of sine oscillations, there will be a smear (which is a hyperbolic curve) around the frequency of the sine/cos signal.

7.4 THE FOURIER TRANSFORM OF PERIODIC FUNCTIONS

We have developed a Fourier transform as a limiting case of the Fourier series by letting the period of a periodic function become infinite. We shall now proceed in the opposite direction and show that the Fourier series is just a limiting case of the Fourier transform. This point of view is very useful, since it permits a unified treatment of both the periodic and the non-periodic functions.

Strictly speaking, the Fourier transform of a periodic function does not exist, since it fails to satisfy the condition of absolute integrability. For any periodic function $f(t)$:

$$\int_{\infty}^{-\infty} |f(t)|\, dt = \infty \quad (7.20(a))$$

But the transform does exist in the limit. We have already found the Fourier transform of cos $\omega_0 t$, sin $\omega_0 t$ in the limit.

Example 1. *A number of impulses of unit strength and separated by T seconds as shown in Fig. 7.11. Find the Fourier transform.*

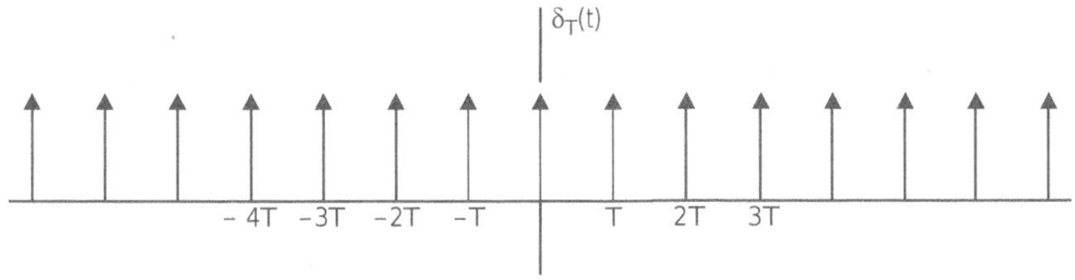

Fig. 7.11. The sequence of a uniform equidistant impulse function.

Solution. This function is very important in sampling theory, and hence it is convenient to denote it by a special symbol $\delta_T(t)$. Thus

$$\delta_T(t) = \delta(t) + \delta(t - T) + \delta(t - 2T) + ... + \delta(t - nT) + ...$$
$$+ \delta(t + T) + \delta(t + 2T) + ... + \delta(t + nT) + ...$$

$$= \sum_{n=-\infty}^{\infty} \delta(t - nT) \quad (7.21)$$

This is obviously a periodic function with period T. We shall first find the Fourier series for this function.

$$\delta_T(t) = \sum_{n=-\infty}^{\infty} F_n e^{jn\omega_0 t}$$

where

$$F_n = \frac{1}{T}\int_{-T/2}^{T/2} \delta_T(t)\, e^{-jn\omega_0 t}\, dt$$

Function $\delta_T(t)$ in the interval (–T/2, T/2) is simply $\delta(t)$. Hence

$$F_n = \frac{1}{T}\int_{-T/2}^{T/2} \delta(t)\, e^{-jn\omega_0 t}\, dt \quad (7.22)$$

From the sampling property of an impulse function as expressed in (Eq. 7.12) the above equation reduces to (note that $\delta(t)$ has value only at $t = 0$).

$$F_n = \frac{1}{T}$$

Consequently, F_n is a constant, 1/T. It therefore follows that the impulse train function of period T contains components of frequencies $\omega = 0, \pm\omega_0, \pm 2\omega_0, ..., \pm n\omega_0, ...,$ etc., ($\omega_0 = 2\pi/T$) in the same amount.

$$\delta_T(t) = \frac{1}{T}\sum_{n=-\infty}^{\infty} e^{jn\omega_0} \quad (7.23)$$

To find the Fourier transform of $\delta_T(t)$, we use Eq. (7.21). Since in this case $F_n = 1/T$, it is evident that

$$F[\delta_T(t)] = 2\pi \sum_{n=-\infty}^{\infty} \frac{1}{T}\delta(\omega - n\omega_0)$$

$$= \frac{2\pi}{T}\sum_{n=-\infty}^{\infty} \delta(\omega - n\omega_0)$$

$$= \omega_0 \sum_{n=-\infty}^{\infty} \delta(\omega - n\omega_0)$$

$$= \omega_0 \delta_{\omega_0}(\omega) \quad (7.24)$$

Relation 7.24 is very significant. It states that the Fourier transform of a unit impulse train of period T is also a train of impulses of strength ω_0 and separated by ω_0 radians

($\omega_0 = 2\pi/T$). Therefore the impulse train function is its own transform. The sequence of impulses with periods T = 1/2 and T = 1 second and their respective transforms are shown in Fig. 7.12. It is

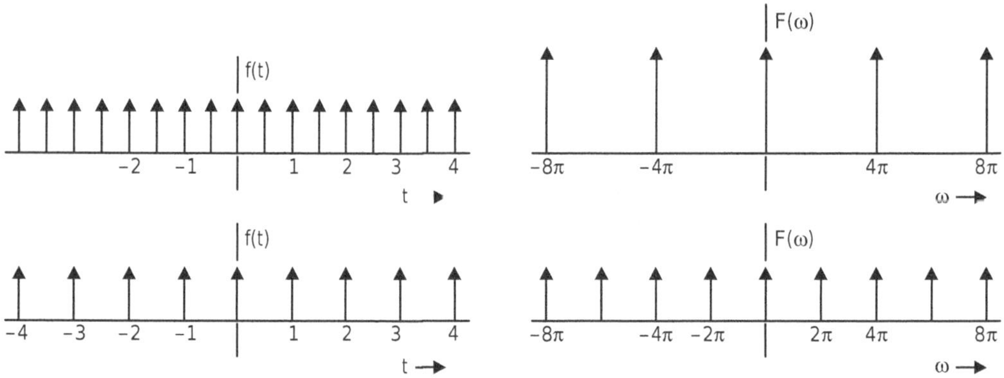

Fig. 7.12. Periodic impulse functions and their transforms.

Note that the coatser function of time (below) gives a fine fourier transform.

7.5 FOURIER TRANSFORMS FOR CERTAIN MATHEMATICAL FUNCTIONS

1. A single rectangular pulse has the F.T. Pulse is G_τ (τ)

$$\tau.\frac{\sin \omega\tau/2}{\omega\tau/2} = \tau.Sa\left(\frac{\omega\tau}{2}\right)$$

$$Sa(x) = \sin x/s, \quad Sa\ (\ \ ) \equiv \text{Sampling Function} \tag{7.25}$$

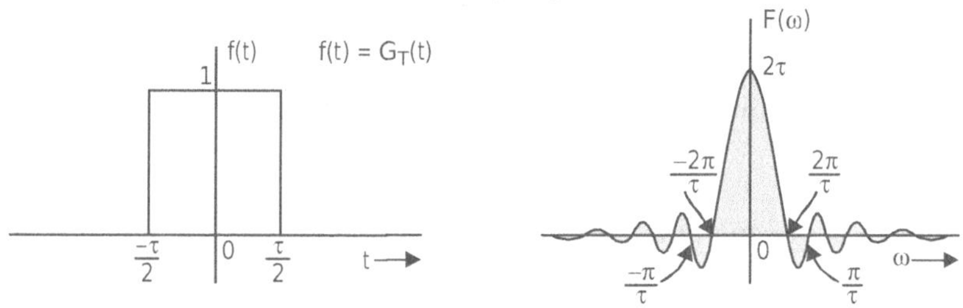

Fig. 7.13

The nonmalised 'sin c' function is given by

$$\sin c\ (x) = \frac{\sin \pi x}{\pi x} \tag{7.25a}$$

So (7.25) can also be written as $\tau \sin c \left(\frac{\omega\tau}{2\Pi}\right)$.

The unnormalised 'sin c' function is however different, it is $\sin x/x$.

2. Triangular pulse gate

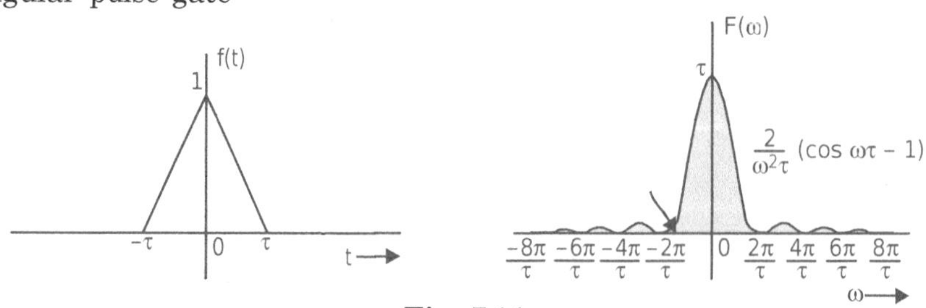

Fig. 7.14

3. Gaussian function. The F.T. is also Gaussian.

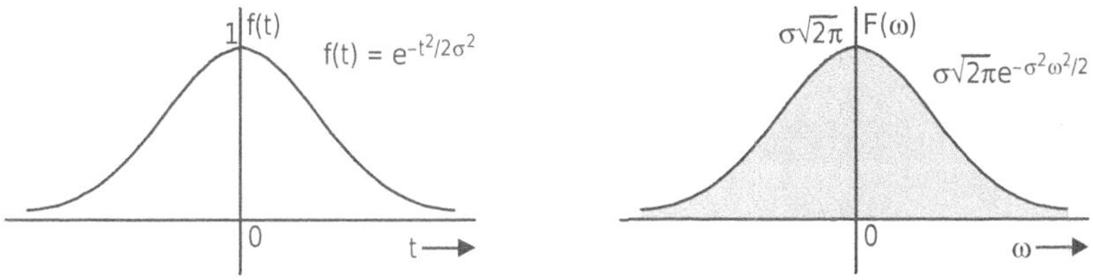

Fig. 7.15

4. Double sided exponential decay.

Note that Phase is zero.

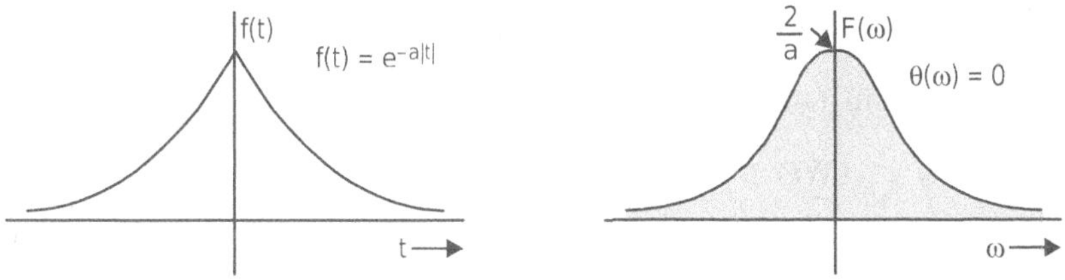

Fig. 7.16

5. Decaying oscillation

Note peak at $\sqrt{a^2 - \omega_0^2}$ and smear on either side. The smear is narrow if decay is fast (α large).

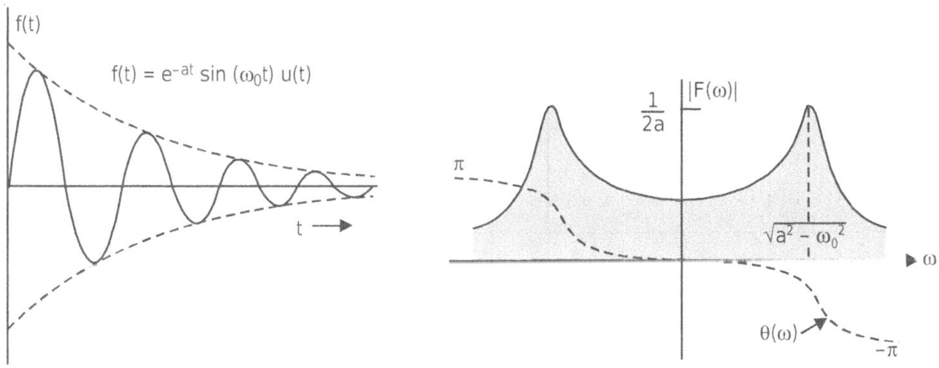

Fig. 7.17

7.6 SOME FOURIER TRANSFORM PAIR

Fourier Transforms

$f(t)$	$F(\omega)$
1. $e^{-at}\, u(t)$	$\dfrac{1}{a + j\omega}$
2. $te^{-at}\, u(t)$	$\dfrac{1}{(a + j\omega)^2}$
3. $\lvert t \rvert$	$\dfrac{-2}{\omega^2}$

4.	$\delta(t)$	1
5.	1	$2\pi\delta(\omega)$
6.	$u(t)$	$\pi\delta(\omega) + \dfrac{1}{j\omega}$
7.	$\cos\omega_0 t\ u(t)$	$\dfrac{\pi}{2}[\delta(\omega-\omega_0)+\delta(\omega+\omega_0)] + \dfrac{j\omega}{\omega_0^2-\omega^2}$
8.	$\sin\omega_0 t\ u(t)$	$\dfrac{\pi}{2j}[\delta(\omega-\omega_0)-\delta(\omega+\omega_0)] + \dfrac{\omega_0}{\omega_0^2-\omega^2}$
9.	$\cos\omega_0 t$	$\pi[\delta(\omega-\omega_0)+\delta(\omega+\omega_0)]$
10.	$\sin\omega_0 t$	$j\pi[\delta(\omega+\omega_0)-\delta(\omega-\omega_0)]$
11.	$e^{-at}\sin\omega_0 t\ u(t)$	$\dfrac{\omega_0}{(a+j\omega)^2+\omega_0^2}$
12.	$\dfrac{W}{2\pi}Sa\dfrac{(Wt)}{2}$	$G_W(\omega)$ (G denotes rect. pulse)
13.	$G_\tau(t)$	$\tau Sa\left(\dfrac{\omega\tau}{2}\right)$
14.	$1-\dfrac{\lvert t\rvert}{\tau} \ldots. \lvert t\rvert<\tau$ $0 \ldots. \lvert t\rvert>\tau$	$\tau\left[Sa\left(\dfrac{\omega\tau}{2}\right)\right]^2$
15.	$e^{-a\lvert t\rvert}$	$\dfrac{2a}{a^2+\omega^2}$
16.	$e^{-t^2/2\sigma^2}$	$\sigma\sqrt{2\pi}e^{-\sigma^2\omega^2/2}$
17.	$\delta_T(t)$	$\omega_0\delta_{\omega_0}(\omega)$ $\left(\omega_0=\dfrac{2\pi}{T}\right)$

7.7 IMPORTANT PROPERTIES OF FOURIER TRANSFORMS

1. Linearity Property

Suppose we have two functions $f_1(t)$ and $f_2(t)$. Let their individual transforms be $F_1(\omega)$ and $F_2(\omega)$. Then for the combined signal $f(t) = f_1(t) \pm f_2(t)$, the Fourier transform is $F(\omega) = F_1(\omega) \pm F_2(\omega)$.

Like this we can generalise for more functions.

2. Scaling Property

If we have $F(\omega)$ for $f(t)$.

To find $f(at)$'s Fourier transform, we use this,

$$f(at) \leftrightarrow \frac{1}{|a|}F\left(\frac{\omega}{a}\right)$$

This means time compression causes frequency expansion and vice versa.

3. Symmetry Property

Suppose we have a signal and its Fourier transform.

Then, if the waveform given by the transform is itself taken a Fourier transform, what would one get?

$f(t)$ gives $F(\omega)$.

Now replace ω variable by x. Then take its F.T.

$$F\{F(x)\} = \int_{-\infty}^{\infty} F(x)e^{-j\omega x}\, dx \tag{7.25i}$$

But
$$f(t) = \frac{1}{2\pi} F(\omega)e^{j\omega t}\, d\omega \tag{7.25ii}$$

Put $(-t)$ in this. Then

$$f(-t) = \frac{1}{2\pi}\int F(\omega)e^{-j\omega t}\, d\omega \tag{7.25iii}$$

In the above, the integrating variable can be replaced as x, without changing the value of the integral, *i.e.*, put $\omega = x$!

$$f(-t) = \frac{1}{2\pi}\int F(x)e^{-jxt}\, dx \tag{7.25iv}$$

Compare (*i*) and (*iv*). The variable is ω in (7.25(*i*)) and t in (7.25(*iv*)). Therefore

$$f(-t) = \frac{1}{2\pi}\int F(x)e^{-jxt}\, dx$$

$$= (1/2\pi)\, F\{F(x)\} \qquad \text{from (7.25}i\text{)}$$

$$= (1/2\pi) \times \text{Fourier transform of } \mathbf{F(\omega)}.$$

Thus when we take another Fourier transform of the F.T. of a signal, we get the time function back as $2\pi\, f(-t)$.

Thus,
$$f(t) \xrightarrow[\text{F.T.}]{} F(\omega) \xrightarrow[\text{F.T.}]{} \frac{1}{2\pi} f(-t) \tag{7.26}$$

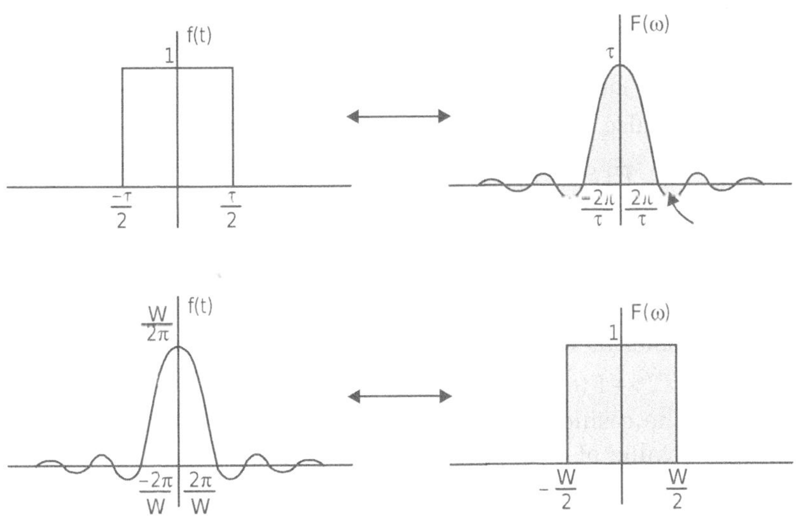

Fig. 7.18. Rectangular time function and its symmetrical transform pair.

Thus a rectangular time function and its F.T. are shown.

For the same function we can take the transform and get the time function as a rectangular time function. The 2π factor comes in.

4. Frequency Shift Theorem

If $f(t)$ has $F(\omega)$ as its transform, then $f(t)\,e^{j\omega_c t}$ has $F(\omega - \omega_c)$ as its transform.

This theorem can be easily proved and is left as an exercise. The meaning of this theorem is applicable in our modulation methods. $e^{j\omega_c t}$ and $f(t)$ multiplied means modulating the signal with the sinusoidal (since $\exp(j\omega_c t) = \cos \omega_c t + j \sin \omega_c t$) carrier frequency ω_c.

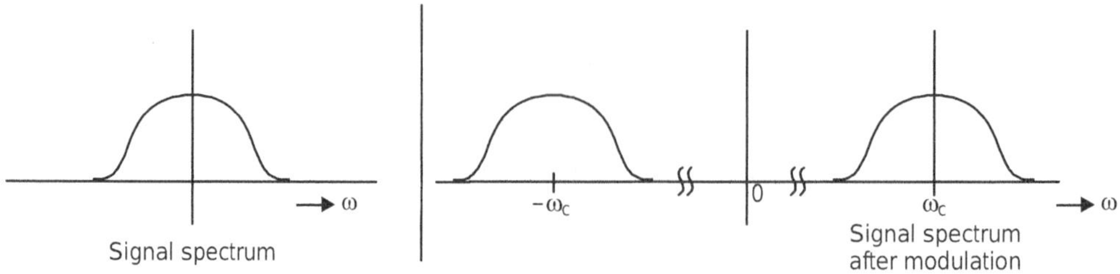

Fig. 7.19 (*a*) modulating signal and (*b*) its spectrum ω_c is carrier.

The modulating signal's spectrum is centered around the $\pm\,\omega_c$ frequencies, as two *side-bands*. The modulated signal $f_m(t)$ is

$$f_m(t) = f(t) \cos \omega_c t = \frac{1}{2}\left\{f(t)\,e^{j\omega_c t} + f(t)e^{-j\omega_c t}\right\} \tag{7.27}$$

Hence the spectrum of $f_m(t)$ is

$$\frac{1}{2}\left[F(\omega + \omega_c) + F(\omega - \omega_c)\right]$$

which are the two side bands shown above.

5. Time Translation Theorem

This says that $f(t - t_0)$ has the F.T. $F(\omega)\,e^{-j\omega t_0}$

So, the spectrum of a known time function, when it is shifted in time can be found by multiplication with $\exp(-j\omega t_0)$.

The proof of this is also very simple.

$$\mathbf{F}\{f(t - t_0)\} = \int f(t - t_0)\,e^{-j\omega t}\,dt = \int f(x)\,e^{-j(t_0 + x)\omega}\,dx \quad (x = t - t_0)$$

$$= F(\omega)\,e^{-j\omega t_0}$$

Suppose we shift a time function $f(t)$ backwards and forwards and add the two.

$$f(t - t_0) + f(t + t_0) \tag{7.28}$$

If we find its transform, we get

$$F(\omega)\,e^{-j\omega t_0} + F(\omega)\,e^{j\omega t_0} = 2\,F(\omega)\cos \omega t_0$$

That means, it is the cosine modulation product which is got as the frequency domain function for the added value of two equally shifted signals $f(t)$.

6. Real, Imaginary, Odd, Even Relation between function *f(t)* and its F.T. F(ω).

See exercise 3. The following table should be memorised.

$f(t)$	$F(\omega)$
Real and even function of t	Real and even function of ω
Real and odd	Imaginary and odd
Imaginary and even	Imaginary and even
Complex and even	Complex and even
Complex and odd	Complex and odd

7.8 CONVOLUTION

Let the response of a system to a single impulse be $h(t)$.

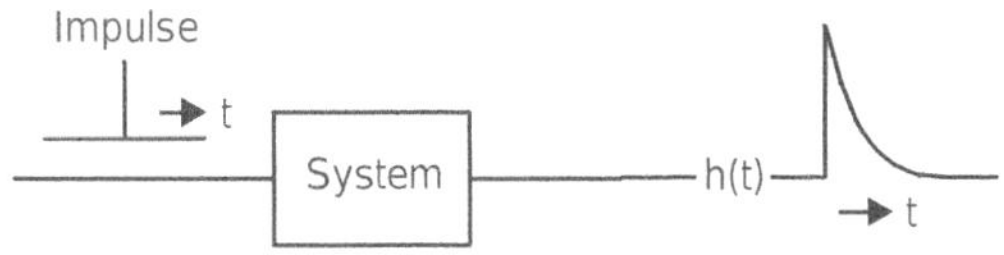

Fig. 7.20

Any other function of time $f(t)$, if applied to this system will give a response that we want to find.

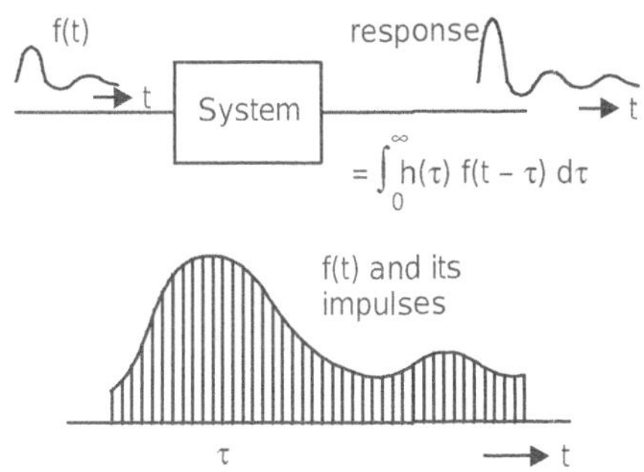

Fig. 7.21

Let $f(t)$ be decomposed into a number of infinitesimal pulses along the t axis, separated by $d\tau$. The variable τ denotes a time that moves on the t-axis. Combining response of each such impulse, we get the system's response to $f(t)$ as

$$r(t) = \int_0^\infty f(t-\tau)\, h(\tau)\, d\tau \equiv f(t) * h(t) \tag{7.29}$$

where $r(t)$ is the net response to $f(t)$, the notation is shown using a * for convolution of two signals.

This integral is called the Duhamel or **Convolution** integral. Note that this has a variable inside the integral, τ, that changes from 0 to ∞ to denote the impulses at all times.

The convolution is defined for two functions $x(t)$ and $h(t)$ as

$$\phi(t) = \int_0^\infty x(\tau)\, h(t-\tau)\, d\tau \tag{7.29a}$$

Note that the variable τ is in both x and h; the time variable is in only one of them. The same integral can also be written as

$$\phi(t) = \int_0^\infty h(\tau)\, x(t-\tau)\, d\tau \tag{7.29b}$$

7.9 TIME DERIVATIVE AND INTEGRAL

Given $f(t)$ and its F.T. as $F(\omega)$, what will be the transform of $df(t)/dt$?

The Fourier transform actually expresses a function $f(t)$ in terms of a continuous sum of exponential functions of the form $e^{j\omega t}$. The derivative of $f(t)$ is therefore equal to the continuous sum of the derivatives of the individual exponential components. But the derivative of an exponential function $e^{j\omega t}$ is equal to $j\omega e^{j\omega t}$. Therefore the process of differentiation of $f(t)$ is equivalent to multiplication by $j\omega$ of each exponential component. Hence

$$\frac{df}{dt} \leftrightarrow j\omega F(\omega)$$

A similar argument applies to the integration.

Similarly $\int_{-\infty}^{t} f(\tau)\,d\tau$ has its F.T. as $\frac{1}{j\omega}F(\omega)$.

Thus multiplication by $j\omega$ is equivalent to differentiation and division by $j\omega$ as equivalent to integration in frequency domain.

7.10 CONVOLUTION PROPERTY

We start with the definition of convolution :

$$y(t) = h(t) * x(t) = \int_{-\infty}^{\infty} x(s)\,h(t-s)\,ds \tag{7.30}$$

By the definition of the Fourier transform,

$$\hat{y}(\omega) = \int_{-\infty}^{\infty}\left[\int_{-\infty}^{\infty} x(s)\,h(t-s)\,ds\right]e^{-j\omega t}\,dt \tag{7.31}$$

Switching the order of integration,

$$\hat{y}(\omega) = \int_{-\infty}^{\infty} x(s)\left[\int_{-\infty}^{\infty} h(t-s)\,e^{-j\omega t}\,dt\right]ds \tag{7.32}$$

Letting $\sigma = t - s$,

$$\hat{y}(\omega) = \int_{-\infty}^{\infty} x(s)\left[\int_{-\infty}^{\infty} h(\sigma)\,e^{-j\omega(\sigma+s)}\,d\sigma\right]ds \tag{7.33}$$

$$= \int_{-\infty}^{\infty} x(s)\,e^{-j\omega s}\left[\int_{-\infty}^{\infty} h(\sigma)\,e^{-j\omega\sigma}\,d\sigma\right]ds$$

Then by the definition of the Fourier transform,

$$\hat{y}(\omega) = \int_{-\infty}^{\infty} x(s)\,e^{-j\omega s}\,\hat{h}(\omega)\,ds \tag{7.34}$$

$$= \hat{h}(\omega)\int_{-\infty}^{\infty} x(s)\,e^{-j\omega s}\,ds$$

$$= \hat{h}(\omega)\hat{x}(\omega)$$

This means that if there are two transform (frequency domain) functions $h(\omega)$ and $X(\omega)$, when we multiply them, they belong to a time signal which is the 'convolution' of their original time signals $h(t)$ and $x(t)$.

That means a product *here* is not a product *there*, but a 'convolution'.

7.11 SIGNUM FUNCTION AND IMPULSE FUNCTION

$$\text{Sgn}(t) = 1 \text{ for } t > 0$$
$$\qquad = -1 \text{ for } t < 0$$

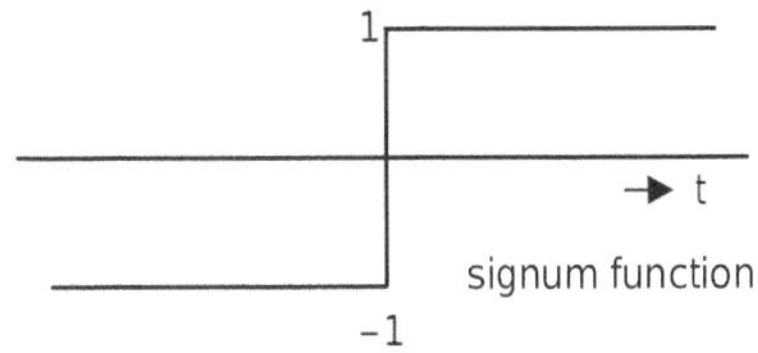

Fig. 7.22

It can be seen that

$$\text{Sgn}(t) = u(t) + u(-t)$$

where $u(t)$ is a step function, starting at $t = 0$.

Since there is a discontinuity at $t = 0$, we write

$$\text{Sgn}(t) = \underset{a \to 0}{\text{Lt}} \left[u(t)\, e^{-at} - e^{at}\, u(-t) \right] \tag{7.35}$$

Then finding the F.T. is easy, as

$$\underset{a \to 0}{\text{Lt}} \int_0^{\infty} e^{-at}\, e^{-j\omega t}\, dt + \int_0^{-\infty} e^{at}\, e^{-j\omega t}\, dt \tag{7.36}$$

$$= \lim_{a \to 0} \frac{-2j\omega}{a^2 + \omega^2}$$

$$= \frac{2}{j\omega}$$

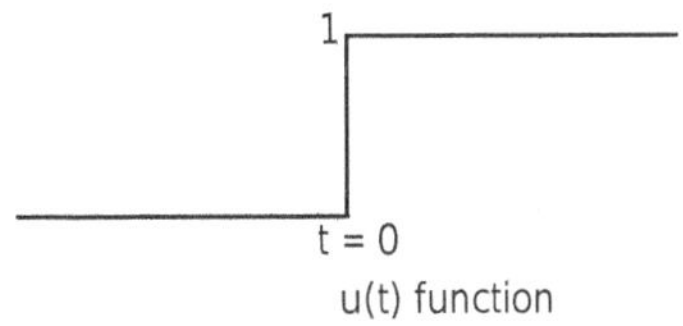

Fig. 7.23

The $u(t)$ function can be written as

$$u(t) = \frac{1}{2}\left[1 + \text{sgn}(t)\right]$$

Hence its F.T. is

$$\frac{1}{2}\text{F}(1) + \frac{1}{2}\cdot\frac{2}{j\omega}$$

Fourier transform of a constant (= 1) is $2\pi \times 1$, so we get

$$\text{F.T. } \{u(t)\} = \pi + \frac{1}{j\omega}. \tag{7.37}$$

Example 2. *Find the F.T. of the Gaussian function*

$$f(t) = e^{-\alpha t^2} \tag{7.38i}$$

Solution. We start with the well-known definite integral

$$\int_{-\infty}^{\infty} e^{-\alpha t^2}\, dt = .\sqrt{\frac{\pi}{\alpha}} \tag{7.38ii}$$

Differentiating n times with respect to α,

$$\int_{-\infty}^{\infty} t^{2n}\, e^{-\alpha t^2}\, dt = \frac{1.3 \,....\, (2n-1)}{2^n} \sqrt{\frac{\pi}{\alpha^{2n+1}}} \tag{7.38iii}$$

Next we break the general Fourier integral into an exponential series :

$$\text{F}(\omega) = \int f(t) \sum \frac{(-j\omega t)^n}{n!}\, dt \left\{ e^{j\omega t} = \sum \frac{(-j\omega t)^n}{n!} \right.$$

$$= \sum_{n=0}^{\infty} (-j)^n . \frac{\omega^n}{n!} \int t^n f(t)\, dt \tag{7.38iv}$$

We can replace n by $2n$ in the above without any loss of generality!

$$F(\omega) = \sum_{n=0}^{\infty} (-j)^{2n} . \frac{\omega^{2n}}{2n!} \int t^{2n} f(t)\, dt \tag{7.38v}$$

If $f(t)$ is given by (*i*), then, using (*iii*),

$$= \sum (-j)^{2n} \frac{\omega^{2n}}{2n!} \frac{1.3 \ldots . (2n-1)}{(2\alpha)^n} \sqrt{\frac{\pi}{\alpha}} \tag{7.38vi}$$

$$= \sqrt{\frac{\pi}{\alpha}}\, e^{-\omega^2/4\alpha} \tag{7.38vii}$$

by a series sum of the exponential series.

So, we get the result in (*vii*). Note this is also a Gaussian function.

7.12 THE HILBERT TRANSFORM*

The Hilbert transform is defined as

$$H(f) = -j \text{ sgn } (f) \tag{7.39}$$

The – sgn (f) is shown. The j term denotes a phase of – 90° or – $\pi/2$ for positive and $+\pi/2$ or 90° for negative frequencies.

Magnitude is constant for all ω.

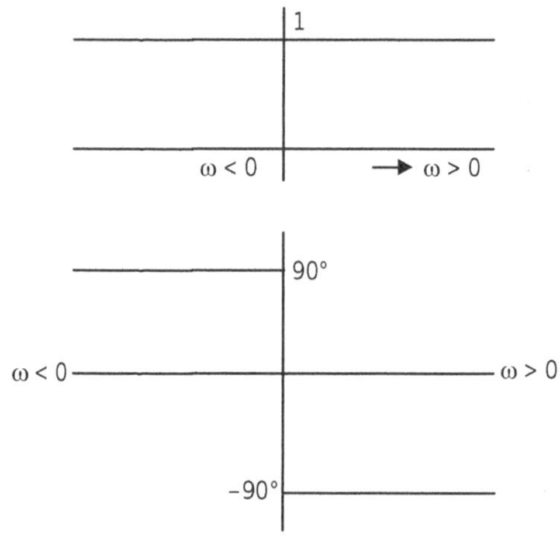

Fig. 7.24. Magnitude and phase of Hilbert transform.

In order to find the corresponding time domain function, for this Hilbert transform, let us define another function

*Not in Syllabus for Ist year ECE Anna University.

$$G(f:\alpha) = \begin{cases} e^{-\alpha f} & \text{for } f < 0 \\ -e^{+\alpha f} & \text{for } f > 0 \end{cases} \tag{7.40}$$

When we put $\alpha \to 0$ in the above, it becomes the signum function. Hence the time function for G is

$$g(t:\alpha) = F^{-1}[G(f:\alpha)]$$

$$= \int_0^{\infty} e^{-\alpha f}\, e^{j2\pi ft}\, df - \int_{-\infty}^{0} e^{\alpha f}\, e^{j2\pi ft}\, df$$

$$= \frac{4\pi\, j.t}{\alpha^2 + (2\pi t)^2} \tag{7.41}$$

When $\alpha \to 0$ in the limit, we get $\frac{j}{\pi t}$.

This is the time function for sgn (f), but the Hilbert transform is $-j$ sgn (f), we get as (h denoting time domain Hilbert transform),

$$h(t) = \frac{-j^2}{\pi t} = \frac{1}{\pi t}.$$

Note that the Hilbert transform is a function which operates on any function of t.

Given an input $m(t)$, the Hilbert transform is got as $\frac{1}{\pi}\int_{-\infty}^{\infty}\frac{m(\tau)}{t-\tau}\,d\tau$ by convolution of $m(t)$ with $h(t)$.

Thus the signal $s(t)$ has the Hilbert transform given by

$$\int_{-\infty}^{\infty} \frac{s(\tau)}{t-\tau}\, d\tau \tag{7.42}$$

The Hilbert transform gives an all-pass effect, since the amplitude of all frequencies is the same; but over the required range, a 90° phase lag is obtained.

Properties of the Hilbert Transform

1. **A sine wave**

   Nothing that we have to given a $-j\pi/2$ (−90°) phase by

$$h(t) = H\{\text{Imag. } e^{j\omega t}\} = \text{Imag.}\left\{\left(-\frac{j\pi}{2}\right) e^{j\omega t}\right\}$$

$$= \sin\left(\omega t - \frac{\pi}{2}\right) = -\cos \omega t.$$

   A *cosine* wave gives, by similar development using real part of $e^{j\omega t}$,

$$h(t) = \sin \omega t$$

   Thus the cos and sine functions, which were called as orthogonal, form a Hilbert transform pair.

2. A number of physical signals based on properties of the medium are known to be Hilbert transforms of each other. The absorption (of light) by a specimen (a liquid, say) gives an absorption signal (over various frequencies of light). The dielectric property of the chemical also varies with frequency. It is known that absorption and dielectric constant are Hilbert transforms of each other.
3. The Hilbert transform of a square wave is shown.

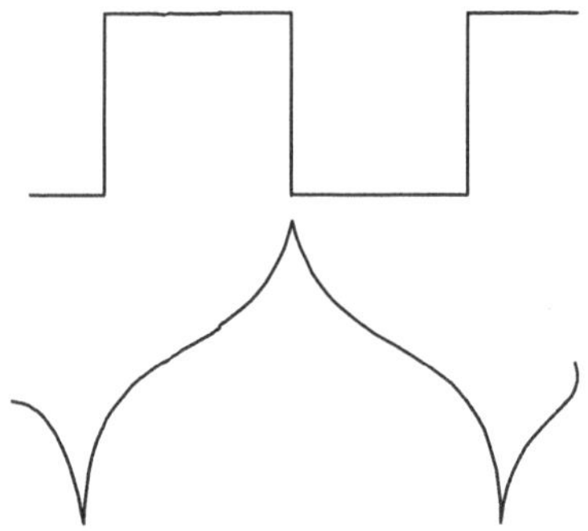

Fig. 7.25. The square wave and its Hilbert transform.

4. Hilbert transform can be realised by a digital filter. Using this, it is used to generate a single side-band amplitude modulated signal.

Example 3. *Show that the Hilbert transform of the square wave*

$$x(t) = A \qquad (-\tau/2 < t < \tau/2)$$
$$= 0$$

for all other t is a logarithmic function.

Solution. Let $\hat{X}(t)$ = Hilbert transform of $x(t)$.

$$\hat{X}(t) = x(t) * \frac{1}{\pi t} \qquad \text{(Convolution)}$$

$$= \frac{1}{\pi}\int_{-\infty}^{\infty} \frac{x(\lambda)}{(t-\lambda)}\, d\lambda = \frac{1}{\pi}\int_{-\tau/2}^{\tau/2} \frac{A}{t-\lambda}\, d\lambda$$

$$= \frac{A}{\pi}\left[+\log_e (t-\lambda)\right]_{t-\tau/2}^{t+\tau/2} = -\frac{A}{\pi}\log_e\left(\frac{t-\tau/2}{t+\tau/2}\right). \qquad (7.43)$$

7.13 PROPERTIES OF H.T.

1. The Hilbert transform of a Hilbert transform is the negative of the original function.

   Since the H.T. corresponds to a phase shift of $\pm \pi/2$, we note that the H.T. of $v(t)$ corresponds to the transfer function

$$[-j\,\text{sgn}\,(f)]^2$$
$$-j^2\,(\text{sgn}\,f)^2 = -1 \text{ or } -\pi. \qquad (7.44)$$

   For example, hence $H[H\{u(t)\}] = -u(t)$.

2. A function and its Hilbert transform are orthogonal over the infinite interval.

i.e., $$\int_{-\infty}^{\infty} u(t)\, v(t)\, dt = 0 \qquad (i)$$

where $u(t)$ and $v(t)$ are H.T.

We note that

$$\int_{-\infty}^{\infty} u(t)\, v^*(t)\, dt = \int_{-\infty}^{\infty} U(f)\, V^*(f)\, df \qquad (ii)$$

This is by the generalised Parseval's theorem, saying that energy products (such as by multiplying two signals) are same whether it is evaluated over infinite time or over the (infinite) frequency range.

Parseval's Theorem

$$\frac{1}{2\pi}\int_{-\infty}^{\infty} V(\omega)\,.\, U^*(\omega)\, d\omega = \frac{1}{2\pi}\left[\int_{-\infty}^{\infty}\int_{-\infty}^{\infty}\left\{v(t)\, e^{-j\omega t}\right\} dt\; U^*(\omega)\, d\omega\right]$$

$$= -\int_{-\infty}^{\infty} v(t) dt \,\frac{1}{2\pi}\int_{-\infty}^{\infty} U^*(\omega)\, e^{-j\omega t}\, d\omega$$

$$= \int_{t=-\infty}^{\infty} v(t)\, u^*(t)\, dt \qquad (7.45)$$

This is Parseval's Theorem.

Now take (i); (To prove it)

$$\int_{-\infty}^{\infty} v(t)\, u^*(t)\, dt = \int_{-\infty}^{\infty} (+\, j\, \text{sgn}\,(f))\, |U(f)|^2\, df \quad \text{(since } v \text{ is } j\text{sgn}(f) \text{ of U).}$$

Note that the R.H.S. integral is an odd function; its $|U(f)|^2$ is always positive which is an *even* function, but j sgn (f) is odd function.

So, any integral which has an odd function and limits $-\infty$ and $+\infty$ would become zero.

This proves that (i) is satisfied; hence orthogonal.

FOURIER TRANSFORM—ADDITIONAL PROBLEM

Example 1. *For each signal, find the Fourier transform, X(ω), and then plot |X(ω)| (note, you may want to use MATLAB for the plot in 3).*

1.

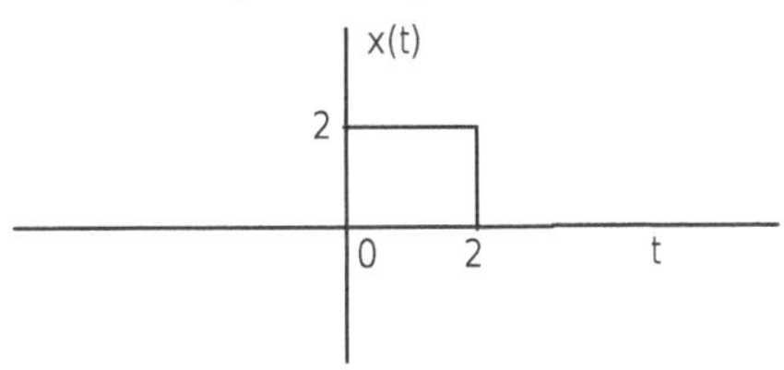

2.

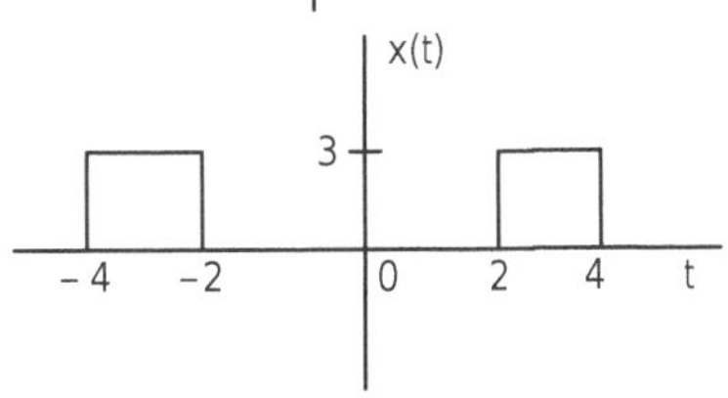

3.

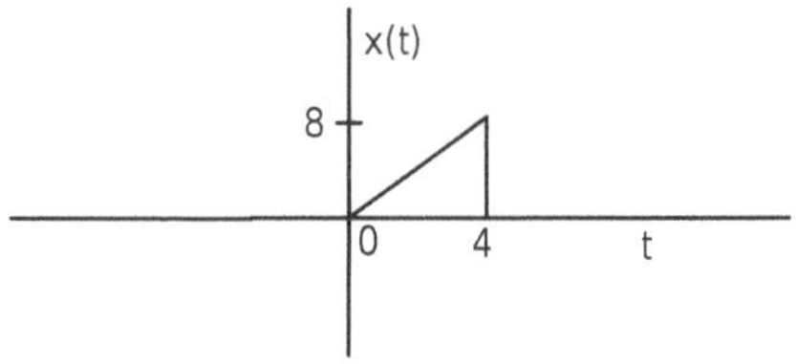

4. $x(t) = cos\ (200t)p_4(t)$

5. $x(t) = e^{-3t}\ cos\ (10t)u(t)$

Solution:

1. The signal is a rectangular pulse of width 2, at $t = 1$ (centre of pulse). Recalling $p(t)$ as a pulse signal and that its F.T. is a 'sinc' function: (eqn. 7.15 and 7.25). Here $\tau = 2$, A = 1 $\omega\tau/ = 2\omega/2$,

$$x(t) = 2\ p_2\ (t - 1)$$

As Sa (α) = sin c (α/π),

$$X(\omega) = 2 \cdot 2 \sin c \left(\frac{2\omega}{2\pi}\right) e^{-j\omega \cdot 1}$$

$$= 4 \sin c \left(\frac{\omega}{\pi}\right) e^{-j\omega}$$

$$|X(\omega)| = 4\,|\sin c\ (\omega/\pi)|$$

The plot is as shown

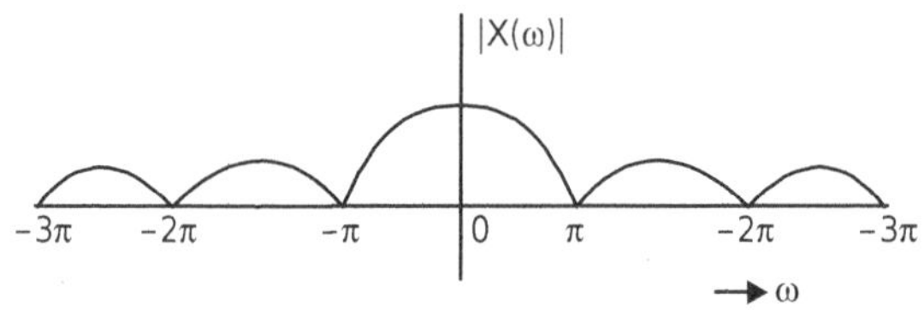

Fig. 7.26

2. Here there are two pulses.

Amplitude = 3 Centre = 3 Width = 2

Amplitude = 3 Centre = –3 Width = 2

$$x(t) = 3p_2(t - 3) + 3p_2(t + 3)$$

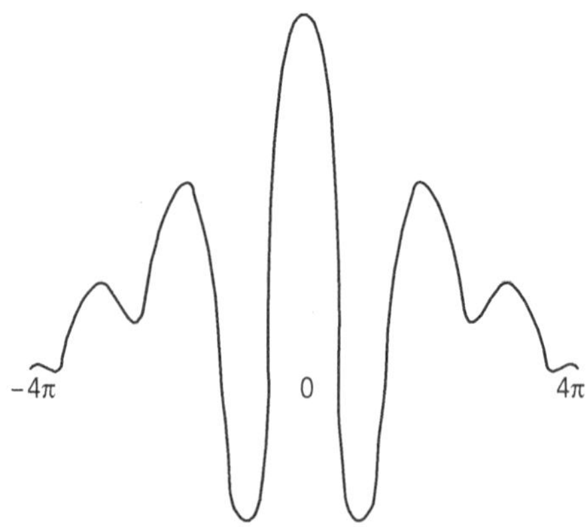

Fig. 7.27

This can be also considered as a wide pulse of width 8 centred at $t = 0$, and subtracted by another of width 4.

$$x(t) = 3p_8(t) - 3p_4(t)$$

This makes for easy evaluation.

$$X(\omega) = 3(8)\,\text{sinc}\left(\frac{8\omega}{2\pi}\right) - 3(4)\,\text{sinc}\left(\frac{4\omega}{2\pi}\right)$$

$$= 24\ \text{sinc}\ (4\omega/\pi) - 12\ \text{sinc}\ (2\omega/\pi)$$

The first has its zero crossing at $\pi/4$.

The second has its zero crossing at $\pi/2$.

The plot is roughly sketched below.

It can be done by a computer program as with simple BASIC.

```
SCREEN 12: PI = 355/113   : I% = 1
FOR OMEGA = –4*PI   TO   4*PI   STEP 0–1
A = 4*OMEGA/PI : B = 2*OMEGA/PI
X = 24*SIN(A/A) : X = X – 12*SIN(B)/B
PSET (2*I%, 200 – 10*X)
I% = I% + 1
NEXT
```

3. In this we have a pulse centred at $t = 2$, but also multiplied by t, giving a ramp width is 4. Amplitude is $2(2 \times 4 = 8)$.

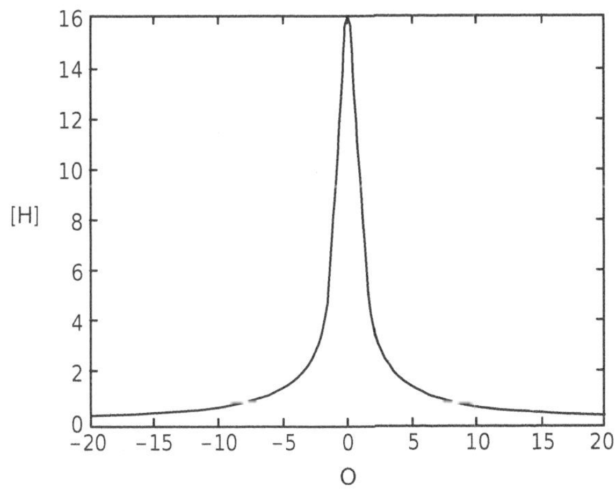

Fig. 7.28

$$x(t) = 2t\ p_4(t - 2)$$

$$p_4(t-2) \longleftrightarrow 4\ \text{sinc}\left(\frac{2\omega}{\pi}\right) e^{-2j\omega}$$

$$2t\ p_4(t-2) \longleftrightarrow j2\frac{d}{d\omega}\left(\frac{4\sin(2\omega)}{2\omega}e^{-2j\omega}\right)$$

$$= 4j\left[-\frac{\omega\,(2\cos(2\omega)\,e^{-2j\omega}-2}{\omega^2}+\frac{\sin(2\omega)\,e^{-2j\omega}-\sin(2\omega)\,e^{-2j\omega}}{\omega^2}\right]$$

$$= \frac{4j}{\omega^2}e^{-2j\omega}\,(\omega 2\cos(2\omega)-2j\omega\sin(2\omega)-\sin(2\omega))$$

$$= \frac{4j}{\omega^2}e^{-2j\omega}\left(\omega 2e^{-2j\omega}-\sin(2\omega)\right)$$

4. $$x(t) = \cos(200t)\,p_4(t)$$

$$p_4(t) \longleftrightarrow 4\,\text{sinc}\left(\frac{2\omega}{\pi}\right)$$

$$\cos(200t)p_4(t) \longleftrightarrow 2\,\text{sinc}\left(\frac{2}{\pi}(\omega-200)\right)+2\,\text{sinc}\left(\frac{2}{\pi}(\omega+200)\right)$$

since $$\cos\theta = (e^{j\theta}+e^{-j\theta})/2.$$

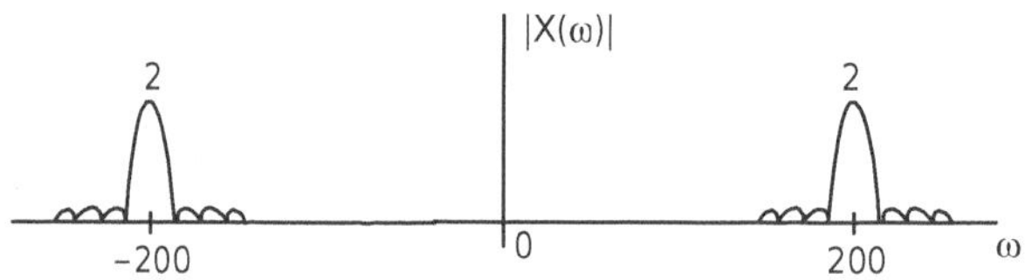

Fig. 7.29

5. $$x(t) = e^{-3t}\cos(10t)\,u(t)$$

From page 181 F.T. table,

$$e^{-3t}u(t) \longleftrightarrow \frac{1}{j\omega+3}$$

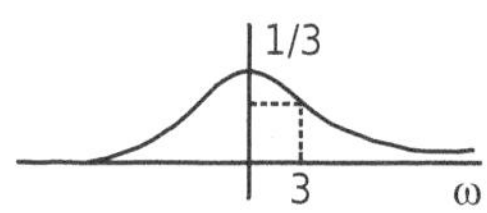

$$e^{-3t}u(t)\cos(10t) \longleftrightarrow \frac{1}{2}\left[\frac{1}{j(\omega-10)+3}+\frac{1}{j(\omega+10)+3}\right]$$

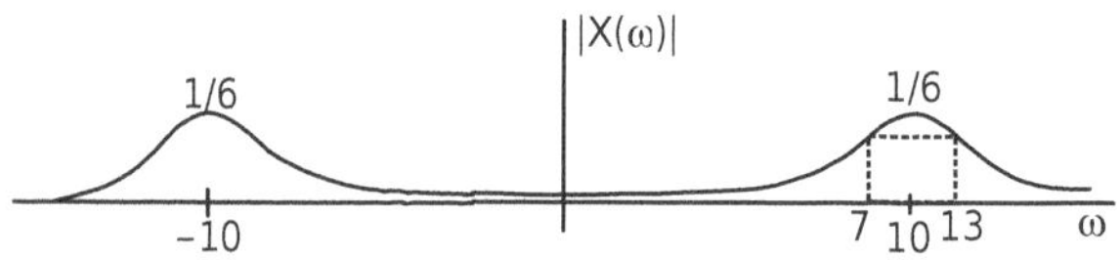

Fig. 7.30 *a*, *b*, *c*

EXERCISES

1. A signal $x(t)$ lasts for 4.096 seconds which is digitised by an ADC at every 8 ms. Find the spacing between the frequency components of X(*f*), its Fourier transform. What is the highest feasible frequency in the spectrum?

   [**Hint :** $4.096/8 \times 10^{-3}$ = 512 points

   8 ms is spacing in time, corresponds to 1000/8 = 125 Hz. in frequency, and the highest frequency is 125/2 = **62.5 Hz.** The steps between frequencies = 125/512 = **0.25 Hz.**]

2. Show that when a signal has only a real component and no imaginary part, the Fourier transform has its real part as an *even* function and imaginary part an *odd* function.

[**Hint :** Use the basic definition of Fourier transform and split into real, imaginary parts. Use the fact that sine function is odd, cosine even.]

3. Show that the Fourier transform of $f(t)$ may also be expressed as

$$F\{f(t)\} = F(\omega) = \int_{-\infty}^{\infty} f(t)\cos\omega t - j\int_{-\infty}^{\infty} f(t)\sin\omega t\, dt$$

Show also that if $f(t)$ is an even function of t, then

$$F(\omega) = 2\int_{0}^{\infty} f(t)\cos\omega t\, dt$$

and if $f(t)$ is odd, then

$$F(\omega) = -2j\int_{0}^{\infty} f(t)\sin\omega t\, dt$$

Thereby show that for the following $f(t)$, the $F(\omega)$ is :

$f(t)$	$F(\omega)$
Real and even function of t	Real and even function of ω
Real and odd	Imaginary and odd
Imaginary and even	Imaginary and even
Complex and even	Complex and even
Complex and odd	Complex and odd

4. A function $f(t)$ can be expressed as a sum of an odd part and even part :

$$f(t) = f_e(t) + f_0(t)$$

Show that $\text{Re}\left[F(\omega)\right]$ is the transform of $f_e(t)$

$j\text{Im}\left[F(\omega)\right]$ is the transform of $f_0(t)$

5. Determine $F(\omega)$ for the following time function.

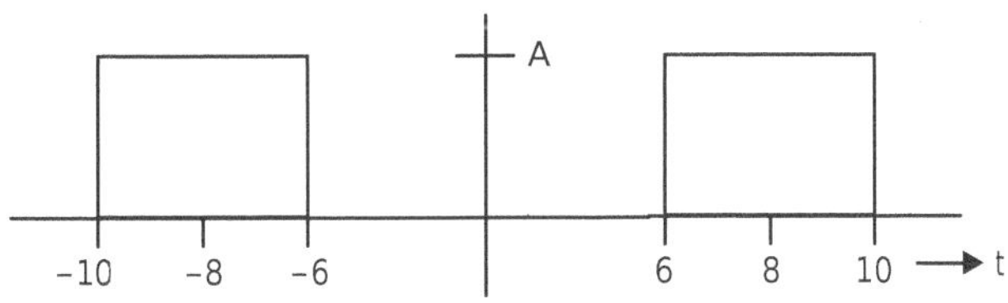

Fig. 1

6. Find the N-point Discrete Fourier Transform (DFT) of

$h(n) = 1/3$ for $n = 0, 1, 2$, and zero otherwise. $\left[\textbf{Ans. } \frac{1}{3} e^{-j2\pi k/N}_{N=3} \left[1 + 2\cos\frac{2\pi k}{N}\right] = H(k)\right]$

7. Find the DFT of $x(n)$ if

$$x(n) = 1, \quad \text{for } n = 2 \text{ to } 6$$

$$x(n) = 0, \quad \text{for } n = 0, 1, 7, 8, 9.$$

Assume $x(n)$ is periodic beyond this interval 0–9.

$$\left[\textbf{Ans. } \exp\left[-j\frac{4\pi k}{5}\right]\sin\left(\frac{\pi k}{2}\right)\Big/\sin\left(\frac{\pi k}{10}\right)\right]$$

8. Find the response of the following system to the input:

$$x[n] = 2 + 2\cos\left(\frac{n\pi}{4}\right) + \cos\left(\frac{2\pi n}{3} + \frac{\pi}{2}\right)$$

System: $H(\omega) = e^{-j\omega}\cos(\omega/2).$

$$\left[\textbf{Ans. } y(n) = 2 + 1.84\cos\left(\frac{n\pi}{4} - \frac{\pi}{4}\right) + \frac{1}{2}\cos\left(\frac{2\pi n}{3} - \frac{\pi}{6}\right)\right]$$

9. The DTFT of a 6 sample signal $x[n]$ is shown below figure. Indicate the values of the DFT on this plot.

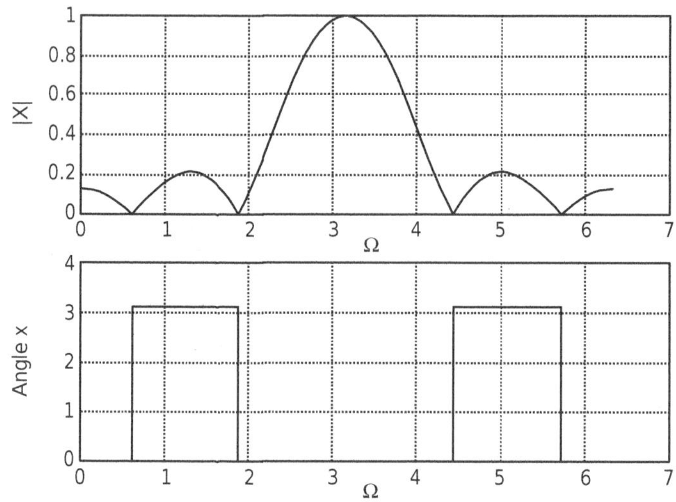

[**Ans.** The DFT is a discretised version of the DTFT at $\omega = \dfrac{2\pi k}{6}$.

$\omega = 0$, $\omega = \pi/3$, $\omega = 2\pi/3$, $\omega = \pi$, $\omega = 4\pi/3$, $\omega = 5\pi/3$ are the points.]

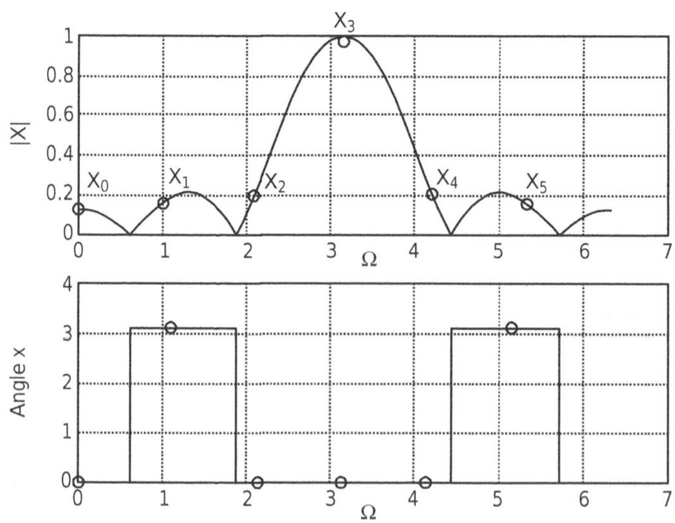

10. Show that the Hilbert Transform of $\exp(j\omega t)$ is $(-\text{ sgn } f) \exp(j\omega t)$

11. Using the formula

$$h(t) = \int_{-\infty}^{\infty} \frac{s(\tau)\, d\tau}{t - \tau}$$

Find the Hilbert transforms of

(*a*) $\sin \omega t$ $[-\cos \omega t]$

(*b*) $\cos \omega t$ $[\sin \omega t]$.

12. Find the Fourier transform of the following signals:

(*a*)

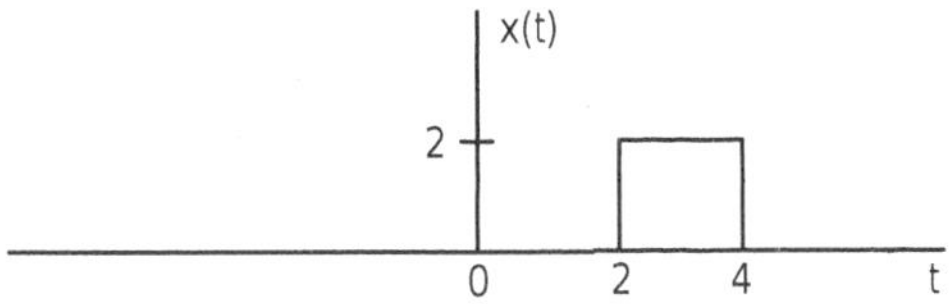

(*b*) $x(t) = 2e^{-2t}\, u(t)$

(*c*) $x(t) = 5e^{-5t}\, u(t)$

(*d*) $x(t) = 2e^{-2t} \cos(4t)\, u(t)$

[**Ans.** (*a*) $X(t) = 2\rho_2(t-3) \longleftrightarrow 4 \text{ sinc}\left(\frac{2\omega}{2\pi}\right) e^{-3\omega j} = X(\omega)$

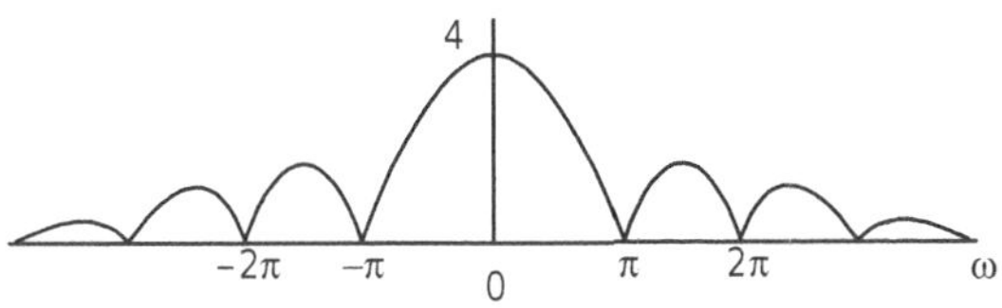

(*b*) $X(\omega) = \dfrac{2}{2 + j\omega}$

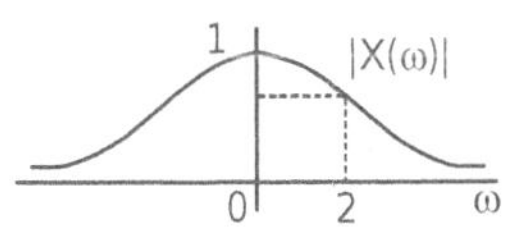

(*c*) $X(\omega) = \dfrac{5}{5 + j\omega}$

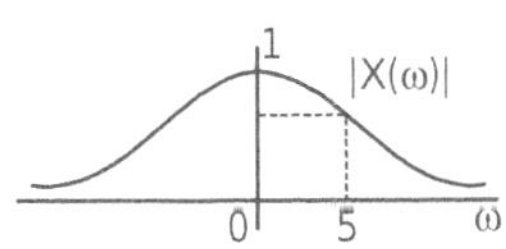

(*d*) $X(\omega) = \dfrac{1}{2}\left[\dfrac{1}{j(\omega-4)+2} + \dfrac{1}{j(\omega+4)+2}\right]$

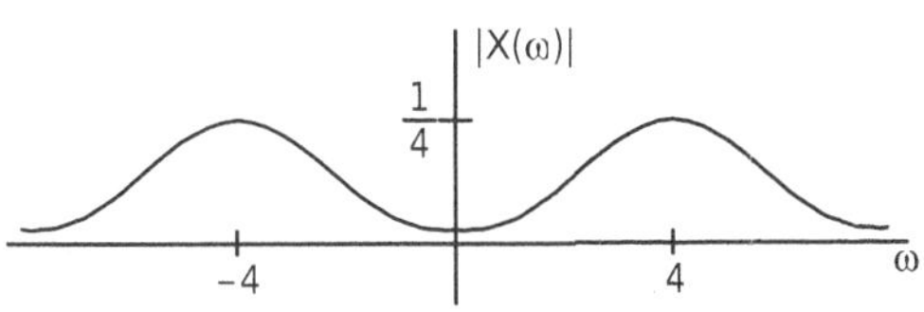

]

13. Match the time responses with the corresponding frequency responses.

1. .... 2. .... 3. .... 4. .... 5. ....

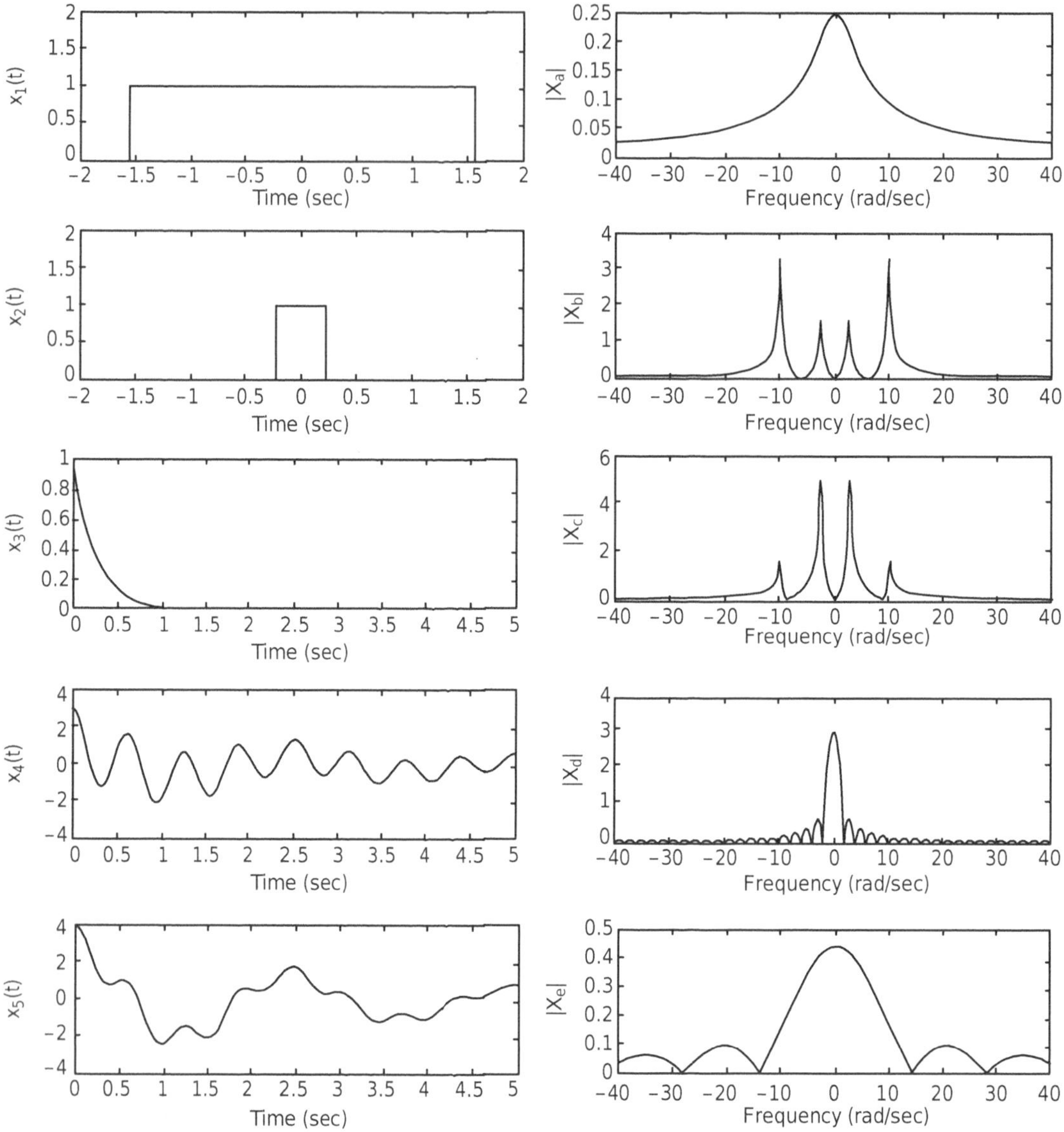

[**Ans.** *d*, *e*, *a*, *b*, *c*.]

14. Compute the F.T. of the following signal (figure).

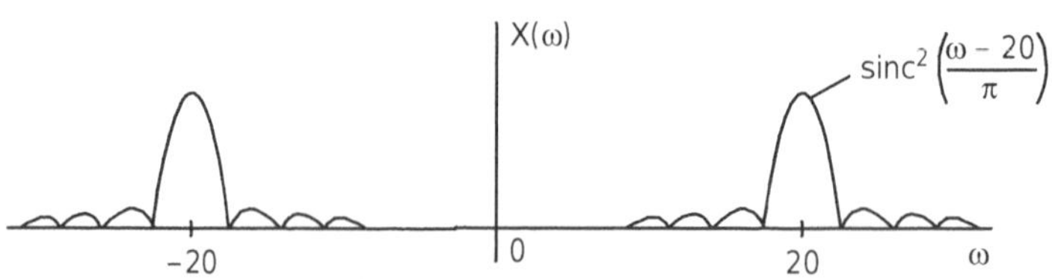

[**Ans.** Consider

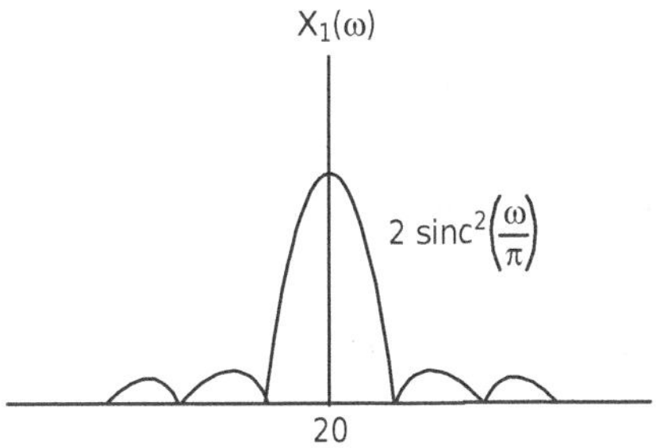

$$x_1(t) = \left(1 - \frac{2|t|}{4}\right) \rho_4(t)$$

Then $$x(t) = x_1(t) \cos (20t)]$$

SIMPLE EXERCISES

1. Write down the relation of the Fourier Inverse transform. Explain the terms in the equation.

[Eqn. 7.5.]

[x is a function of time t. The frequency variable as angular frequency is radians/sec and is ω. j is the $\sqrt{-1}$.]

2. From the inverse Fourier Transform, derive the direct transform.

[Equation 7.5 to 7.7 and the integral I including the δ function]

3. Draw the Fourier transform plot of a single sine wave, which (a) starts at $t = 0$, (b) which starts at $t = \pi/\omega$ and (c) at any other t somewhere in between the two.

[Fig 7.9, 7.8 and Fig. 7.10.]

4. Derive the relation of the Fourier transform of a unit step function. Explain the smear on the function and give its mathematical form.

5. For a decaying sinusoidal oscillation, determine the Fourier Transform and sketch the same. Infer upon the same, marking the peak amplitude and the phase change.

[Fig. 7.17.]

6. Write down the F.T. of the following time functions:

$|t|$, $\exp(-t^2/2)$ $[-2/\omega^2$, $\sqrt{2\pi} \exp(-\omega^2/2)$

7. What is meant by "Scaling property" of the Fourier Transform? Explain.

8. Derive the F.T. of a single square pulse of width τ. Then, using the symmetry property find its inverse and show that it is the same square pulse, except for 2π factor.

[Eqn. 7.25 to 7.27 and Fig. 7.18.]

9. Explain amplitude modulation of a carrier signal. Given the F.T.of the signal, sketch how the amplitude modulated signal spectrum would look like.

[Eqn. 7.27a and Fig. 7.19.]

10. A time shift (a transport lag as in a measurement system for *e.g.*, of temperature in a heated vessel obtained by the outlet temperature from a pipe of length L) of t_a for a function $f(t)$ will have a different Fourier Transform than the signal itself. Is it true?

What is the effect of this with regard to the Fourier Transform's amplitude? Of its phase?

[Equation 7.28. Amplitude not affected. Phase is linearly increasing as $-\omega t_a$. Thus spectrum looks same, but phase lag is more.

11. Explain the inferences relating the Fourier transform and the signal itself for (*a*) odd and even functions (*b*) real and imaginary functions.

[See exer. 3. Also Table after equation (7.28).]

12. Explain convolution of two functions of the same variable. What is its significance? State the Duhamel integral and its limits. What are its two forms? [Eqn. 7.29.]

13. Derive the Fourier transform relation of the convolution of two signals.

[Eqn. 7.30-7.34.]

14. Explain Hilbert transform and derive its time relation from the frequency relation. What is the Hilbert transform of a square wave? [Eqn. 7.42, 7.43.]

15. Derive Parseval's theorem and state is significance to orthogonal functions, [Eqn. 7.45.]

8

LAPLACE TRANSFORMS AND TRANSIENT ANALYSIS

8.1 CONCEPT OF LAPLACE TRANSFORM

An exponentially decaying oscillation, as in a series R-L-C circuit after an excitation, can be represented as

$$A_m e^{-\sigma t} \sin \omega t$$

or in exponential complex form as

$$A_m e^{-\sigma t} e^{-j\omega t}$$

since sin ωt is part of exp $(-j\omega t)$.

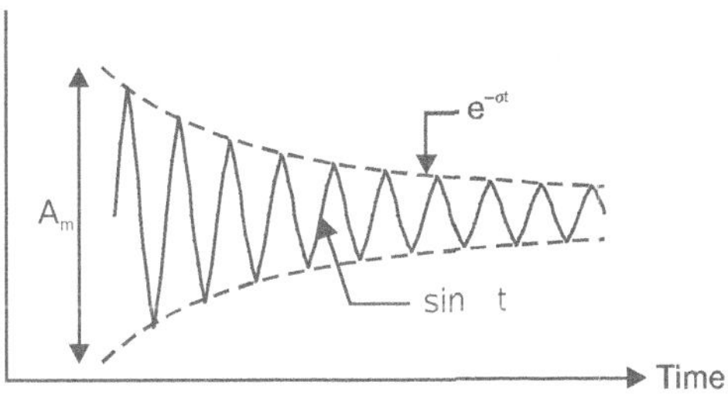

Fig. 8.1(*a*)

Combining the exponents, we get it as $A_m e^{-(\sigma + j\omega)t}$. This $\sigma + j\omega$ is considered as a complex frequency. Thus a concept of general frequency is brought out as having a real part σ and the imaginary part ω. It is represented by 's'.

The Fourier transform gives, for a signal $f(t)$, its frequency components for various ω values as

$$F(\omega) = \int_{-\infty}^{\infty} f(t)\, e^{-j\omega t}\, dt,$$

but for some functions ($f(t) = e^{\alpha t}$ for example, the integral does not converge. So, a convergence factor $e^{-\sigma t}$ was included; σ could take any positive value so that the integral converges. Also the integral limits were changed from $-\infty$ to $+\infty$, to 0 to ∞. This gave the Laplace Transform.

8.1.1 Laplace Transforms

Any function $f(t)$ can be transformed into a function of the complex variable $s = \sigma + j\omega$ by the Laplace transformation

$$F(s) = \int_0^\infty f(t)\, e^{-st} dt \qquad (8.0)$$

As an example, transform $\sin\omega t$ into a function of s.

$$F(s) = \int_0^\infty \sin \omega t\, e^{-st} dt$$

$$= \int \frac{e^{j\omega t} - e^{-j\omega t}}{2j} e^{-st} dt$$

$$= \frac{1}{2j} \int e^{-(s-j\omega)t} - e^{-(s+j\omega)t}\, dt$$

$$= \frac{1}{2j}\left(\frac{1}{s-j\omega} - \frac{1}{s+j\omega}\right) = \frac{\omega}{s^2+\omega^2}$$

Laplace transform of sin ωt is :

$$L\sin \omega t = \omega/(s^2 + \omega^2)$$

Similarly $\quad L\cos \omega t = s/(s^2 + \omega^2)$

For a unit step function, which is unity for $t > 0$,

$$\int_0^\infty 1e^{-st} dt = \frac{1}{s} \text{ or } L(1) = \frac{1}{s}.$$

For an exponential function e^{-at}.

$$\int_0^\infty e^{-at} e^{-st} dt = \int e^{-(s+a)t} dt = \frac{1}{s+a}$$

8.1.2. Laplace Transforms of Step, Ramp Inputs

Step Input $\quad t > 0, f(t) = 1,$

$$Lf(t) = \int_0^\infty 1.e^{-st}\, dt = \frac{1}{s}.$$

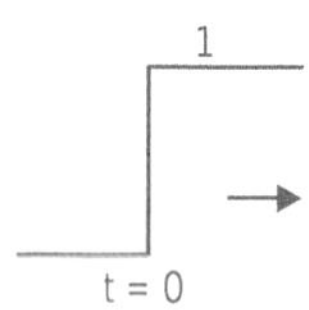

Ramp Input $\quad t > 0, f(t) = t.$

$$Lf(t) = \int_0^\infty te^{-st}\, dt = \frac{1}{s^2}.$$

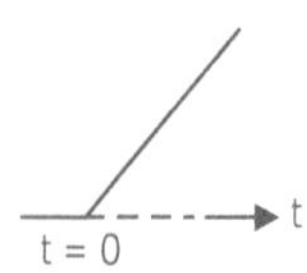

Like this the transforms for several functions are evaluated.

Thus a table of Laplace transforms may be written.

$f(t)$	F(s)
1 (step function)	$1/s$
A	A/s
e^{-at}	$1/(s + a)$
$\sin at$	$a/(s^2 + a^2)$
$\cos at$	$s/(s^2 + a^2)$
$e^{-at} \sin bt$	$b/[(s + a)^2 + b^2]$
$e^{-at} \cos bt$	$(s + a)/[(s + a)^2 + b^2]$
t	$1/s^2$
t^n	$n!/s^{n+1}$
te^{-at}	$1/(s + a)^2$
$(t - a)$	e^{-as}/s

8.2 SOME THEOREMS

1. Laplace transform of $d/dt\ [f(t)]$ is

$$\mathrm{L}\ f'(t) = s\ \mathrm{L}\, f(t) - f(0)$$

$$\mathrm{L}\ f'(t) = e^{-st}(df/dt)dt$$

$$= \left[e^{-st} f(t)\right]_0^{\infty} + \int s e^{-st} f(t) dt$$

$$= s\mathrm{F}(s) - f(0) \tag{8.1}$$

since e^{-st} at $t = 0$ is 1 and at $t \to \infty$, it is zero.

Likewise, for $f''(t)$ we get:

$$\mathrm{L} f''(t) = s^2\mathrm{F}(s) - sf(0) - f'(0) \tag{8.2}$$

Relation (8.1) is useful in connection with an inductor.

$$V_L = L\ di/dt$$

The Laplace transform V_L is got as

$$V_L(s) = L s\ I(s) - Li(0) \tag{8.3}$$

where V and I are Laplace transforms of $v(t)$ and $i(t)$.

2. Laplace transform of $\int f(t)dt$

$$\mathrm{L}\, f'(t) = s\mathrm{L}\, f(t) - f(0)$$

put $\qquad f'(t) = \theta(t).$

Then $\qquad f(t) = \int \theta(t) = \theta^{-1}(t)$

$\therefore \qquad \mathrm{L}\theta(t) = s\ \mathrm{L}\theta^{-1}(t) - \theta^{-1}(0)$

$\therefore \qquad \mathrm{L}\theta^{-1} = \mathrm{L}\theta(t)/s + \theta^{-1}(0)/s$

or $$Lf^{-1}(t) = \frac{Lf(t)}{s} + \frac{f^{-1}(0)}{s} \quad (8.4)$$

This is useful for the capacitor relation of voltage v across it as

$$v_c = 1/C \int i\, dt$$

$$v_c = I(s)/Cs + v(0)/s \quad (8.5)$$

where $v(0)$ is the initial voltage on the capacitor C due to residual charge.

3. Initial Value Theorem

$$\lim_{s\to\infty} s\,F(s) = \lim_{t\to 0} f(t) \quad (8.6)$$

To prove this, consider:

$$\int_0^\infty e^{-st} f'(t)dt = sF(s) - f(0) \quad (8.6a)$$

Put $s \to \infty$ in the left integral; we get zero. So the right side gives the result.

4. Final Value Theorem

$$\lim_{s\to 0} s\,F(s) = \lim_{t\to\infty} f(t) \quad (8.7)$$

Consider (8.6 a) again and putting $s = 0$, we get the integral as

$$f(\infty) - f(0)$$

and the right side as

$$\lim_{s\to 0} s\,F(s) - f(0)$$

Equating both, we get the relation (8.7).

5. Time Translation Theorem

Given $f(t)$, a function of time as shown in Fig. 8.1 (b). We call the function of Fig. 8.1 (c) as $f(t - a)$. The difference between $f(t)$ and $f(t - a)$ is that the latter starts from $t = a$ and is zero for $t < a$.

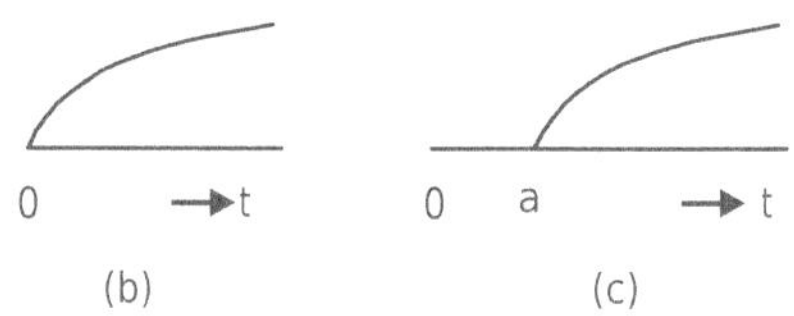

Fig. 8.1

Now consider the theorem:

$$L\,f(t - a) = e^{-as}\,L\,f(t) \quad (8.8)$$

$$L\,f(t - a) = \int e^{-st} f(t - a)dt$$

Putting $t - a = u$ and integrating, we get the result.

6. Unit Impulse-transform

The impulse function $\delta(t)$ is defined as shown. Its value is unity at $t = 0$ and zero at all other t.

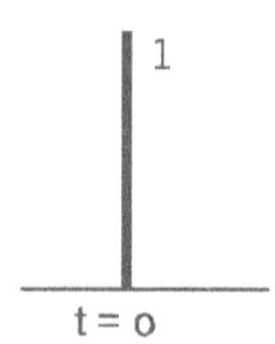

Fig. 8.2 Impulse function.

$$\int e^{-st}\delta(t)dt = [e^{-s0} \times 1] = 1$$

The Laplace transform of $\delta(t)$ is unity. Hence, the inverse, transform of unity is $\delta(t)$, the impulse function.

7. Convolution (vide equation 7.30)

The Laplace transform of the convolution of $h(t)$ and $f(t)$ is

$$L\,(f(t) * h(t)) = H(s)\,.\,F(s)$$

The product of the two transforms. Its proof is similar to the derivation given for equation (7.30) earlier.

8.2.1 Advantage of Laplace Transform Over Fourier Transform

1. Fourier transform is defined for stable systems or signals that taper off at infinity.

   But Laplace transform is defined for unstable signal or system, including an unbounded signal.
2. The Laplace transform can be used for systems with even non-zero initial conditions.

Examples.

We have (1)
$$L\,(1) = \frac{1}{s}$$

$\therefore$
$$L(e^{-at}) = \frac{1}{s+a}$$

(2)
$$L\,(\cos bt) = \frac{s}{s^2+b^2}$$

$\therefore$
$$L(e^{-at}\cos bt) = \frac{s+a}{(s+a^2)+b^2}$$

and
$$L(e^{at}\cos bt) = \frac{s-a}{(s+a^2)+b^2}$$

(3)
$$L\,(\sin bt) = \frac{b}{s^2+b^2}$$

$\therefore$
$$L(e^{-at}\sin bt) = \frac{b}{(s+a)^2+b^2}$$

and $L(e^{at}\sin bt) = \dfrac{b}{(s-a)^2+b^2}$

(4)
$$L(t^n) = \frac{n!}{s^{n+1}} \text{ if } n \text{ is a positive integer.}$$

$\therefore$
$$L(e^{-at}t^n) = \frac{n!}{(s+a)^{n+1}}$$

$$L(e^{at}t^n) = \frac{n!}{(s-a)^{n+1}}$$

Theorem (*i*) If
$$L\,[f(t)] = F(s), \text{ then}$$

$$L\,[t\,f\,(t)] = -\frac{d}{ds}F(s)$$

$$F(s) = \int_0^\infty e^{-st} f(t)dt$$

$$\frac{d}{ds}F(s) = \frac{d}{ds}\int_0^\infty e^{-st} f(t)dt$$

$$= \int_0^\infty \frac{d}{ds}[e^{-st} f(t)]dt = \int_0^\infty -t e^{-st} f(t)dt$$

$$= -\int_0^\infty e^{-st}\, t f(t)dt = -L[t f(t)]$$

$\therefore$
$$L[t f(t)] = -\frac{d}{ds}F(s)$$

Cor.
$$L[t^n f(t)] = (-1)^n \frac{d^n}{ds^n}\{L[f(t)]\}$$

We have
$$L[t^2 f(t)] = -\frac{d}{ds} L\,[tf(t)]$$

$$= -\frac{d}{ds}\left\{-\frac{d}{ds}L[f(t)]\right\}$$

$$= (-1)^2 \frac{d^2}{ds^2}L[f(t)].$$

Continuing the process, we get the result.

This result can also be written as follows :

If
$$L[f(t)] = F(s), \text{ then}$$

$$F'(s) = L[-t f(t)]$$

$$F''(s) = L[(-t)^2 f(t)]$$

$$F^n(s) = L[(-t)^n f(t)].$$

8.3 PARTIAL FRACTION EXPANSIONS

This is necessary when we solve problems using Laplace transform.

Splitting up a ratio of large polynomials into a sum of ratios of small polynomials can be a useful tool, especially for many problems involving Laplace-like transforms. This technique is known as partial fraction expansion. Here's an example of one ratio being split into a sum of three simpler ratios :

$$\frac{8x^2+3x-21}{x^3-7x-6} = \frac{1}{x+2}+\frac{3}{x-3}+\frac{4}{x+1} \quad (1)$$

There are several methods for expanding a rational function via partial fractions. These include the method of clearing fractions, the Heaviside "cover-up" method, and different combinations of these two. For many cases, the Heaviside "cover-up" method is the easiest, and is therefore the method that we will introduce here. For a more complete discussion, see *Signal Processing and Linear Systems* by B. P. Lathi, Berkeley-Cambridge Press, 1998, pp. 24-33. Some of the material below is based upon this book.

Non-Repeated Roots

Let's say we have a proper function $G(x) = \frac{N(x)}{D(x)}$ (by proper we mean that the degree m of the numerator $N(x)$ is less than the degree p of denominator $D(x)$. In this section we assume that there are no repeated roots of the polynomial $D(x)$.

The first step is to factor the denominator $D(x)$:

$$G(x) = \frac{N(x)}{(x-a_1)(x-a_2)\ldots(x-a_p)} \tag{2}$$

where $a_1 \ldots a_p$ are the roots of $D(x)$. We can then rewrite $G(x)$ as a sum of partial fractions :

$$G(x) = \frac{\alpha_1}{x-a_1} + \frac{\alpha_2}{x-a_2} + \ldots + \frac{\alpha_p}{x-a_p} \tag{3}$$

where $a_1 \ldots a_p$ are constants. Now, to complete the process, we must determine the values of these α coefficients. Let's look at how to find α_1. If we multiply both sides of the equation of $G(x)$ as a sum of partial fractions Eq. (3) by $x - a_1$ and then let $x = a_1$, all of the terms on the right-hand side will go to zero except for α_1. Therefore, we will be left over with :

$$\alpha_1 = (x-a_1)G(x)\big|_{x = a_1} \tag{4}$$

We can easily generalize this to a solution for any one of the unknown coefficients :

$$\alpha_r = (x-a_r)G(x)\big|_{x = a_r} \tag{5}$$

This method is called the "cover-up" method because multiplying both sides by $x - a_r$ can be thought of as simply using one's finger to cover up this term in the denominator of $G(x)$. With a finger over the term that would be canceled by the multiplication, you can plug in the value $x = a_r$ and find the solution for α_r.

Example. In this example, we'll work through the partial fraction expansion of the ratio of polynomials introduced above. Before doing a partial fraction expansion, we must make sure that the ratio we are expanding is proper. If it is not, we should do long division to turn it into the sum of a proper fraction and polynomial. Once this is done, the first step is to factor the denominator of the function :

$$\frac{8x^2+3x-21}{x^3-7x-6} = \frac{8x^2+3x-21}{(x+2)(x-3)(x+1)} \tag{6}$$

Now, we set this factored function equal to a sum of smaller fractions, each of which has one of the factored terms for a denominator.

$$\frac{8x^2+3x-21}{(x+2)(x-3)(x+1)} = \frac{\alpha_1}{x+2}+\frac{\alpha_2}{x-3}+\frac{\alpha_3}{x+1} \tag{7}$$

To find the alpha terms, we just cover up the corresponding denominator terms in G(*x*) and plug in the root associated with the alpha :

$$\alpha_1 = (x+2)G(x)\Big|_{x=-2}$$

$$= \frac{8x^2+3x-21}{(x-3)(x+1)}\Bigg|_{x=-2} \tag{8}$$

$$= 1$$

$$\alpha_2 = (x-3)G(x)\Big|_{x=3} = \frac{8x^2+3x-21}{(x+2)(x+1)}\Bigg|_{x=3} \tag{9}$$

$$= 3$$

$$\alpha_3 = (x+3)G(x)\Big|_{x=-1} = \frac{8x^2+3x-21}{(x+2)(x-3)}\Bigg|_{x=-1} \tag{10}$$

$$= 4$$

We now have our completed partial fraction expansion :

$$\frac{8x^2+3x-21}{(x+2)(x-3)(x+1)} = \frac{1}{x+2}+\frac{3}{x-3}+\frac{4}{x+1} \tag{11}$$

Repeated Roots

When the function G(*x*) has a repeated root in its denominator, as in

$$G(x) = \frac{N(x)}{(x-b)^r(x-a_1)(x-a_2)\dots(x-a_j)} \tag{12}$$

Somewhat more special care must be taken to find the partial fraction expansion. The non-repeated terms are expanded as before, but for the repeated root, an extra fraction is added for each instance of the repeated root :

$$G(x) = \frac{\beta_0}{(x-b)^r}+\frac{\beta_1}{(x-b)^{r-1}}+\dots+\frac{\beta_{r-1}}{x-b}+\frac{\alpha_1}{x-a_1}+\frac{\alpha_2}{x-a_2}+\dots+\frac{\alpha_j}{x-a_j} \tag{13}$$

All of the alpha consants can be found using the non-repeated roots method above. Finding the beta coefficients (which are due to the repeated root) has the same Heaviside feel to it, except that this time we will add a twist by using the derivative to eliminate some unwanted terms.

Starting off directly with the cover-up method, we can find β_0. By multiplying both sides by $(x-b)^r$, we'll get :

$$(x-b)^r G(x) = \beta_0+\beta_1(x-b)+\dots+\beta_{r-1}(x-b)^{r-1}+\alpha_1\frac{(x-b)^r}{x-a_1}+\alpha_2\frac{(x-b)^r}{x-a_2}+\dots+\alpha_j\frac{(x-b)^r}{x-a_j} \tag{14}$$

Now that we have "covered up" the $(x-b)^r$ term in the denominator of $G(x)$, we plug in $x = b$ to each side; this cancels every term on the right-hand side except for β_0, leaving the formula

$$\beta_0 = (x-b)^r G(x)\Big|_{x=b} \tag{15}$$

To find the other values of the beta coefficients, we can take advantage of the derivative. By taking the derivative of the *equation after cover up (Eq. 14)* (with respect to x the right-hand side becomes β_1 plus terms containing an $x - b$ in the numerator. Again, plugging in $x = b$ eliminates everything on the right-hand side except for β_1, leaving us with a formula for β_1 :

$$\beta_1 = \frac{d}{dx}\left((x-b)^r G(x)\right)\Big|_{x=b} \tag{16}$$

Generalizing over this pattern, we can continue to take derivatives to find the other beta terms. The solution for all beta terms is

$$\beta_k = \frac{1}{k!}\frac{d^k}{dx^k}\left((x-b)^r G(x)\right)\Bigg|_{x=b} \tag{17}$$

[**Hint :** To check if you've done the partial fraction expansion correctly, just add all of the partial fractions together to see if their sum equals the original ratio of polynomials.]

Example 1. *Find L* $(t\sin at)$.

Solution.

$$L(t\sin at) = -\frac{d}{ds}L(\sin at)$$

$$= -\frac{d}{ds}\frac{a}{s^2+a^2} = \frac{2as}{(s^2+a^2)^2}.$$

Example 2. *Find L* $(t^2 e^{-3t})$.

Solution.

$$L(t^2 e^{-3t}) = (-1)^2\frac{d^2}{ds^2}L(e^{-3t})$$

$$= (-1)^2\frac{d^2}{ds^2}\left(\frac{1}{s+3}\right) = \frac{2}{(s+3)^3}.$$

Example 3. *Find L* $(t\, e^{-t}\sin t)$.

Solution.

$$L(t\, e^{-t}\sin t) = -\frac{d}{ds}L(e^{-t}\sin t)$$

$$= -\frac{d}{ds}\left[\frac{1}{(s+1)^2+1}\right]$$

since

$$L(\sin t) = \frac{1}{s^2+1}.$$

$$\therefore \qquad L(e^{-t}\sin t) = \frac{1}{(s+1)^2+1}.$$

$$\therefore \qquad L(te^{-t}\sin t) = \frac{2(s+1)}{(s^2+2s+2)^2}.$$

Theorem (*ii*) If L[*f (t)*] = F *(s)* and if $\frac{f(t)}{t}$ has a limit as $t \to 0$, then

$$L\left[\frac{f(t)}{t}\right] = \int_s^\infty F(s)\,ds.$$

$$F(s) = L\,[f(t)] = \int_0^\infty e^{-st} f(t)\,dt$$

$$\therefore \qquad \int_s^\infty F(s)\,ds = \int_s^\infty ds \int_0^\infty e^{-st} f(t)\,dt$$

$$= \int_0^\infty f(t)\left\{\int_s^\infty e^{-st}\,ds\right\}dt$$

on interchanging the order of integration

$$= \int_0^\infty f(t)\left[\frac{e^{-st}}{-t}\right]_s^\infty dt$$

$$= \int_0^\infty e^{-st}\frac{f(t)}{t}dt = L\left[\frac{f(t)}{t}\right].$$

Example 4. *Find* $L\left(\frac{1-e^t}{t}\right)$.

Solution.

$$L\left(\frac{1-e^t}{t}\right) = \int_s^\infty L\,(1-e^t)\,ds$$

$$= \int_s^\infty \{L\,(1) - L\,(e^t)\}ds$$

$$= \int_s^\infty \left(\frac{1}{s} - \frac{1}{s-1}\right)ds$$

$$= \left[\log\frac{s}{s-1}\right]_s^\infty = -\log\frac{s}{s-1}$$

$$= \log\frac{s-1}{s}.$$

Example 5. *Find* $L^{-1}\left[\dfrac{s}{(s^2+a^2)^2}\right]$.

Solution.

$$F'(s) = \frac{s}{(s^2+a^2)^2};\ \therefore\ F(s) = \int \frac{s}{(s^2+a^2)^2}ds$$

$$= -\frac{1}{2(s^2+a^2)}$$

$$\therefore\quad L^{-1}\left[\frac{s}{(s^2+a^2)^2}\right] = -t\,L^{-1}\left[-\frac{1}{2(s^2+a^2)}\right]$$

$$= \frac{t}{2}L^{-1}\left(\frac{1}{s^2+a^2}\right)$$

$$= \frac{t}{2a}\sin at.$$

Example 6. *Find* $L^{-1}\left[\dfrac{s}{(s^2-1)^2}\right]$.

Solution.

Here
$$F'(s) = \frac{s}{(s^2-1)^2};$$

$$\therefore\quad F(s) = \int \frac{s}{(s^2-1)^2}ds = -\frac{1}{2(s^2-1)}$$

$$\therefore\quad L^{-1}\left[\frac{s}{(s^2-1)^2}\right] = -t\,L^{-1}\left[-\frac{1}{2(s^2-1)}\right]$$

$$= \frac{t}{2}L^{-1}\left(\frac{1}{s^2-1}\right) = \frac{t}{2}\sinh t.$$

8.3.1 Inverse Transforms by Direct Transform Result Modification

We can modify the results we have obtained in finding the Laplace transforms of functions to get the inverse transforms of functions.

1. If L $[f(t)]$ = F (s), then L e^{-at} $[f(t)]$ = F $(s + a)$.

Hence we get the result

$$L^{-1}[F(s+a)] = e^{-t}f(t)$$

$$= e^{-at}L^{-1}[F(s)]$$

Thus for example

1. $$L^{-1}\left[\frac{1}{(s+a)^2}\right] = e^{-at}L^{-1}\left[\frac{1}{s^2}\right]$$

$$= e^{-at}t.$$

2. $$L^{-1}\left[\frac{1}{(s+2)^2+16}\right] = e^{-2t}L^{-1}\left[\frac{1}{s^2+4^2}\right]$$

$$= \frac{e^{-2t}\sin 4t}{4}.$$

3. $$L^{-1}\left[\frac{s-3}{(s-3)^2+4}\right] = e^{3t}L^{-1}\left[\frac{s}{s^2+2^2}\right]$$

$$= e^{3t}\cos 2t.$$

4. $$L^{-1}\left[\frac{1}{(s+a)^n}\right] = e^{-at}L^{-1}\left[\frac{1}{s^n}\right]$$

$$= \frac{e^{-at}\,t^{n-1}}{(n-1)!}.$$

Theorem *(iii)* If $L[f(t)] = F(s)$, then $L[f(at)] = \frac{1}{a}F\left(\frac{s}{a}\right)$.

This result can be written in the form

$$L^{-1}\left[\frac{1}{a}F\left(\frac{s}{a}\right)\right] = f(at), \text{ where } f(t) = L^{-1}[F(s)].$$

Putting $\frac{1}{a} = k$, we have

$$L^{-1}[F(ks)] = \frac{1}{k}f\left(\frac{t}{k}\right), \text{ where } f(t) = L^{-1}[F(s)].$$

Example 7. *Find* $L^{-1}\left[\frac{s}{s^2a^2+b^2}\right]$.

Solution. $$\frac{s}{s^2a^2+b^2} = \frac{1}{a}\left(\frac{sa}{s^2a^2+b^2}\right) = \frac{1}{a}F(sa).$$

$\therefore$ $$F(s) = \left(\frac{s}{s^2+b^2}\right).$$

$\therefore$ $$L^{-1}\left(\frac{s}{s^2a^2+b^2}\right) = \frac{1}{a}L^{-1}\left(\frac{sa}{s^2a^2+b^2}\right)$$

$$= \frac{1}{a}L^{-1}[F(sa)] = \frac{1}{a}\frac{1}{a}.f\left(\frac{t}{a}\right)$$

where $$f(t) = L^{-1}\left(\frac{s}{s^2+b^2}\right)$$

$$= \cos(bt).$$

$$\therefore \qquad L^{-1}\left(\frac{s}{s^2a^2+b^2}\right) = \frac{1}{a^2}\cos\left(\frac{bt}{a}\right).$$

Theorem (*iv*)

If $\quad L[f(t)] = F(s)$, then $L[t\,f(t)] = -F'(s)$.

Hence we get the result

$$L^{-1}[F'(s)] = -t\,f(t)$$
$$= -t\,L^{-1}[F(s)].$$

Example 8. *Find* $L\left[\dfrac{sinat}{t}\right]$.

Solution.
$$L\left[\frac{\sin at}{t}\right] = \int_s^\infty L(\sin at)\,ds$$
$$= \int_s^\infty \frac{a}{s^2+a^2}\,ds = \left[\tan^{-1}\frac{s}{a}\right]_s^\infty = \tan^{-1}\infty - \tan^{-1}\left(\frac{s}{a}\right)$$
$$= \frac{\pi}{2} - \tan^{-1}\frac{s}{a} = \cot^{-1}\left(\frac{s}{a}\right).$$

Example 9. Find $L^{-1}\left[\dfrac{s+2}{(s^2+4s+5)^2}\right]$.

Solution.
$$F(s) = \frac{s+2}{(s^2+4s+5)^2};$$

$$\therefore \qquad F(s) = -\frac{1}{2(s^2+4s+5)}.$$

$$\therefore \qquad L^{-1}\left[\frac{s+2}{(s^2+4s+5)^2}\right] = -t\,L^{-1}\left[\frac{1}{-2(s^2+4s+5)}\right]$$
$$= \frac{t}{2}L^{-1}\frac{1}{(s+2)^2+1^2}$$
$$= \frac{t}{2}e^{-2t}L^{-1}\frac{1}{s^2+1^2} = \frac{t\,e^{-2t}\sin t}{2}.$$

Theorem (*v*) If $L[f(t)] = F(s)$, then $L[t\,f(t)] = -F'(s)$.

This theorem can be used in the following way to get inverse transforms of certain functions :

Take for example $L^{-1}\left(\log\dfrac{s+1}{s-1}\right)$.

Let this be equal to $f(t)$.

Then $$\mathrm{L}\,[t\,f\,(t)] = -\frac{d}{ds}\log\left(\frac{s+1}{s-1}\right)$$

$$= -\frac{d}{ds}[\log(s+1)-\log(s-1)]$$

$$= -\frac{1}{s+1}+\frac{1}{s-1}$$

$\therefore$ $$t\,f\,(t) = \mathrm{L}^{-1}\left(-\frac{1}{s+1}+\frac{1}{s-1}\right)$$

$$= -\mathrm{L}^{-1}\frac{1}{s+1}+\mathrm{L}^{-1}\frac{1}{s-1}$$

$$= -e^{-t}+e^{t}$$

$$= 2\sinh t.$$

$\therefore$ $$f\,(t) = \frac{2\sinh t}{t}.$$

Theorem (*vi*) If L $[f\,(t)] = s$ F (s) and if $\phi\,(t)$ is a function such that L $[\phi\,(t)]$ = F (s) and $\phi\,(0) = 0$, then $f\,(t) = \phi'(t)$.

We have $$\mathrm{L}\,[\phi'\,(t)] = s\,\mathrm{L}\,[\phi\,(t)]-\phi(0)$$

$$= s\,\mathrm{L}\,[\phi\,(t)]$$

$$= s\,\mathrm{F}\,(s)$$

$$= \mathrm{L}\,[f\,(t)]$$

$\therefore$ $$f\,(t) = \phi'\,(t).$$

This result for finding inverse transforms can be put in the form

$$\mathrm{L}^{-1}\,[s\,\mathrm{F}\,(s)] = \frac{d}{dt}\mathrm{L}^{-1}[\mathrm{F}\,(s)]$$

provided $$\mathrm{L}^{-1}[\mathrm{F}\,(s)] = 0 \text{ when } t = 0.$$

Example 10. *Find* $L^{-1}\left(\dfrac{s}{s^2+b^2}\right)$.

Solution. $$\mathrm{L}^{-1}\left(\frac{s}{s^2+b^2}\right) = \frac{d}{dt}\mathrm{L}^{-1}\left(\frac{1}{s^2+b^2}\right)$$

$$= \frac{d}{dt}\left(\frac{1}{b}\sin bt\right)$$

$$= \cos bt.$$

Example 11. *Find* $L^{-1}\left[\dfrac{s}{(s+3)^2+4}\right]$

Solution.

$$L^{-1}\left[\frac{s}{(s+3)^2+4}\right] = \frac{d}{dt}L^{-1}\left[\frac{1}{(s+3)^2+4}\right]$$

$$= \frac{d}{dt}\left\{e^{-3t}\,L^{-1}\left(\frac{1}{s^2+4}\right)\right\}$$

$$= \frac{d}{dt}\left(e^{-3t}\frac{\sin 2t}{2}\right)$$

$$= e^{-3t}\cos 2t - \frac{3}{2}e^{-3t}\sin 2t$$

$$= \frac{e^{-3t}}{2}(2\cos 2t - 3\sin 2t).$$

Example 12. *Find* $L^{-1}\left(\dfrac{s-3}{s^2+4s+13}\right)$.

Solution.

$$L^{-1}\left(\frac{s-3}{s^2+4s+13}\right) = L^{-1}\left(\frac{s}{s^2+4s+13}\right) - 3\,L^{-1}\left(\frac{1}{s^2+4s+13}\right)$$

$$= \frac{d}{dt}L^{-1}\left(\frac{1}{s^2+4s+13}\right) - 3\,L^{-1}\left(\frac{1}{s^2+4s+13}\right)$$

$$= \frac{d}{dt}L^{-1}\left[\frac{1}{(s+2)^2+3^2}\right] - 3\,L^{-1}\left[\frac{1}{(s+2)^2+3^2}\right]$$

$$= \frac{d}{dt}\left(\frac{e^{-2t}\sin 3t}{3}\right) - 3.\frac{e^{-2t}\sin 3t}{3}$$

$$= \frac{e^{-2t}}{3}(3\cos 3t - 5\sin 2t).$$

Example 13. *Find* $L^{-1}\left[\dfrac{s}{(s+2)^2}\right]$.

Solution.

$$L^{-1}\left[\frac{s}{(s+2)^2}\right] = \frac{d}{dt}L^{-1}\left[\frac{1}{(s+2)^2}\right]$$

$$= \frac{d}{dt}\left\{e^{-2t}\,L^{-1}\left(\frac{1}{s^2}\right)\right\}$$

$$= \frac{d}{dt}(e^{-2t}t) = e^{-2t}(1-2t).$$

Example 14. *Find* $L^{-1}\left[\dfrac{s^2}{(s-1)^3}\right]$.

Solution.

$$L^{-1}\left[\frac{s^2}{(s-1)^3}\right] = \frac{d}{dt} L^{-1}\left[\frac{s}{(s-1)^3}\right]$$

$$= \frac{d^2}{dt^2} L^{-1}\left[\frac{1}{(s-1)^3}\right] = \frac{d^2}{dt^2} e^t L^{-1}\left(\frac{1}{s^3}\right) = \frac{d^2}{dt^2}\left(\frac{e^t . t^2}{2}\right)$$

$$= \frac{e^t}{2}(t^2+4t+2).$$

Theorem (*vii*) $L\left[\int_0^t f(x)\,dx\right] = \frac{1}{s} L\,[f(t)].$

Hence $$L^{-1}\left\{\frac{1}{s} L\,[f(t)]\right\}_0^t = \int_0^t f(x)\,dx. \qquad (a)$$

$\therefore$ If $L[f(t)] = f(s)$, we have

Proof: $$L^{-1}\left[\frac{1}{s}\phi(s)\right] = \int_0^t L^{-1}[\phi(s)]\,dt. \qquad (b)$$

Let $$\int_0^t f(x)\,dx = F(t)$$

Then $$F(0) = \int_0^0 f(x)\,dx = 0.$$

Also $$F'(t) = f(t)$$

Hence $$L[F'(t)] = L[f(t)].$$

But $$L[F'(t)] = s\,L[F(t)] - F(0)$$

$$= s\,L[F(t)] \text{ as } F(0) = 0.$$

$\therefore$ $$L[f(t)] = s\,L[F(t)]$$

$$= s\,L\left[\int_0^t f(x)\,dx\right]$$

$\therefore$ $$L^{-1}\left\{\frac{1}{s} L\,[f(t)]\right\} = \int_0^t f(x)\,dx. \qquad (a)$$

$\therefore$ If $L[f(t)] = \phi(s)$, we have

$$L^{-1}\left[\frac{1}{s}\phi(s)\right] = \int_0^t L^{-1}[\phi(s)]\,dt. \qquad (b)$$

Example 15. *Find* $L^{-1}\left[\dfrac{1}{s(s+a)}\right]$.

Solution.

$$L^{-1}\left[\frac{1}{s(s+a)}\right] = \int_0^t L^{-1}\left[\frac{1}{s+a}\right]dt$$

$$= \int_0^t e^{-at}dt = \left[\frac{e^{-at}}{-a}\right]_0^t = \frac{1}{a}(1-e^{-at}).$$

Example 16. *Find* $L^{-1}\left[\dfrac{1}{s(s^2+a^2)}\right]$.

Solution.

$$L^{-1}\left[\frac{1}{s(s^2+a^2)}\right] = \int_0^t L^{-1}\left(\frac{1}{s^2+a^2}\right)dt$$

$$= \int_0^t \frac{\sin at}{a}\,dt = \frac{1}{a}\left[\frac{-\cos at}{a}\right]_0^t$$

$$= \frac{1}{a^2}(1-\cos at).$$

Example 17. *Find* $L^{-1}\left[\dfrac{1}{(s^2+a^2)^2}\right]$.

Solution.

$$L^{-1}\left[\frac{1}{(s^2+a^2)^2}\right] = L^{-1}\left[\frac{1}{s}\,\frac{s}{(s^2+a^2)^2}\right]$$

$$= \int_0^t L^{-1}\left[\frac{s}{(s^2+a^2)^2}\right]dt$$

$$= \int_0^t \frac{t\sin at}{2a}\,dt$$

$$= \frac{1}{2a}\left[\frac{-t\cos at}{a} + \frac{\sin at}{a^2}\right]_0^t$$

$$= \frac{1}{2a^3}(\sin at - at\cos at).$$

The method of partial fractions can be used to find the inverse transforms of certain functions.

The method is illustrated in the following examples :

Example 18. *Find* $L^{-1}\left[\dfrac{1}{s(s+1)(s+2)}\right]$.

Solution. We can split $\dfrac{1}{s(s+1)(s+2)}$ into partial fractions and we have

$$\frac{1}{s(s+1)(s+2)} = \frac{1}{2}\frac{1}{s} - \frac{1}{s+1} + \frac{1}{2}\frac{1}{s+2}.$$

$$\therefore \qquad L^{-1}\left\{\frac{1}{s\ (s+1)\ (s+2)}\right\} = \frac{1}{2}L^{-1}\left(\frac{1}{s}\right) - L^{-1}\left(\frac{1}{s+1}\right) + \frac{1}{2}L^{-1}\left(\frac{1}{s+2}\right)$$

$$= \frac{1}{2} - e^{-t} + \frac{1}{2}e^{-2t}.$$

Example 19. *Find* $L^{-1}\left[\dfrac{1}{(s+1)(s^2+2s+2)}\right]$.

Solution. Splitting into partial fractions we have

$$\frac{1}{(s+1)(s^2+2s+2)} = \frac{1}{s+1} - \frac{s+1}{s^2+2s+2}.$$

$$\therefore \qquad L^{-1}\left[\frac{1}{(s+1)\ (s^2+2s+2)}\right] = L^{-1}\left(\frac{1}{s+1}\right) - L^{-1}\left(\frac{s+1}{s^2+2s+2}\right)$$

$$= e^{-t} - L^{-1}\left\{\frac{s+1}{(s+1)^2+1^2}\right\}$$

$$= e^{-t} - e^{-t}L^{-1}\left(\frac{s}{s^2+1^2}\right)$$

$$= e^{-t} - e^{-t}\cos t$$

$$= e^{-t}(1-\cos t).$$

Example 20. *Find* $L^{-1}\left[\dfrac{1+2s}{(s+2)^2(s-1)^2}\right]$.

Solution.
$$\frac{1+2s}{(s+2)^2(s-1)^2} = \frac{1}{3}\frac{(s+2)^2-(s-1)^2}{(s+2)^2(s-1)^2}$$

$$= \frac{1}{3}\left[\frac{1}{(s-1)^2} - \frac{1}{(s+2)^2}\right]$$

Hence $$L^{-1}\left[\frac{1+2s}{(s+2)^2\,(s-1)^2}\right] = \frac{1}{3}L^{-1}\left[\frac{1}{(s-1)^2}\right]-\frac{1}{3}L^{-1}\left[\frac{1}{(s+2)^2}\right]$$

$$= \frac{1}{3}t\,e^{t} - \frac{1}{3}t\,e^{-2t} = \frac{t}{3}\,(e^{t}-e^{-2t}).$$

Solving Differential Equations Using L.T. Additional Problems

Example 1. *Solve the following differential equations using the Laplace Transform method:*

$\dot{y} + 4y = 3x;\ x(t) = \sin(2t),\ y(0) = 1$

Solution. $\dot{y} + 4y = 3x;\ x = \sin(2t),\ y(0) = 1$

$$sY(s) - y(0) + 4Y(s) = 3X(s)$$

$$(s+4)\,Y(s) = \frac{6}{s^2+4} + 1$$

$$Y(s) = \frac{6}{(s^2+4)(s+4)} + \frac{1}{s+4}$$

$$= \frac{C_1 s + C_2}{s^2+4} + \frac{C_3}{s+4} + \frac{1}{s+4}$$

$$C_3 = 3/10$$

$$6 = (C_1 s + C_2)(s+4) + 3/10(s^2+4)$$

$$\Rightarrow \quad C_1 = -3/10$$

$$6 = 4C_2 + 6/5$$

$$6/5 = C_2$$

$$Y(s) = \frac{-3/10s + 6/5}{s^2+4} + \frac{13/10}{s+4}$$

$$= \frac{-3/10s}{s^2+4} + \frac{6}{10}\frac{2}{s^2+4} + \frac{13/10}{s+4}$$

$$y(t) = \left(-3/10\cos(2t) + 6/10\sin(2t) + 13/10e^{-4t}\right)u(t)$$

8.4 LAPLACE TRANSFORMS FOR SIGNAL FUNCTIONS

Example 1. *Compute the Laplace transforms of the following functions :*

(a) $x(t) = 4\sin(100t)u(t)$

(b) $x(t) = 4\sin(100t - 10)u(t - 0.1)$

(c) $x(t) = 2u(t) + \delta(t-4) - \cos(5t)u(t)$

(d) $x(t) = tu(t) - 2(t-2)u(t-2) + (t-3)u(t-3)$

(e) $x(t) = u(t) - e^{-2t}\cos(10t)u(t)$

Solution. (*a*) $$4\sin(100t)u(t) \leftrightarrow \frac{400}{s^2+100^2}$$

(*b*) $$4\sin(100t-10)u(t-0.1) = 4\sin(100(t-0.1))$$

$$4\sin(100t) \leftrightarrow \frac{400}{s^2+100^2}$$

$$4\sin(100(t-.1)u(t-.1) \leftrightarrow \frac{400}{s^2+100^2}e^{-0.1s}$$

(*c*) $$2u(t)+(t-4)-\cos\,(5t)\,u(t) \leftrightarrow \frac{2}{s}+e^{-4s}-\frac{s}{s^2+25}$$

(*d*) $$tu(t)-2(t-2)u(t-2)+(t-3)u(t-3) \leftrightarrow \frac{1}{s^2}-\frac{2e^{-2s}}{s^2}+\frac{1}{s^2}e^{-3s}$$

(*e*) $$u(t)-e^{-2t}\cos(10t) \leftrightarrow \frac{1}{s}-\frac{s+2}{(s+2)^2+10^2}$$

Example 2. *Compute the inverse Laplace Transforms of the following functions :*

(*a*) $$X(s) = \frac{10(s+1)}{s^2+4s+3}$$

(*b*) $$X(s) = \frac{10(s+1)}{s^2+4s+8}$$

(*c*) $$X(s) = \frac{2s+100}{(s+1)(s+8)(s+10)}$$

(*d*) $$X(s) = \frac{10(s+1)}{s^2+4s+3}e^{-2s}$$

(*e*) $$X(s) = \frac{20}{s(s^2+10s+16)}$$

(*f*) $$X(s) = \frac{10(s+1)}{(s^2+4s+8)s}$$

Solution. (*a*) $$X(s) = \frac{10(s+1)}{s^2+4s+3} = \frac{10(s+1)}{(s+1)(s+3)} = \frac{10}{s+3}$$

$$X(t) = 10e^{-3t}\,u(t)$$

(*b*) $$X(s) = \frac{10(s+1)}{s^2+4s+8} = \frac{10(s+1)}{(s+2)^2+4} = \frac{10(s+2)}{(s+2)^2+4} - \frac{10}{(s+2)^2+4}$$

$$X(t) = (10e^{-2t}\cos(4t)-5e^{-2t}\sin(4t))\,u(t)$$

(*c*) $$X(s) = \frac{2s+100}{(s+1)(s+8)(s+10)} = \frac{C_1}{s+1}+\frac{C_2}{s+8}+\frac{C_3}{s+10}$$

$$C_1 = X(s)(s+1)\big|_{s=-1} = 1.555$$

$$C_2 = -4.667$$

$$C_3 = 4.44$$

$$X(t) = (1.555e^{-t} - 4.667e^{-8t} + 4.44e^{-10t})\,u(t)$$

(*d*) $$X(s) = \frac{10(s+1)}{s^2+4s+3}e^{-2s} = X_1(s)e^{-2s}$$

From (*a*) $$X_1(t) = 10e^{-3t}\,u(t)$$

so $$x(t) = 10e^{-3(t-2)}\,u(t-2)$$

(*e*) $$X(s) = \frac{20}{s(s^2+10s+16)} = \frac{C_1}{s} + \frac{C_2}{s+2} + \frac{C_3}{s+8}$$

$$C_1 = \frac{5}{4},$$

$$C_2 = \frac{20}{s(s+8)}\bigg|_{s=-2} = \frac{-20}{12} = \frac{-5}{3}, \quad C_3 = \frac{20}{s(s+2)}\bigg|_{s=-8} = \frac{5}{12}$$

$$X(t) = \left(\frac{5}{4} - \frac{5}{3}e^{-2t} + \frac{5}{12}e^{-8t}\right)u(t)$$

(*f*) $$X(s) = \frac{10(s+1)}{s(s^2+4s+8)} = \frac{10(s+1)}{s\left((s+2)^2+4\right)} = \frac{C_1}{s} + \frac{C_2 s + C_3}{(s+2)^2+4}$$

$$C_1 = 5/4$$

$$10(s+1) = \frac{5}{4}(s^2+4s+8) + C_2 s^2 + C_3 s \Rightarrow C_2 = \frac{-5}{4}, C_3 = 5$$

$$X(s) = \frac{5/4}{5} + \frac{-5/4s+5}{(5+2)^2+4} = \frac{5/4}{s} - \frac{5/4(s+2)}{(s+2)^2+4} + \frac{15/2}{(s+2)^2+4}$$

$$X(t) = \frac{5}{4}u(t) - \frac{5}{4}e^{-2t}\cos(2t)\,u(t) + \frac{15}{4}e^{-2t}\sin(2t)\,u(t)$$

Example 3. *Find the limit as $t \to \infty$ of x(t) (if the limit exists)*

(*a*) $$X(s) = \frac{10(s+1)}{s(s^2+4s+3)}$$

(*b*) $$X(s) = \frac{10(s+1)}{s(s^2+4s+8)}$$

(*c*) $$X(s) = \frac{10(s+1)}{s(s^2+2s-3)}$$

Solution. (*a*) $$X(s) = \frac{10(s+1)}{s(s^2+4s+3)}$$

poles of $sX(s)$ are –3, –1 $\Rightarrow$ limit exists

$$\lim_{t\to\infty} x(t) = \lim_{s\to 0} sX(s) = \frac{10}{3}$$

(*b*) $$X(s) = \frac{10(s+1)}{s(s^2+4s+8)}$$

poles of $sX(s)$ are $-2 \pm 2j$

$< 0 \Rightarrow$ limit exists

$$\lim_{t\to\infty} x(t) = \lim_{s\to 0} sX(s) = \frac{5}{4}$$

(*c*) $$X(s) = \frac{10(s+1)}{s(s^2+2s-3)}$$

poles of $sX(s)$ are 3, –1 ; 3 > 0 so no limit exists

Example 4. *Give the general form of x(t) (do not solve for the coefficients explicity).*

(*a*) $$X(s) = \frac{2s+100}{(s+2)(s+6)(s+10)}$$

(*b*) $$X(s) = \frac{2s+100}{s(s+1)(s+8)(s-4)}$$

(*c*) $$X(s) = \frac{s-40}{(s+1)(s+8)(s+10)}$$

(*d*) $$X(s) = \frac{10(s+1)}{s(s^2+4s+3)}$$

(*e*) $$X(s) = \frac{10(s+1)}{s(s^2+4s+8)}$$

(*f*) $$X(s) = \frac{s+1}{s(s^2+4)(s+8)}$$

(*g*) $$X(s) = \frac{20(s+1)}{(s^2+16)((s+4)^2+25)(s+1)}$$

Solution. (*a*) $$X(s) = \frac{2s+100}{(s+2)(s+6)(s+10)}$$

$$x(t) = C_1e^{-2t} + C_2e^{-6t} + C_3e^{-10t},\ t \geq 0$$

(*b*) $$X(s) = \frac{2s+100}{s(s+1)(s+8)(s-4)}$$

$$x(t) = C_1 + C_2e^{-t} + C_3e^{-8t} + C_4e^{4t},\ t \geq 0$$

(c)
$$X(s) = \frac{s-40}{(s+1)(s+8)(s+10)},$$
$$x(t) = C_1e^{-t} + C_2e^{-8t} + C_3e^{-10t},\ t \geq 0$$

(d)
$$X(s) = \frac{10(s+1)}{s(s^2+4s+3)} = \frac{10(s+1)}{s(s+1)(s+3)}$$
$$x(t) = C_1 + C_2e^{-t} + C_3e^{-3t},\ t \geq 0$$

(e)
$$X(s) = \frac{10(s+1)}{s(s^2+4s+8)} = \frac{10(s+1)}{s\left((s+2)^2+4\right)}$$
$$x(t) = C_1 + C_2e^{-2t}\cos(2t+\theta),\ t \geq 0$$

(f)
$$X(s) = \frac{s+1}{s(s^2+4)(s+8)},$$
$$x(t) = C_1 + C_2\cos(2t+\theta) + C_3e^{-8t},\ t \geq 0$$

(g)
$$X(s) = \frac{20(s+1)}{(s^2+16)(s+4)^2+25)(s+1)},$$
$$x(t) = C_1\cos(4t+\theta_1) + C_2e^{-4t}\cos(5t+\theta_2) + C_3e^{-t},\ t \geq 0$$

8.5 NODE AND MESH ANALYSIS IN THE LAPLACE TRANSFORM VARIABLE 'S'

The circuit elements of inductors and capacitors have an impedance given by (2.10) and (2.14) as

$$X_L = j\omega L \qquad X_C = \frac{1}{j\omega C} \tag{8.9}$$

which is the ratio of voltage by current through the element.

Now, if the complex frequency s is used, we get them as

$$X_L = sL \quad \text{and } X_C = \frac{1}{sC} \tag{8.10}$$

In this way, the current through a series circuit is got by

$$I(s) = \frac{E(s)}{R + sL + \frac{1}{sC}}$$

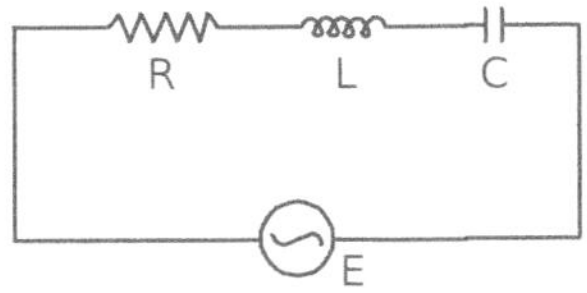

Fig. 8.3

Note that E(s) is the transform of the actual time-function of E, which could be E sin ωt.

Replacing s with $j\omega$, we get the current I($j\omega$) as

$$I(j\omega) = \frac{E(j\omega)}{R + j\omega L + \dfrac{1}{j\omega C}} \quad (8.11)$$

Example. Consider the loop equation solution of Fig. 2.38. First transform into 's' variable.

The impedances of 2H, $\frac{1}{2}$ F for ω = 1 (ω = 1 for volts E) are $2s$, $\frac{2}{s}$ respectively.

So the node equations for node 'd' as common can be written as

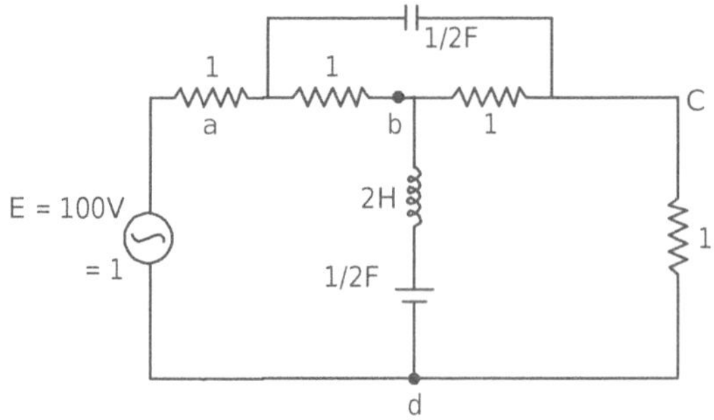

Fig. 8.4 (Fig. 2.38 redrawn)

$$\begin{vmatrix} \left(2 + 2s + \dfrac{2}{s}\right) & -\left(2s + \dfrac{2}{s}\right) & -1 \\ -\left(2s + \dfrac{2}{s}\right) & 2 + 2s + \dfrac{2}{s} & -1 \\ -1 & -1 & 2 + \dfrac{2}{s} \end{vmatrix} \begin{vmatrix} I_1(s) \\ I_2(s) \\ I_3(s) \end{vmatrix} = \begin{vmatrix} E(s) \\ 0 \\ 0 \end{vmatrix}$$

Here E(s) is the transform of 100 sin ωt.

We can solve for $I_1(s)$, $I_2(s)$ and $I_3(s)$ and then take the inverse transforms. Then we would get the currents, along with the transients.

To get merely the steady-state sinusoidal currents, we put $s = j\omega$ and solve the above set of equations, which is already given in section 2.23. This is easy.

So, only for finding the transient part of the currents, we need to use 's' variable.

8.5.1 Certain Networks used in Communication Circuits

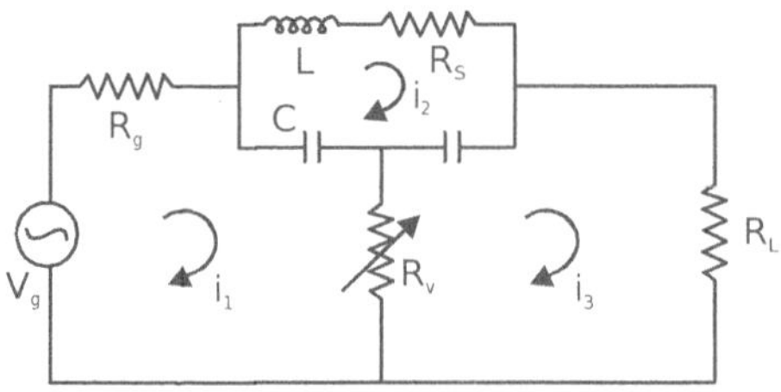

Fig. 8.4.1 Bridged – T Notch filter

The network equations are written as (V_g is the generator voltage).

$$\begin{vmatrix} R_g + \frac{1}{sC} + R_v & \frac{-1}{sC} & -R_v \\ -\frac{1}{sC} & sL + Rs + \frac{2}{sC} & -\frac{1}{sC} \\ -R_v & -\frac{1}{sC} & R_v + \frac{1}{sC} + R_L \end{vmatrix} \begin{vmatrix} i_1 \\ i_2 \\ i_3 \end{vmatrix} = \begin{vmatrix} V_g \\ 0 \\ 0 \end{vmatrix}$$

$$i_3 = \frac{V}{\Delta}\left(\frac{1}{s^2C^2} + sR_vL + R_LR_S + \frac{2R_v}{sC}\right)$$

where Δ is the determinant of the above impedance matrix.

We want the condition for which the numerator is zero.

$$s^3LC^2R_v + s^2C^2R_sR_v + CR_v + 1 = 0$$

Now for sinusoidal steady state condition, $s = j\omega$.

$$(1 - \omega^2C^2R_sR_v) + j\omega C.R_v(2 - \omega^2LC) = 0$$

Only if the above is zero in both the real and imaginary parts, the network will have a transmission Null.

Using imaginary part of the above,

$$\omega_r = \sqrt{\frac{2}{LC}}$$

Use this ω in the real part of it.

$$R_v = \frac{L}{2CR_s}$$

Thus for this R_v, the network will reject the ω_r frequency and act as a Notch filter.

8.5.2 RC-Phase Shift Circuit

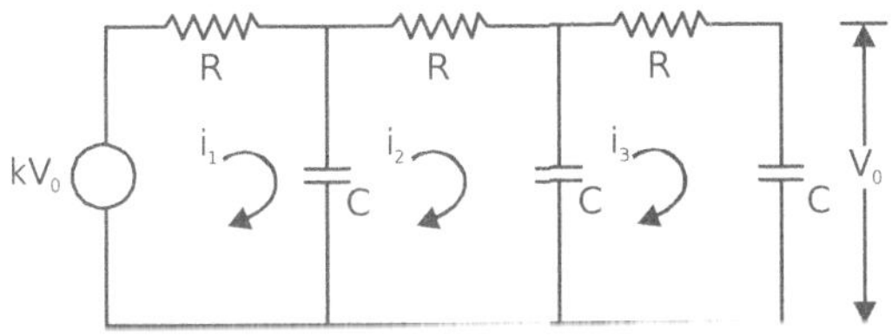

Fig. 8.4.2

The network equation in matrix form is: V_0 is the input voltage.

$$\begin{vmatrix} R + \frac{1}{sC} & -\frac{1}{sC} & 0 \\ -\frac{1}{sC} & R + \frac{2}{sC} & -\frac{1}{sC} \\ 0 & -\frac{1}{sC} & R + \frac{2}{sC} \end{vmatrix} \begin{vmatrix} i_1 \\ i_2 \\ i_3 \end{vmatrix} = \begin{vmatrix} kV_0 \\ 0 \\ 0 \end{vmatrix}$$

This network will produce a phase-lag for any ω. We want to find at what ω it is –180°!

Solve for i_3:

$$i_3 = \frac{1}{\Delta}\begin{vmatrix} R+\frac{1}{sC} & -\frac{1}{sC} & kV_0 \\ -\frac{1}{sC} & R+\frac{2}{sC} & 0 \\ 0 & -\frac{1}{sC} & 0 \end{vmatrix}$$

$$= \frac{kV_0}{s^2C^2\Delta} \qquad (\Delta = \text{Matrix determinant})$$

But $i_3 = V_0sC$ from figure, being the current through the final C. So,

$$s^3C^3\Delta = k$$

$$s^3R^3C^3 + 5s^2C^2R^2 + 6sRC + 1 = k.$$

Substituting $s = j\omega$ and separating real-imag. parts,

$$(1 - 5\omega^2R^2C^2) + j\omega CR\,(6 - \omega^2C^2R^2) = k$$

Because the Imag. part is zero and solving for ω,

$$\omega = \frac{\sqrt{6}}{RC}$$

Substitute this in the real part,

$$k = -29.$$

So, if an amplifier with a gain of 29 (at least) is used with this network in feedback, the same will work as a sine wave oscillator of freq. ω. Such oscillators are used in communication.

8.5.3 Twin-T Notch Filter

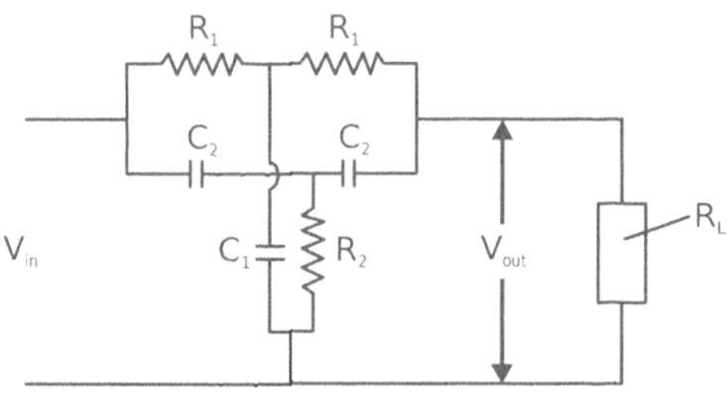

Fig. 8.4.3(*a*)

This acts as a Notch filter and is typically used with signals having 50 Hz hum (as added to signal). It will reject the 50 Hz hum.

The circuit has to be redrawn as below to enable loop analysis.

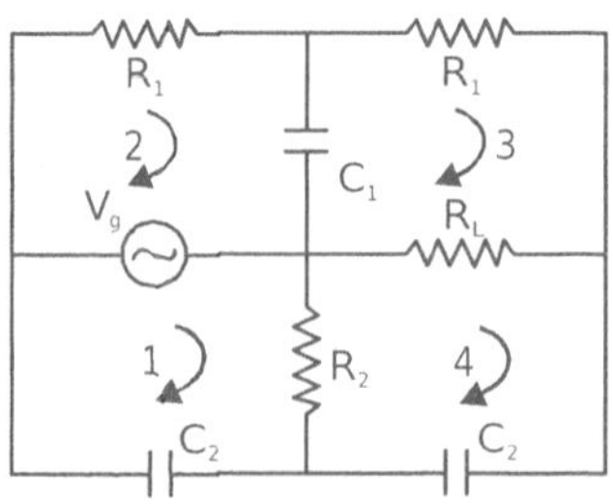

Fig. 8.4.3 (*b*)

The matrix equations for loop currents 1 to 4 is:

$$\begin{vmatrix} R_2 + \dfrac{1}{sC_2} & 0 & 0 & -R_2 \\ 0 & R_1 + \dfrac{1}{sC_1} & -\dfrac{1}{sC_1} & 0 \\ 0 & -\dfrac{1}{sC_1} & R_1 + R_L + \dfrac{1}{sC_1} & -R_L \\ -R_2 & 0 & -R_L & R_2 + R_L + \dfrac{1}{sC_2} \end{vmatrix} \begin{vmatrix} i_1 \\ i_2 \\ i_3 \\ i_4 \end{vmatrix} = \begin{vmatrix} V_g \\ -V_g \\ 0 \\ 0 \end{vmatrix}$$

From this, if Δ is the matrix determinant,

$$i_3 = \frac{1}{\Delta}\begin{vmatrix} R_2 + \dfrac{1}{sC_2} & 0 & V_g & -R_2 \\ 0 & R_1 + \dfrac{1}{sC_1} & -V_g & 0 \\ 0 & -\dfrac{1}{sC_1} & 0 & -R_L \\ -R_2 & 0 & 0 & R_2 + R_L + \dfrac{1}{sC_2} \end{vmatrix}$$

Likewise,

$$i_4 = \frac{1}{\Delta}\begin{vmatrix} R_2 + \dfrac{1}{sC_2} & 0 & 0 & V_g \\ 0 & R_1 + \dfrac{1}{sC_1} & -\dfrac{1}{sC_1} & -V_g \\ 0 & -\dfrac{1}{sC_1} & R_1 + R_L + \dfrac{1}{sC_1} & 0 \\ -R_2 & 0 & R_L & 0 \end{vmatrix}$$

The null conditions are got by the numerators of the above equations and after simplifying,

$$\left(R_1^2R_2 - \frac{2R_2}{\omega^2C_1C_2}\right) + j\left(\frac{1}{\omega^3C_1C_2^2} - \frac{2R_1R_2}{\omega C_1}\right) = 0$$

From which,

$$\omega^2 = \frac{1}{2R_1R_2C_2^2}$$

and $$R_1C_1 = 4R_2C_2$$

But in practice, $R_1 = 2R_2$, $C_2 = 2C_1$; then

$$f = \frac{\omega}{2\pi} = \frac{1}{2\pi RC}.$$

Thus this frequency will be 'notched' or rejected if R and C are chosen correctly for 50 Hz hum.

8.6 USE OF LAPLACE TRANSFORMS FOR TRANSIENT PROBLEMS

Consider the differential equation of the L–R transient

$$L\, di/dt + R\, i = E$$

Transforming i into its Laplace transform and denoting the transform as $I(s)$,

$$Ls\, I(s) - L\, i(0) + R\, I(s) = E/s.$$

$\because$ di/dt transforms into $sI(s) - i(0)$

E transforms into E/s for step input at $t = 0$.

$$\therefore \qquad I(s) = \frac{E}{s(Ls+R)} + \frac{L\,i\,(0)}{L\,s+R}$$

Since $\qquad i(0) = 0,\ I(s) = E/s(Ls + R)$

Let $$I(s) = \frac{E}{s(Ls+R)} = \frac{A}{s} + \frac{B}{sL+R} \qquad (8.12)$$

The above expansion is the *partial fraction* expansion.

A and B are to be found. Cross multiply by $s\,(Ls + R)$.

$$E = A\,(Ls + R) + Bs$$

Put $s = 0$ in this identity. We get $A = E/R$.

Put $s = -R/L$ and then $B = -LE/R$.

The relation (8.12) is a function of s. To transform into the function of time, we find the inverse transform of $I(s)$. From the table of transforms, the inverse of $1/s$ is unity; that for $1/(sL + R)$ is $(1/L) \exp(-R/L)t$. So

$$i(t) = A + B/L\, e^{-Rt/L}$$
$$= E/R\,(1 - e^{-Rt/L})$$

Advantages of Laplace Transform Method

(1) Initial conditions of transient problems can be easily incorporated.

(2) The particular integral and the complimentary function (the steady-state and transient solutions) are obtained together simultaneously.

(3) The writing of differential equations for meshy circuits is difficult without Laplace transforms.

As an example suppose the voltage across R_2 is required.

A mesh equation problem using Laplace Transform Variable 's'

Method (a) $$R_1 i_1 = 1/C \int i_2\, dt$$

$E = R_1\, i_1 + R_2\,(i_1 + i_2)$ are the two mesh equations, having currents i_1, i_2 as marked.

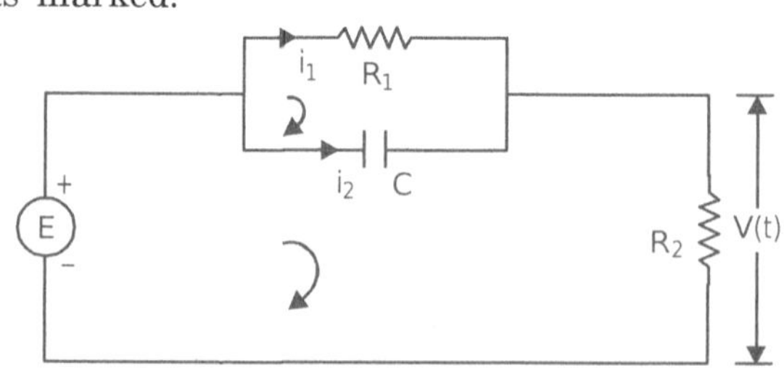

Fig. 8.5

$$E = \frac{R_1 + R_2}{R_1 C} \int_0^t i_2\, dt + R_2 i_2 \qquad (a)$$

The required voltage across R_2 is

$$v = R_2(i_1 + i_2) = (R_2 / R_1)(1/C)\int i_2\, dt + R_2\, i_2 \qquad (b)$$

Transform relations (*a*) and (*b*).

$$E\,(s) = \frac{R_1 + R_2}{R_1 C}\, \frac{I_2(s)}{s} + R_2 I_2(s) \qquad (a')$$

The above assumes that there is no initial charge on C.

$$v(s) = \frac{R_2}{R_1\, C}\, \frac{I_2\,(s)}{s} + R_2 I_2(s) \qquad (b')$$

Eliminate I_2 from a' and b' to get

$$v(s) = \frac{R_2(1 + R_1 Cs)}{R_1 + R_2 + R_1 R_2 Cs}\, E(s) \qquad (8.13)$$

This can be transformed back to get $v(t)$, the transient solution.

Method (*b*) Suppose we find the impedance of the whole circuit using the symbolic method for a frequency ω and compute $v(t)$ for that input.

$$\text{Impedance } Z = R_2 + \frac{R_1 (1/j\omega C)}{R_1 + (1/j\omega C)}$$

The voltage $$V = [E/Z] \times R_2$$

$$= \frac{E\, R_2 [R_1 + (1/j\omega C)]}{R_2 [R_1 + 1/j\omega C] + R_1/j\omega C}$$

Now, in this, if we replace $j\omega$ by s we find that, we get relation (8.13) above. So we see this method as simpler. In method (*b*), we are generalising the result which we know as true for a frequency ω to be true also for a complex frequency $s = \sigma + j\omega$. We discuss this further in the next article.

Example 1. *Find the transient currents i_1 and i_2 and voltages v_1, v_2 after closure of switch. The capacitor is initially uncharged.*

Solution. From (8.13), which was derived for a similar network.

$$V(s) = \frac{R_2(1 + R_1 C_1 s)}{R_1 + R_2 + R_1 R_2 Cs} E(s)$$

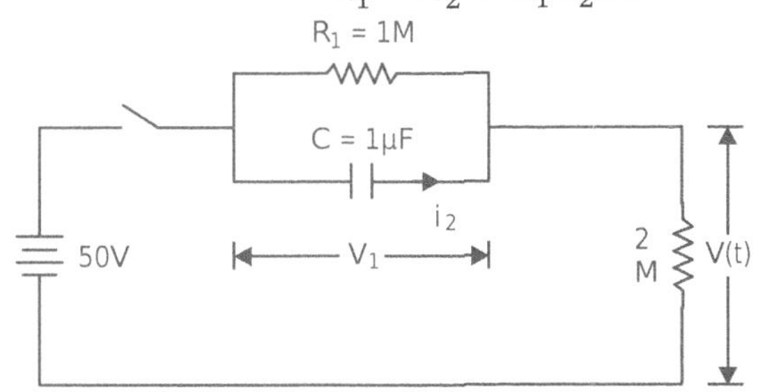

Fig. 8.6

For a step input (*i.e.*, switch suddenly closing at $t = 0$), $E(s) = E/s$.

$$V(s) = \frac{E}{s}\frac{(1 + R_1C_1s)}{(s + 1/\tau)R_1C} \quad \text{where } \tau = \frac{R_1R_2C}{R_1 + R_2} \qquad (a)$$

$$= \frac{A}{s} + \frac{B}{(s + 1/\tau)} \qquad \left\{= \frac{1\times 2\times 1}{1+2} = \frac{2}{3}\right\}$$

To get A, put $s = 0$ in expression (*a*), omitting the denominator term s.

$$A = \frac{E(1 + R_1C_1 \times 0)}{(0 + 1/\tau)\,R_1C} = \frac{E\tau}{R_1C} = \frac{50 \times 2/3}{1 \times 1} = \frac{100}{3}$$

To get B, put $s = -1/\tau$ in expression (*a*), omitting $(s + 1/\tau)$.

$$B = \frac{E[1 + R_1C_1\,(-1/\tau)]}{R_1C(-1/\tau)} = \frac{50[1 - (3/2)]}{-3/2} = \frac{50}{3}$$

Taking the inverse Laplace transform of V(*s*), we get

$$v(t) = A + B\,e^{-t/\tau}$$

$$= 50/3\ (2 + e^{-3t/2})$$

Hence $v(t) + v_1(t) = 50$.

$$v_1(t) = 50 - v(t) = 50/3\ (1 - e^{-3t/2}).$$

Example 2. *In the figure shown switch 1 closes at t = 0 and switch 2 follows 1 millisec later. Find the current through the switch S_2 after 2 milliseconds.*

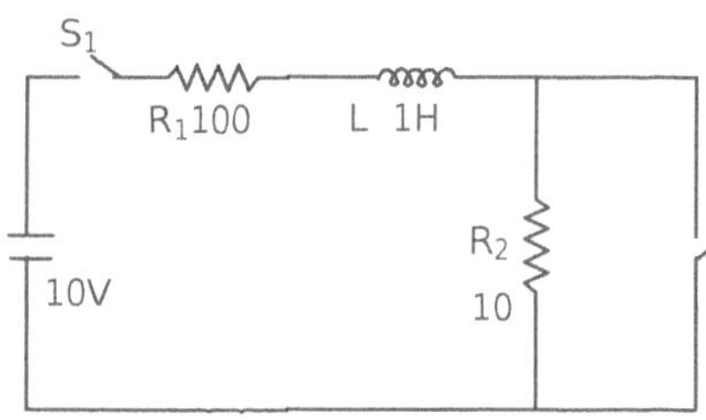

Fig. 8.7

Solution. In this problem there are 2 transient states. When S_1 is closed with S_2 open, it is in the first state for which the current is easily found. Initially current is 0 and final current would be E/(100 + 10). Hence, the current in the first state (S_1 alone closed)

$$= E/110 - [E/110 - 0]e^{-Rt/L}$$

$$= E/110\ [1 - e^{-110t}]$$

Now, when S_2 is closed at $t = 1$ millisecond, the initial current in L is E/110 $(1 - e^{-0.11})$ and final current would be E/100.

The time constant is now $L/R_1 = 1/100$

So, the transient current after 1 millisecond (use t' to denote the time after S_2 closes)

$$i^{(t')} = \text{Final current} - [\text{Final-Initial}]\ \exp(-t'/\tau)$$

$$= E/100 - [E/100 - E/100(1-e^{-0.11})]\ \exp(-100t')$$

Note that $t' = t - 0.001$.

If E = 10, after 2 milliseconds, we get

$$i(t) = 0.1 - [0.1 - .091\ (1 - \exp(-0.11))]\ \exp\ (-0.1)$$
$$= 0.018 \text{ amperes.}$$

Laplace Transform method. Upto t = 1ms, we have the current in the circuit as

$$I_1(s) = \frac{E/s}{110 + 1s}$$

where E/s is the transform of the step input E and Ls + R is the transform of L and R. (L = 1, R = 110)

$$i_1(t) = E/110\ [1 - e^{-110t}]$$
$$t = 1\text{ms},\ i_t(1) = 10/110(1 - e^{-.11}) = 0.091(1 - e^{-.11})$$

When the switch 2 is closed, let i_2 be the new current.

$$L\ s\ I_2(s) - L\ i_1(0) + 100\ I_2(s) = E/s$$

$$I_2(s) = \frac{E}{s(100+s)} + \frac{L\,i_1(0)}{s+100}$$

Putting L = 1 and taking the inverse Laplace transform,

$$i_2(t') = E/100\ [1 - e^{-100t}] + i_1(0)e^{-100t}$$
$$= 0.1(1 - e^{-100t}) + .091(1 - e^{-.11})\ e^{-100t}$$

At t = 2ms, from start, the current can be got by putting t = 1 millisec in the above, as i_2 starts 1ms later. Substituting, t = 1ms in the above, we get the same result as before.

8.7 R–L CIRCUIT RESPONSE TO SINUSOIDAL INPUT

A sine wave voltage V sin ($\omega t + \alpha$) is applied at t = 0.

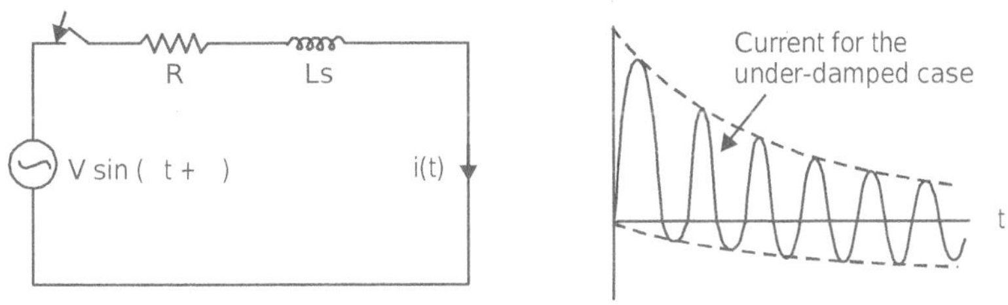

Fig. 8.8

The transformed differential equation is:

$$(Ls + R)\ I(s) = V \times \text{Laplace transform of } \sin\ (\omega t + \alpha)$$
$$= V[\omega \cos \alpha + s \sin \alpha] \div (s^2 + \omega^2)$$

$$I(s) = \frac{V\omega \cos\alpha}{(Ls+R)(s^2+\omega^2)} + \frac{Vs \sin\alpha}{(Ls+R)(s^2+\omega^2)} \quad (8.14)$$

We have to separate the right-hand side function into partial fractions before we can find the inverse transform.

$$\frac{1}{(Ls+R)(s^2+\omega^2)} = \frac{As+B}{s^2+\omega^2}+\frac{C}{Ls+R} \tag{8.15}$$

$$1 = (As + B)(Ls + R) + C(s^2 + \omega^2)$$

Put $s = -R/L$ in the above identity.

$$C = 1 \div (R^2/L^2 + \omega^2)$$

Equating co-efficients of s^2

$$AL + C = 0$$

$$A = -1/(R^2 + L^2\omega^2)$$

Equating co-efficients of s,

$$B = -AR/L = R/(R^2 + L^2\omega^2)$$

Next, write the partial fraction expansion of the second term.

$$\frac{s}{(Ls + R)\,(s^2 + \omega^2)} = \frac{A's + B'}{s^2 + \omega^2}+\frac{C'}{Ls + R} \tag{8.16}$$

$$s = C'\,(s^2 + \omega^2) + (A's + B')\,(L\,s + R)$$

As before, we get, from the above.

$$A' = R/(R^2 + L^2\,\omega^2)$$

$$B' = \omega^2/L\,(R^2/L^2 + \omega^2)$$

$$C' = -RL/(R^2 + L^2\,\omega^2)$$

Next, we can write the inverse Laplace transform of (8.16).

Using (8.16), (8.15) and the table of transforms,

$$i(t) = V\omega\cos\alpha\,[A\cos\omega t + (B/\omega)\sin\omega t + (C/L)\,e^{-Rt/L}] + V\sin\alpha\,[A'\cos\omega t + (B'/\omega)\sin\omega t + (C'/L)\,e^{-Rt/L}]$$

Since the transform of $s/(s^2 + \omega^2)$ is $\cos\omega t$,

of $1/(s^2 + \omega^2)$ is $(1/\omega)\sin\omega t$,

of $1/(sL + R)$ is $e^{-Rt/L}$.

Substituting for A, B, C, A′, B′, C′ and re-arranging,

$$i(t) = \frac{V\omega\cos\omega t}{R^2 + L^2\omega^2}(R\sin\alpha - L\omega\cos\alpha) + \frac{V\sin\omega t}{R^2 + (L\omega)^2}(R\cos\alpha + L\omega\sin\alpha) + \frac{Ve^{-Rt/L}}{R^2 + (L\omega)^2}(L\omega\cos\alpha - R\sin\alpha)$$

Putting $\tan\varphi = L\omega/R$, the angle φ recognised as the power factor angle φ, the above can be written :

$$i(t) = \frac{Ve^{-Rt/L}}{Z}\left(\frac{L\omega}{Z}\cos\alpha - \frac{R}{Z}\sin\alpha)\right)$$

$$+\frac{V}{Z}\cos\omega t\left(\frac{R}{Z}\sin\alpha - \frac{L\omega}{Z}\cos\alpha\right)$$

$$+\frac{V}{Z}\sin\omega t\left(\frac{R}{Z}\cos\alpha + \frac{L\omega}{Z}\sin\alpha\right)$$

$$= V/Z\,[e^{-Rt/L}(\sin\varphi\cos\alpha - \cos\varphi\sin\alpha)$$

$$+\cos\omega t(\cos\varphi\sin\alpha - \sin\varphi\cos\alpha)$$

$$+\sin\omega t(\cos\varphi\cos\alpha + \sin\varphi\sin\alpha)]$$

$$= V/Z\,[\sin(\omega t + \alpha - \varphi) + \sin(\varphi - \alpha)\,e^{-Rt/L}] \quad (8.17)$$

Though the above expression appears to have been so difficult to derive it has a simple explanation. The first term is easily recognised as the steady-state value, which we have studied in A.C. Circuit theory, *viz.*, the voltage V divided by impedance Z, but lagging the voltage sin (ωt + α) by φ. The second term is the transient term. It decays as t increases. The initial current at $t = 0$ is evidently zero (when the switch was just closed) and it is the second term which makes the current equal to zero at $t = 0$. For, at $t = 0$, the first term gives V/Z sin ($\alpha - \varphi$) and the second term is

$$\frac{V}{Z}\sin(\varphi - \alpha)\,e^{0} = -\frac{V}{Z}\sin(\alpha - \varphi),$$

thereby cancelling the first and satisfying the initial requirement. The final value of the negative exponent (second term) is zero, when the transient has decayed and steady-state has set in.

The significance of the angle α. We initially took a term sin (ωt + α) instead of sin ωt, because, for a most general case, we may close the switch at any unknown instant t. Suppose we close the switch at that time when the A.C. voltage has passed an angle φ from its zero in the positive direction, (point P). Then $t = 0$ starts at P. Now the waveform has a value V sin φ at t =0. So the general term V sin (ωt + α) becomes, at $t = 0$, equal to V sin α and in the case of the switch closing at P, it is V sin φ (see Fig.) Hence $\alpha = \varphi$ for this case.

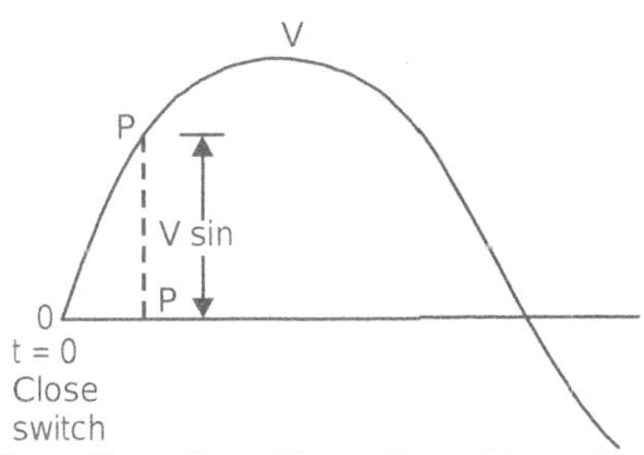

Fig. 8.9

Then the transient current is

$$i(t) = V/Z\sin(\omega t + \varphi - \varphi) + V/Z\sin(\varphi - \varphi)\exp(-Rt/L)$$

$$= V/Z\sin\omega t + 0.$$

Thus, there is *no* transient term If we close the switch at a specific point in the waveform as it passes !

Example 3. *A voltage v = V sin 100 πt is applied to a series circuit of resistance 25 Ω and 1 henry inductance. Calculate the ratio of the maximum value to which the current rises, to the steady-state maximum value, if the voltage is applied at the instant t = 0.002 second.*

Solution. The steady-state value of the current is first found ωL = 1 × 100π.

$$Z \approx \sqrt{25^2 + (1 \times 100\pi)^2} = 102.3$$

$$\tan \varphi = 100\pi/25 = 12.56,\ \varphi = 85°\ 26' \text{ or } 1.491 \text{ radians.}$$

So,
$$i_{ss} = (V/102.3) \sin (100\pi t - 1.491) \quad (a)$$

At $t = 0.002$, when the switch is just closed, the current ought to be zero, but the value given by (*a*) will be

$$V/102.3 \sin (100\pi \times .002 - 1.491) = -(V/102.3) \sin (0.863)$$

which must therefore be counteracted by a transient term. Hence the transient current is

$$(V/102.3) \times 0.76 \exp (-Rt/L) = (V/102.3) \times 0.76 \exp (-25t)$$

Hence the total $i(t)$ is

$$V/Z \sin (100\pi t - 1.491) + (V/Z) \times 0.76\, e^{-25t}$$

The maximum value of the current $i(t)$ is got by differentiating $i(t)$ and equating it to zero.

$$\cos (100\pi t - 1.49) = 0.76/4\pi \exp (-25t)$$

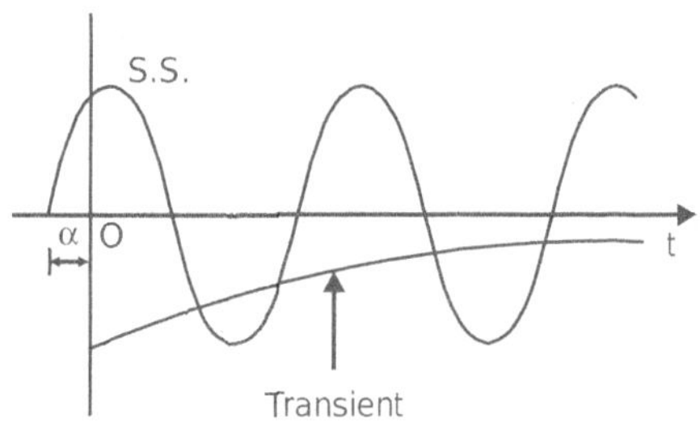

Fig. 8.10

It is difficult to find t from this transcendental equation. It can be approximately determined by noting that the maximum value will occur nearly at the time of the first peak of the steady-state part of the current. Thus at the first peak,

$$\sin(100\pi t - 1.491) = 1,$$

i.e.,
$$100\pi t = 1.491 + (\pi/2) \qquad (\text{as } \sin \pi/2 = 1)$$

from which
$$25t = 0.244$$

$$i \text{ peak} = V/Z\,[1 + 0.76 \exp(-.244)] = 1.585\ V/Z.$$

The ratio $i_p : i_{SSm}$ is thus 1.585 : 1.

8.8 APPLICATION OF THE R.L TRANSIENT PROBLEM

In the above problem the maximum value of the transient current is about 58% more than the steady-state peak. Suppose we closed the switch at such time that $\alpha = \varphi + \pi/2$. Then

$$i(t) = V/Z\{(\sin(\omega t + \varphi + \pi/2 - \varphi) + \sin (\varphi - \varphi - \pi/2) \times \exp (-Rt/L)\}$$
$$= V/Z\ ((\cos \omega t) - e^{-Rt/L})$$

The maximum peak will occur nearly when $\cos \omega t = 1$

or $\omega t = \pi (\omega t = 0 \text{ is trivial})$

and $i \text{ peak} = V/Z[-1 - \exp(-R\pi/\omega L)]$ (8.18)

If a circuit is highly inductive, R is small and $R\pi/\omega L \simeq 0$, giving

$$i_{peak} : i_{sp} = \{1 + \exp(-Rp/wL)\} : 1 = 1 + 1 : 1 \simeq 2 : 1$$

So, in a highly inductive circuit, the maximum transient current is nearly double the steady-state peak; it occurs during the first oscillation, and decays with a time constant R/L.

The increase of current immediately after switching is common to the alternating current generator when it is short circuited (by fault). The alternator winding has a large inductance and low resistance. When it is short circuited, the problem is similar to the R-L transient.

The likely maximum transient current is twice the steady-state fault current.

8.9 SINUSOIDAL INPUT TO AN R-C CIRCUIT

In this case, the steady-state current is known from circuit theory as

$V/Z \sin (\omega t + \alpha + \varphi)$ (8.19*a*)

At $t = 0$, this expression has the value

$V/Z \sin (\alpha + \varphi)$

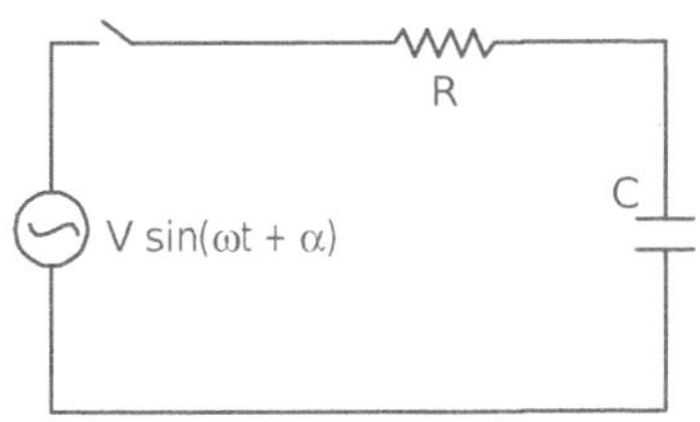

Fig. 8.11

and hence there should be an opposing transient current with a time constant CR, so that at $t = 0$, the current is zero. Hence the transient component is

$$V/Z \sin (\alpha + \varphi)\, e^{-t/CR}$$

Thus $i(t) = V/Z \sin (\omega t + \alpha + \varphi) - V/Z \sin (\alpha + \varphi) e^{-t/CR}$

Hence $Z = \sqrt{R^2 + (1/\omega C)^2}$ and $\varphi = \tan^{-1} (1/\omega CR)$ (8.19*b*)

The mathematical derivation on the lines of the R-L circuit is left to the reader.

8.10 R-L-C CIRCUIT WITH SINUSOIDAL INPUT

At $t = 0$, the sinusoidal voltage is suddenly applied to the R-L-C series circuit. We know that the steady-state current I_{ss} is

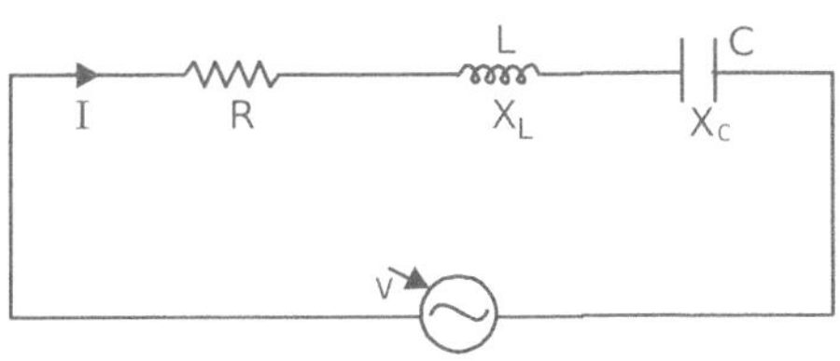

Fig. 8.12

$$I_{ss} = \frac{E}{[R^2 + (L\omega - 1/C\omega)]^{1/2}} \sin(\omega t + \alpha - \varphi) \tag{8.20}$$

where $\tan \varphi = [\omega L - (1/\omega C)] \div R$

As per this expression, at $t = 0$, we have a current

$$i_{ss} = E/Z \sin (0 + \alpha - \varphi)$$

in the circuit which must necessarily be absent at $t = 0$.

So, we apply an opposite current of $I_0 = -i_{ss}(0)$ at $t = 0$ and this will cause the transient part of the current that we want. Also, at $t = 0$, the steady-state expression gives a voltage across C which actually is not present. Therefore we have to apply an opposite voltage V_0 at $t = 0$ and this dies away in the transient period slowly.

$$V_C = -IX_C = -I/\omega C \sin (\omega t + \alpha - \varphi - 90°)$$

$$V_C(0) = I_{max} X_C \cos (\alpha - \varphi)$$

So, at $t = 0$, the current I_0 in L and voltage V_0 in C are present which die away as per the equation

$$L\, di/dt + Ri + 1/C \int i\, dt = 0$$

and the solution of this gives the transient part of the current in the R–L–C circuit with the sinusoidal input. Transforming and inserting initial constraints,

$$Ls\, I(s) - LI_0 + RI(s) + I(s)/Cs + V_0/s = 0$$

$$I(s) = \frac{LI_0 - V_0/s}{Ls + R + (1/Cs)}$$

$$= \frac{sI_0 - V_0/L}{(s + R/2L)^2 + (1/LC - R^2/4L^2)} \tag{8.21}$$

The inverse transform gives the transient part of $i(t)$, using the table of transforms. Three cases arise.

Over damped Case : $1/LC < R^2/4L^2$

$$I(s) = (sI_0 - V_0/L) \div [(s + a)^2 - b^2]$$

$$a = R/2L \text{ and } b = R^2/4L^2 - 1/LC.$$

$$i(t) = I_0 e^{-at} [\cosh bt - (a/b) \sinh bt] - (V_0/Lb) \sinh bt \tag{8.22}$$

For critical damping : $R^2/4L^2 = 1/LC$

$$I(s) = (sI_0 - V_0/L) \div (s + R/2L)^2$$

$$i(t) = I_0 e^{-at} [1 - (V_0/LI_0 + a)t] \tag{8.23}$$

where $a = R/2L$.

For underdamped cases : $R^2/4L^2 < 1/LC$

$$I(s) = \frac{I_0(s + a)}{(s + a)^2 + b^2} - \frac{I_0 a + V_0/L}{(s + a)^2 + b^2}$$

where $a = R/2L$, $b = 1/LC - R^2/4L^2$

$$i(t) = I_0 e^{-at} \cos bt - [(I_0 a + V_0/L)/b] \sin bt\, e^{-at} \tag{8.24}$$

The actual transient current is the steady-state current (8.20) together with one of 8.22 to 8.24. The current waveform for the underdamped case contains transient oscillations of frequency $b/2\pi$ (see cos bt and sin bt terms in 8.22). Usually, for power frequency transients; b is generally more than ω and these oscillations appear to modulate the applied frequency sin ωt.

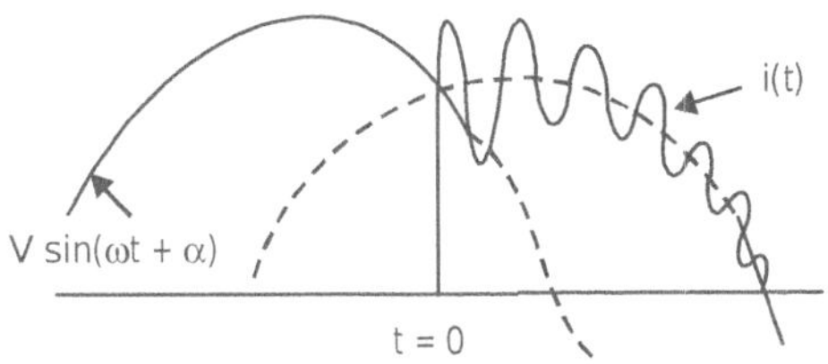

Fig. 8.13

Example 4. *In an R–L–C series circuit having R = 50, L = 0.1 and C = 100μ F, a 100V 50 c/s voltage is suddenly applied at the instant of maximum voltage. Find the current if capacitor is already charged to 50.51 volts.*

Solution. The steady-state current is given by

$$i_{ss} = E/Z \sin(\omega t + \alpha - \varphi)$$

$$\omega L = 0.1 \times 314,\ 1/\omega C = 10^6/314 \times 100 = 31.8$$

$$Z^2 = 50^2 + (31.4 - 31.8)^2 \text{ or } Z = 50\angle\tan^{-1}(0.4/50)$$

$$= 50\angle 0°27'$$

The circuit is capacitive and φ is positive

$$i_{ss} = 100/50 \sin(\omega t + \alpha + 0°27')$$

At $t = 0$, the voltage is maximum; $\omega t + \alpha = \pi/2$ at $t = 0$.

So $\alpha = \pi/2$ So

$$i_{ss} = 2\cos(\omega t + 0°27')$$

$I(0) = -2 \cos 27' \simeq -2$ must be applied at $t = -0$ to satisfy the initial condition.

(as V_C lags I_C) $\quad V_C = IX_C = 31.8 \times 2 \cos(\omega t + \alpha + 27' - 90°)$

$$V_C(0) = 31.8 \times 2 \times \sin 27' = 0.51 \text{ volts}$$

So, in order to satisfy the initial condition of 50.51V upon C, we have to apply a voltage V_0 at $t = 0$ suddenly across C, so that $V_0 + V_{C0} = 50.51$; So, $V_0 = 50$ V.

Now the transient part of the current can be found.

$$R/2L = 50/0.2 = 250 = a$$

$1/LC - R^2/4L^2 = (10^6/0.1 \times 100) - 250^2 = 37500 = b^2$

Hence this is an undamped case, giving oscillatory response.

The solution is got from

$$(Ls + R + 1/Cs)\, I(s) = LI_0 - V_0/s$$

where the I_0, V_0 are the initial condition requirements.

$$I(s) = [sI_0 - V_0/L] \div (s + a)^2 + b^2$$

$$i(t) = I_0\, e^{-at} \cos bt - (1/b)\,(I_0\, a + V_0/L)\, e^{-at} \sin bt$$

$$I_0 = -2,\ (I_0 a + V_0/L)/b = (-250 \times 2 + 50/0.1)/b = 0$$

Hence the transient part of the current is

$$I_0\, e^{-at} \cos bt \;=\; 2e^{-250t} \cos 193.7t$$

and this together with i_{ss} = 2cos (ωt + 0°27′) is the total current.

Example 5. *The circuit shown has no stored energy in it. Find the Laplace transform of the current supplied by the battery upon closure of switch at t = 0. Hence find the initial and final values of this current. The battery is a 10V battery.*

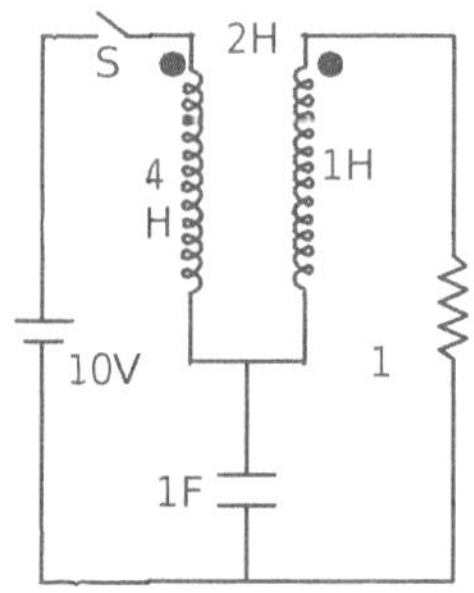

Fig. 8.14

Solution. Assume loop currents i_1, i_2 in the two loops. Current i_2 leaves the dot and so the induced emf M di_2/dt is negative at the dot of the primary. Writing loop equations using operational impedances,

$$10/s \;=\; 4s\, I_1 + (1/s)\,(I_1 - I_2) - (2s)I_2$$

Here 10/s is the transform of the 10 volt step input; –2s I_2 represents the mutual term. This equation can be rewritten as

$$I_1[4s + (1/s)] - I_2[2s + (1/s)] \;=\; 10/s$$

The second loop has the equation

$$I_2[1 + s + (1/s)] - I_1[2s + (1/s)] \;=\; 0$$

Substituting for I_2 in the first equation,

$$I_1(s) \;=\; \frac{10(1+s+s^2)}{(4s^2+1)(s^2+s+1)-(2s^2+1)^2}$$

$$= \frac{10(1+s+s^2)}{s(4s^2+s+1)}$$

To find initial and final values, evaluate the limits of $sI_1(s)$ as $s \to \infty$ and $s \to 0$ respectively.

$$\text{Initial value} \;:\; \frac{10(1+s+s^2)}{1+s+4s^2}\;(\text{as } s\to\infty) = \frac{10}{4}$$

$$\text{Final value} \;:\; \frac{10(1+s+s^2)}{1+s+4s^2}\;(\text{as } s\to 0) = 10.$$

(Inductors are shorts capacitor is open).

The above five examples are rather tough but standard problem.

8.11 STEADY STATE ERROR CALCULATION IS CONTROL SYSTEMS

Control system theory uses Laplace transforms only for all its development. A typical control system is given by a function relating the output of the system to its reference input as

$$G(s) = \frac{k(1 + T_1 s)}{(1 + T_2 s)(1 + T_3 s)}$$

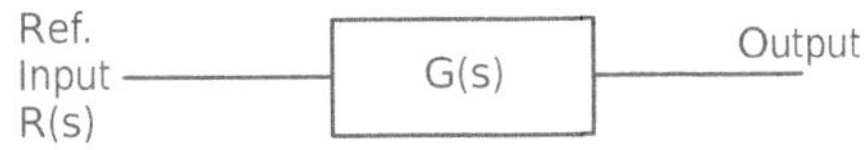

(Open-loop control)

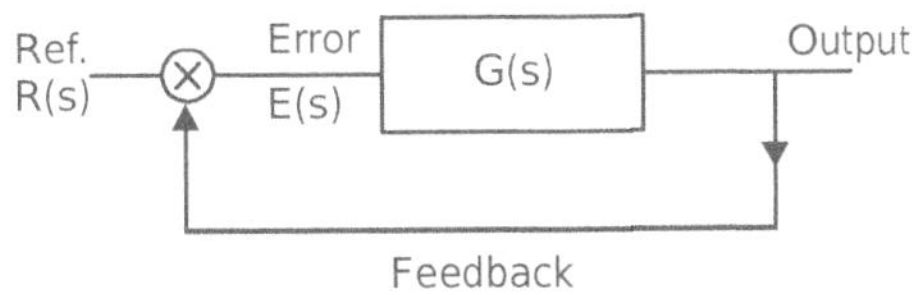

(Feedback control)

Fig. 8.15

The above is the transfer function of the open loop system. Here T_1, T_2, T_3 are system time constants.

Suppose we give a step or ramp input. For *e.g.,* in a temperature controller, we may give a step input from 40°C to 50°C to heat up. This can be also uniformly increased from 40°C at 1°C/sec, as a ramp input.

Now for such input, what will be the error in the output (*i.e.,* with reference to input) at the final time ? (at $t \to \infty$?).

This can be found from the use of the Final Value Theorm (8.24). The error function is

$$E(s) = R(s)\left[1 - \frac{G(s)}{1 + G(s)}\right] = \frac{R(s)}{1 + G(s)}$$

For a step input $\qquad R(s) = 1/s.$

The error function

$$E(s) = \frac{1}{s\left[1 + \dfrac{k(1 + T_1 s)}{(1 + T_2 s)(1 + T_3 s)}\right]}$$

The steady state error is

$$\underset{t\to\infty}{\text{Lt}}\; e(t) = \underset{s\to 0}{\text{Lt}}\; sE(s) = \frac{1}{1 + k}.$$

Thus the error is found for step input as $1/(1 + k)$. (Higher the k, less to error).

Additional Problems—Transients Using *s*-domain

Example 1. *In the figure below, find the current in L after closing the switch, given* $V = 200V$, $v_C(0) = 100$ V, $R_1 = 30$, $R_2 = 10$ *and* $L = 0.1$ H, $C = 1000\ \mu F$.

Solution. The loop equations are written as

$$40I_1 + 0.1sI_1 - 0.1i_1(0) - 10I_2 = \frac{200}{s}$$

$$-10I_1 + 10I_2 + \frac{1000I_2}{s} - \frac{V_C(0)}{s} = 0$$

I_1 is initially $\frac{200V}{30 + 10} = 5A.$ So 0.1 I(0) = 0.5

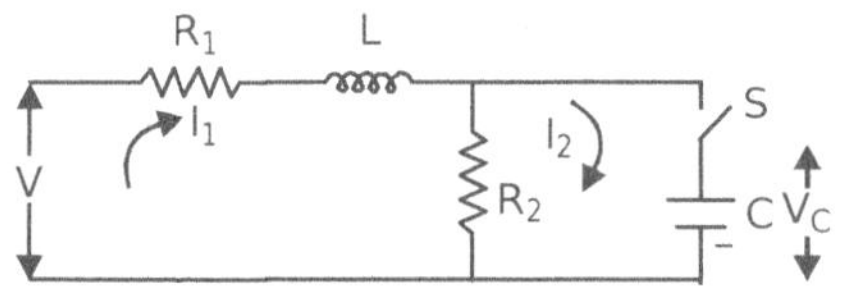

Fig. 8.16

Note:

$$Ls = 0.1s$$

$$\frac{1}{Cs} = \frac{10^6}{1000s} = \frac{1000}{s}.$$

$$I_1 = \frac{1}{\Delta}\begin{vmatrix} 200/s + 0.5 & -10 \\ 100/s & 10 + \frac{1000}{s} \end{vmatrix}$$

where

$$\Delta = \begin{vmatrix} 40 + 0.1s & -10 \\ -10 & 10 + 1000/s \end{vmatrix}$$

Hence

$$I_1 = 5\left(\frac{1}{s} + \frac{300}{s^2 + 400s + 40000}\right)$$

The roots of $(s^2 + 400s + 40000)$ are $p_1 = p_2 = P = -200$, a multiple root on writing the second term as

$$\frac{5 \times 300}{s^2 + 400s + 40000} = \frac{1500}{(s + 200)^2}$$

We find its inverse transform as

$$1500t\, e^{-200t}$$

Hence the

$$i_1(t) = 5 + 1500t\, e^{-200t}$$

EXERCISES A (MATHEMATICS)

1. Find the Laplace transforms of the following functions :

1. $t\, e^{-5t}$. **[Ans.** $(1/(s + 5)^2]$

2. $t^2\, e^{3t}$. **[Ans.** $2/(s - 3)^2]$

3. $t \cos 2t$. [**Ans.** $(s^2 - 4)/(s^2 + 4)^2$]

4. $t \cos^2 t$. [**Ans.** $1/2s^2 + 2s/\ (s^2 + 4)^2$]

5. $t^2 \sin 2t$. [**Ans.** $12s^2 - 16/(s^2 + 4)^3$]

6. $t^2 e^{-3t}$. [**Ans.** $2/(s + 3)^2$]

7. $t \sinh at$. [**Ans.** $2as/(s^2 - a^2)^2$]

8. $t \cosh at$. [**Ans.** $s^2 + a^2/(s^2 - a^2)^2$]

9. $\sin at - at \cos at$. [**Ans.** $2a^3/(s^2 + a^2)^2$]

10. $t\, e^{-t} \cos t$. [**Ans.** $s^2 + 2s/(s^2 + 2s + 2)^2$]

2. Find the inverse transforms :

1. $\dfrac{1}{s^2\,(s^2 + a^2)}$.

2. $\dfrac{s^2}{(s^2 + a^2)^2}$. $\left[\textbf{Ans.}\ \dfrac{1}{2a}\sin at + \dfrac{t}{2}\cos at\right]$

3. $\dfrac{s}{(s^2 + 4)^2}$. [**Ans.** $t/4 \sin 2t$]

4. $\dfrac{1}{(s^2 + 9)^2}$.

5. $\dfrac{3a^2}{s^3 + a^3}$.

6. $\dfrac{s^2 - a^2}{(s^2 + a^2)^2}$.

7. $\dfrac{1}{s^2(s^2 + 1)(s^2 + 9)}$.

8. $\dfrac{1}{s^4 - a^4}$.

9. $\dfrac{6s + 3}{(s - 1)\,(s^2 + 2s + 5)}$.

10. $\dfrac{s + 1}{s^2 + 2s}$. [**Ans.** $(1/2 + 1/2e^{-2t})$]

EXERCISES B (CIRCUITS)

1. A capacitor C_1 is charged to V_1 volts and is switched across a parallel combination of R and C_2 at $t = 0$. C_2 is charged to V_2 volts intitially ($V_2 < V_1$). Calculate the current through R. Using the initial and final value theorems, calculate the initial value of current and its rate of change and the final value of voltage across C_1 (or C_2).

$$\left\{\frac{V_1C_1 + V_2C_2}{R(C_1 + C_2)} \exp\,[-t/(RC_1 + RC_2)];\ \frac{V_1C_1 + V_2C_2}{R(C_1 + C_2)},\ \infty,\ 0.\right\}$$

2. Find the currents through and the voltages across the resistance, inductance and capacitance in Fig. 1 below, one second after closing the switch.

[i_R = 12.6 = i_C, i_L = 0, V_R = 6.3, V_L = V_C = 3.7V.

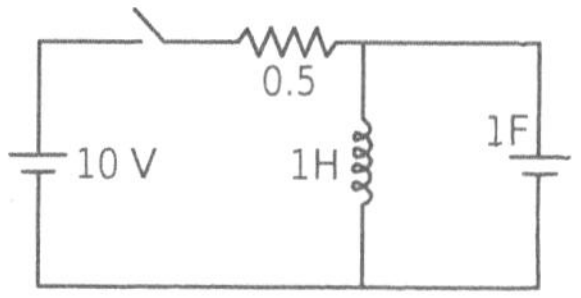

Fig. 1

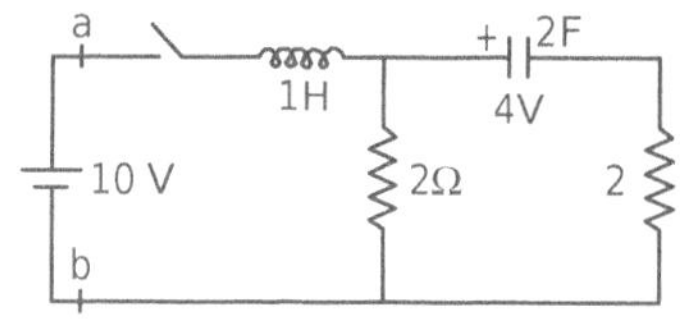

Fig. 2

3. Find the transient current through the capacitor after switch closure if capacitor is initially charged to 4V (in Fig. 2 above).

[$i_1 = 5 + 13.1\, e^{-1.125t} \sin(1.572t + 0.702^C)$, $i_2 = 12.72\, e^{-1.125t} \cos 1.572t$.]

4. A. voltage of 10 Sin $10^4 t$ is switched at $t = 0$ into a series RC circuit comprising R = 10 kiloohms and C = 0.01 μF. Determine an expression for the current in terms of the initial capacitor voltage.

Hence deduce the value of the initial capacitor voltage for which the circuit attains the steady state immediately after switching, *i.e.*, without any transient. [–5V]

Hint: By superposition, the transient current $V_0/R \exp(-t/CR)$ due to initial charge will be added to the a.c. transient current given in (8.19). So

$$i = 0.707 \sin(10^4 t + 45°) - (½ \quad + V_0/10) \exp(-t/CR)$$

5. A voltage $v = V \sin 100\pi t$ is applied to a circuit having 25 Ω and 1.0 H and the switch is closed at t = 0.01 sec. What is the ratio between the peak transient current and the steady state a.c. current? [1.78 : 1]

6. A 100V (max) 50c/s source is connected to $Z_1 = 3 + j4$ and $Z_2 = 2 + j8$ in series and the current in them have settled. Suddenly at the instant when the current is maximum and is just decreasing, the Z_2 is short circuited by a switch. Find the current through Z_1 thereafter.

[20 cos ($314t$ + 0.16) – 12.05 exp ($-235.5t$).]

Hint:

Initial current	=	$10\,(100/13) \cos \omega t$.
Voltage	=	$100 \cos(\omega t + \alpha)$ [$\alpha = \tan^{-1}(12/5) = 67.4°$
Final current	=	$20 \cos(\omega t + \alpha - \varphi)$
$\tan \varphi$	=	$4/3$, $\varphi = 58{,}2°$

Current at (t = 0) is 100/13.

∴ Transient part of current is

$-[20 \cos(\omega \times 0 + \alpha - \varphi) - 100/13] \exp(-Rt/L)$

7. A parallel circuit having an R-L branch in parallel to a capacitor C is supplied from an A.C. source of 100V max , of ω = 1. The values are R = 3, L = 4, C = 1/5. The circuit (Fig. 3) is opened at the instant of current zero. Find the transient voltage across the capacitor and its rate of rise at t = 0. Also find the rate of change of current at t = 0 in the inductor.

[14 exp ($-3t/16$) sin ($1.053t$ – 1.43), 69.3 V/sec, ∞]

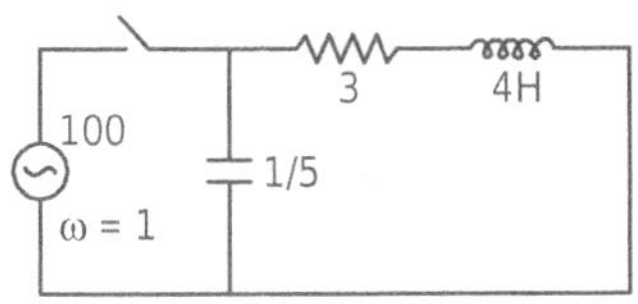

Fig. 3

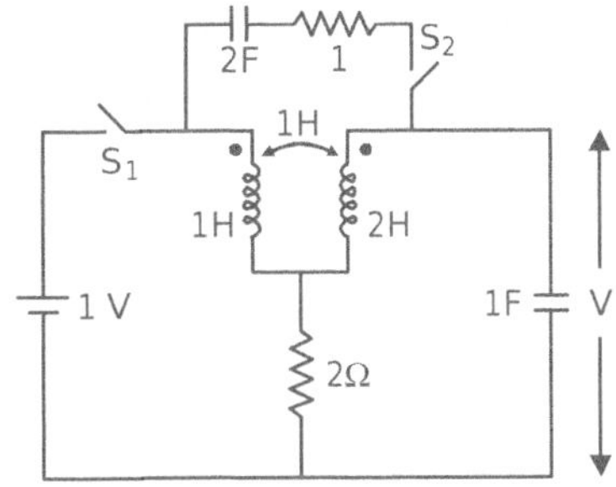

Fig. 4

8. S_1 is initially kept closed in Fig. 4. S_2 is suddenly closed and S_1 opened. Find the voltage across the 1F capacitor. [$1 + 0.189\, e^{-t/4} \sin 0.66t - ½\; e^{-t}$]

SIMPLE EXERCISES

1. Relate the Fourier transform with the Laplace transform, explaining the significance of the variable '*s*'. [See 8.1, 8.1.1, 8.1.2]. Find the transform for a step and a Ramp signal.

2. Write down the time function for the inverse of 1, $1/s$, $1/s^2$, $1/s^3$, $1/s^4$.

$$\left[\delta(t),\ u(t),\ t,\ \frac{t^2}{2!},\ \frac{t^3}{3!}\right]$$

3. Explain the Laplace transform's Initial value theorem. Give an example for use in an inductance circuit. [8.2.1]

4. In what manner does the Laplace transform consider into account the initial charge on a capacitor ? [$V_C = I/Cs + V(0)/s$]

5. If you are given the transfer function of a system and if a step input is given to it, how will you find the steady state final value of a current or voltage output of the system ?

$$\left[\text{Let } \frac{A(s)}{B(s)} \text{ be the transfer function of the system with a step input, it is } \frac{1}{s}\text{; so output is } \frac{A(s)}{sB(s)}\right]$$

Evaluate $\underset{s\to 0}{\text{Lt}}\, sF(s) = \underset{s\to 0}{\text{Lt}}\, \dfrac{sA(s)}{sB(s)}$ to get the final value.

6. Evaluate the time function for V_2 obtained when a step voltage is applied to an RC circuit shown.

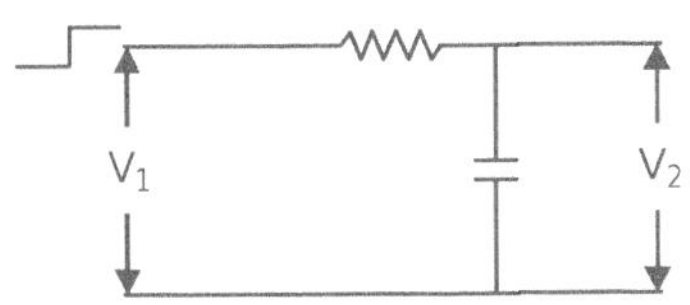

$$\left[\textbf{Ans. } \frac{V_1}{s}\,\frac{1/Cs}{R + 1/Cs} = V_2(s).\ v_2(t) = V_1(1 - e^{-t/CR})\right]$$

7. Explain the method of partial fraction expansion for the case of multiple roots. [See Eq. 13, on page 208]

8. Explain scaling theorem for Laplace transform. Use it to evaluate the inverse of

$$\frac{s}{a^2 s^2 + f^2}$$ [See Theorem (V) and *e.g.,* 1]

9. If the inverse of X(s) is $x(t)$, what are the inverse of X(s)/s and sX(s) ? [$\int x\, dt$ and dx/dt.]

10. What are the advantages of solving differential equations of systems using the Laplace Transform?

[1. Initial condition are directly included. 2. Looking up tables of transforms makes it easy. 3. Differentials become algebraic.]

11. Write down the impedance of a parallel circuit containing R_1 and L in one branch and R_2 and C in another.

$$\left[\frac{(R_1 + sL)(R_2 + 1/sC)}{R_1 + R_2 + s(L) + 1/sC}\right]$$

12. Write down the loop equations of the standard R.C phase shift network and find the condition for 180° phase shift and the corresponding gain and frequency. [See 8.5.2]

13. Write down s-domain loop equations for the R-C Notch filter and find the frequency of Notch. [See 8.5.3]

14. What are the advantages of Laplace Transform in finding the transient currents in any circuit ? [See 8.6]

15. Solve the transient current in an R-L series circuit subject to a sine wave input V sin ($\omega t + \alpha$).

$$\left[V/\sqrt{R^2 + \omega^2 L^2}\left\{\sin(\omega t + \alpha - \phi) + \sin(\phi - \alpha)\, e^{-Rt/L}\right\}\right]$$

16. Explain the phenomenon of doubling effect with reference to the closing of a switch in an R-L transient with sine input. [See 8.8]

17. In an R-L-C series circuit if an excitation of V sin ($\omega t + \alpha$) is applied at $t = 0$, find the current at any t. [Eqn. 8.22]

18. Explain the basic principle is solving a Mutual Inductive circuit transient problem using Laplace Transforms.

[In a Mutual Inductor, Voltage 1 = M di_2/dt and V_2 = M di_1/dt. Taking L-Transforms, $V_1(s) = MsI_2(s)$ and $V_2(s) = MsI_1(s)$. In this way, the KCL and KVL equations are written and finally solved by taking inverse transform. See example 3, See 8.10]

19. Find the steady state error for a type 1 system transfer function

$$G(s) = \frac{k(1 + 2s)}{s(1 + 3s)(1 + 4s)}$$

in a control loop, for a velocity input $\theta(t) = 2t$.

$$\left[\text{S.S. error} = \underset{s\to 0}{\text{Lt}}\; sE(s), \text{ where } E(s) = \frac{R(s)}{1 + G(s)}, \text{ with } R(s) = 2/s^2\right]$$

[**Ans.** $2/(1 + k)$]

20. Find the current $i(t)$ for figure when switch closes at t = 0.

[Anna University, May 2007]

[$5 - 3e^{-t/8}$]

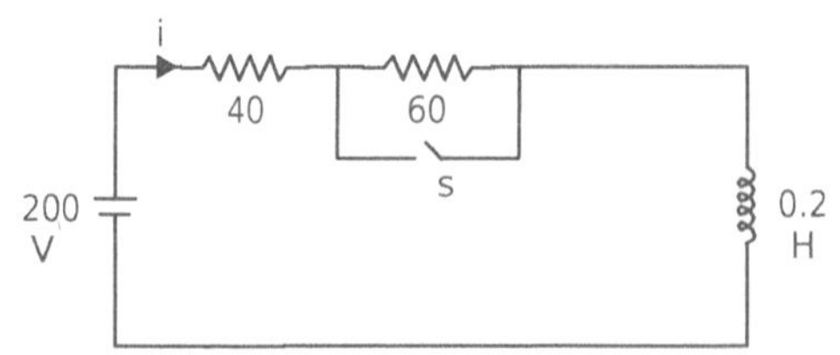

9

FREQUENCY DOMAIN AND TIME DOMAIN

9.1 THE NATURAL FREQUENCIES OF NETWORKS

Suppose we have an R-L-C circuit to which we apply a step voltage and observe the transient current that passes. Let R be a variable resistance. Let us vary R from zero upto a high value. The current will first be oscillatory for low values of R, of the form $e^{-at} \sin (bt + \theta)$. Then at a critical R value, the response becomes non-oscillatory when $a = b$. Further increase of R makes the transient current exponential in nature.

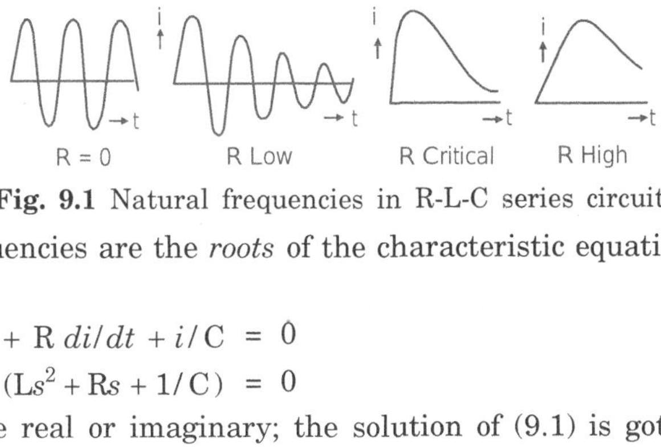

Fig. 9.1 Natural frequencies in R-L-C series circuit.

The natural frequencies are the *roots* of the characteristic equation of the R-L-C circuit differential equation.

$$L\, d^2i/dt^2 + R\, di/dt + i/C = 0 \tag{9.1}$$

viz.,

$$(Ls^2 + Rs + 1/C) = 0 \tag{9.2}$$

The roots may be real or imaginary; the solution of (9.1) is got as

$$A \exp (s_1 t) + B \exp (s_2 t) \tag{9.3}$$

by Differential equation theory, s_1 and s_2 may be real or imaginary. They are natural frequencies, which are, in general, complex numbers. Thus the roots s_1, s_2 may be written is general as

$$s = \sigma + j\omega \tag{9.4}$$

where σ is the real part and ω the imaginary part. In this case, the values of σ and ω are got as:

$$LCs^2 + RCs + 1 = 0$$

roots are $\dfrac{-RC \pm \sqrt{(RC^2) - 4LC}}{2LC}, = \dfrac{-R}{2L} \pm j\sqrt{\dfrac{1}{LC} - \dfrac{R^2}{4L^2}}$

so that $\sigma = -R/2L$; $\omega_0 = [1/LC - R^2/4L^2]^{1/2}$.

If the response is oscillatory, ω will be the frequency of oscillation. If the response is exponential, the imaginary part will be zero. This too can be considered as an oscillation, if we generalise the frequency variable to s, a complex quantity.

9.2 NETWORK FUNCTIONS AND LAPLACE TRANSFORMS

The response of a network to a complex frequency can be found in the same manner as we did in chapter 2 for real frequencies, *viz.*, voltages of the form $e_i = \text{V}_{e}{}^{j\omega t}$.

$$\text{L}\, di/dt + \text{R}i + 1/\text{C}\int i\, dt = e_i(t) = \text{E}e^{st} \tag{9.5}$$

is the R-L-C circuit differential equation, and the input is $e_i(t)=\text{E}e^{st}$. The particular integral, as we know, gives the steady state current. Let us assume that the steady state current is $\text{I}e^{st}$. Substituting in (9.5),

$$[\text{L}\, s + \text{R} + 1/\text{C}s]\,\text{I}e^{st} = \text{E}e^{st}$$

giving

$$\text{E/I} = \text{L}s + \text{R} + (1/\text{C}s) \tag{9.6}$$

In chapter 2.14, when we considered the input of $e^{j\omega t}$, we got the impedance Z as

$$\text{Z} = \text{L}j\omega + \text{R} + (1/\text{C}j\omega) \tag{9.7}$$

Eq. (9.6) is a general impedance function for a complex excitation exp (st) while (9.7) is the same obtained for the case of $s = j\omega$.

If we take the Laplace transform of (9.5), we get,

$$\text{L}s\text{I}(s) + \text{RI}(s) + (1/\text{C}s)\,\text{I}(s) = \text{E}_t(s) \tag{9.8}$$

where it is assumed that initial values are zero and $\text{E}_t(s)$ is the transform of $e_i\,(t)$. From (9.8) we get

$$\text{E}(s)/\text{I}(s) = \text{L}s + \text{R} + (1/\text{C}s)$$

which is the same as (9.6). Thus we see that the Laplace transform when s is substituted for $j\omega$ gives the already familiar relation (9.7) for a frequency ω. Functions such the input impedance, transfer function etc., of networks expressed in terms of s are called network functions in terms of the complex frequency s. Such functions are said to be the frequency domain representations of the circuits. The corresponding time functions can be obtained by the aid of the Inverse Laplace transform.

Example 1. *Find the current in the network after the switch is closed.*

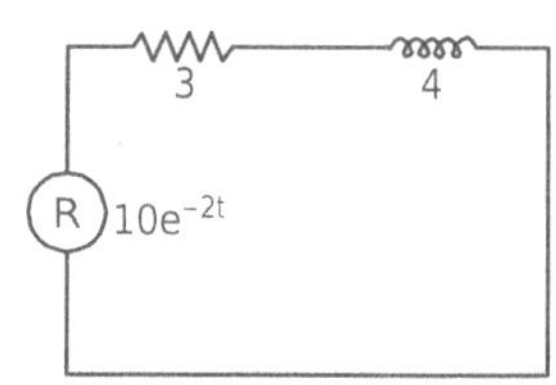

Fig. 9.2

Solution. The circuit differential equation

$$4\, di/dt + 3i = 10e^{-2t}$$

can be transformed to give

$$4s\text{I}(s) - 4i(0) + 3\text{I}(s) = 10/(s+2)$$

$$\text{If } i(0) = 0,$$

$$\text{I}(s) = \frac{10}{(s+2)(4s+3)} = \frac{-2}{s+2} + \frac{8}{4s+3}$$

The inverse transform may be found as

$$i(t) = -2e^{-2t} + 2e^{-3/4t}.$$

Example 2. *The Laplace transform I(s) of a certain time-varying current i(t) is given by*

$$I(s) = \frac{3s^2 + 1}{(s^2 + 4)(2s + 3)}$$

Find the value of the current at t = 0. Find the voltage and charge acquired by a 2 farad capacitor at the end of a long period of time, if this i(t) passes through it. Express the current as a function of time.

Solution. By the theorem of initial value (9.6),

$$\lim_{s\to\infty} s\,\mathrm{I}(s) = i(0)$$

$$= \lim[s(3s^2+1)/(s^2+4)(2s+3)]$$

$$= 3/2 \text{ as } s \to \infty$$

By the final value theorem,

$$\lim_{s\to 0} s\,\mathrm{F}(s) = f(t) \text{ for } t \to \infty$$

The capacitor voltage will be $\mathrm{I}(s)/s\mathrm{C} = \mathrm{I}(s)/2s = \mathrm{F}(s)$

Hence we get lim $s\mathrm{F}(s) \to \frac{1}{2}$ (1/12) = 1/24 volts. Charge = CV = 2 × 1/24 = 1/12 Coulombs.

The current can be expressed as a function of time by taking the inverse transform. Expand in partial fractions :

$$\frac{3s^2+1}{(s^2+4)(2s+3)} = \frac{\mathrm{A}s+\mathrm{B}}{s^2+4} + \frac{\mathrm{C}}{2s+3} \tag{9.9}$$

$$3s^2+1 = (\mathrm{A}s+\mathrm{B})(2s+3) + \mathrm{C}(s^2+4)$$

$$\mathrm{A} = -23/13,\ \mathrm{B} = -69/26,\ \mathrm{C} = 31/13.$$

Then $$i(t) = \mathrm{A}\cos 2t + \tfrac{1}{2}\,\mathrm{B}\sin 2t + \tfrac{1}{2}\,\mathrm{C}e^{-3/2t}. \tag{9.10}$$

Example 3. *A certain impedance has the following admittance function :*

$$Y(s) = \frac{4s + 8}{3s^2 + 12s + 8}$$

Find the steady state current for a sinusoidal input of 10 sin 2t.

Solution. Since I = VY, we can find the current. The admittance for the given frequency ω = 2 can be got by substituting $s = j\omega$ in the above.

$$\mathrm{Y}(j\omega) = (4j\omega + 8) \div [(3j\omega)^2 + 12(j\omega) + 8]$$

$$= (8j + 8) \div [-36 - 24j + 8] = (j+1)/(j-1) = 1\angle -90°.$$

Hence current is got from A.C. theory as

$$\mathrm{I} = 10 \times 1\angle -90°$$

or in time function as $$i(t) = 10\sin(2t - \pi/2).$$

Example 4. *A passive one-port (two terminal) network has the following impedance Z(s).*

$$Z(s) = \frac{2s^2 + 2\times 10^5 s + 10^4}{8s^2 + 0.1s + 10^4}$$

Find the frequency at which this one-port behaves as a pure reactance. Find the value of this reactance.

Solution. We have seen that by putting $s = j\omega$, we obtain the symbolic form of the impedance at frequency ω.

$$Z(j\omega) = \frac{2(j\omega)^2 + 2.10^5 j\omega + 10^4}{8(j\omega)^2 + 0.1 j\omega + 10^4} = \frac{(-2\omega^2 + 10^4) + 2.10^5 j\omega}{(-8\omega^2 + 10^4) + 0.1 j\omega}$$

$$= \frac{[(10^4 - 2\omega^2) + 2.10^5 j\omega]\,[(10^4 - 8\omega^2) - 0.1 j\omega]}{(10^4 - 8\omega^2)^2 + (0.1\omega)^2}$$

The frequency at which it is a pure reactance implies that the real part of $Z(j\omega) = 0$.

$$\text{Re}\,[Z(j\omega)] = 0$$

$$\therefore\ (10^4 - 2\omega^2)(10^4 - 8\omega^2) + 0.2 \times 10^5 \omega^2 = 0$$

$$16\omega^4 - 8\omega^2.10^4 + 10^8 = 0$$

giving ω = 500 rad/sec upon solving the above equation.

9.2.1 Transfer Functions : Additional Problems

Example 1. *Find the transfer functions of the following systems :*

(a) $$\dot{y} + 4y = 3x$$

(b) $$\ddot{y} + 4\dot{y} + 20y = 2\dot{x} - x$$

(c) $$\dddot{y} - 3\ddot{y} + 4\dot{y} + 8y = 4\ddot{x} - 2\dot{x} + x$$

Solution. (a) $$\dot{y} + 4y = 3x$$

$$(s + 4)\,Y(s) = 3X(s)$$

$$H(s) = \frac{3}{s+4}$$

(b) $$\ddot{y} + 4\dot{y} + 20y = 2\dot{x} - x$$

$$(s^2 + 4s + 20)\,Y(s) = (2s - 1)\,X(s)$$

$$H(s) = \frac{2s-1}{s^2 + 4s + 20}$$

(c) $$\dddot{y} - 3\ddot{y} + 4\dot{y} + 8y = 4\ddot{x} - 2\dot{x} + x$$

$$(s^3 - 3s^2 + 4s + 8)\,Y(s) = (4s^2 - 2s + 1)\,X(s)$$

$$H(s) = \frac{4s^2 - 2s + 1}{s^3 - 3s^2 + 4s + 8}$$

Example 2. *Find the transfer function of*

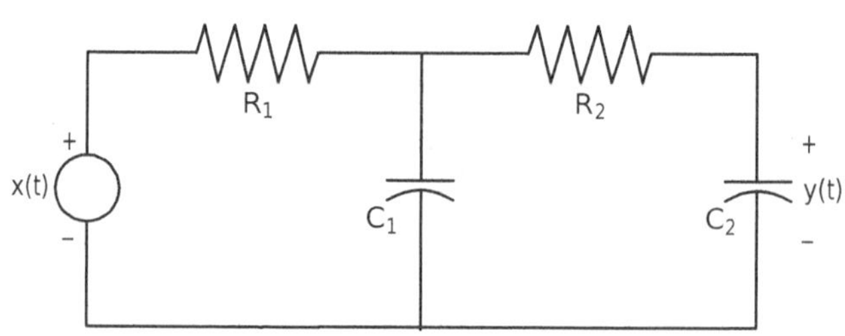

Give the result for $C_1 = C_2 = 100\mu f$, $R_1 = R_2 = 2000\ \Omega$.

Solution.

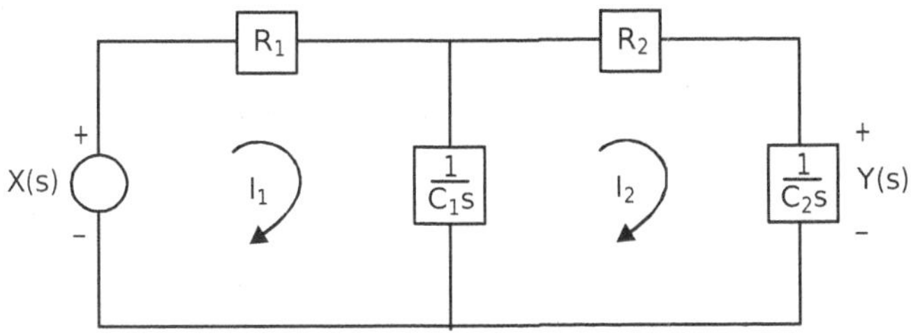

(1) $$X(s) = I_1R_1 + (I_1 - I_2)\frac{1}{C_1s}$$ using Mesh and analysis

(2) $$0 = R_2I_2 + \frac{1}{C_2s}I_2 + (I_2 - I_1)\frac{1}{C_1s}$$

$$0 = \left(R_2 + \frac{1}{C_2s} + \frac{1}{C_1s}\right)I_2 - I_1\frac{1}{C_1s}$$

$\Rightarrow$ $$I_1 = C_1s\left(R_2 + \frac{1}{C_2s} + \frac{1}{C_1s}\right)I_2$$

into (1) : $$X(s) = \left[\left(R_1 + \frac{1}{C_1s}\right)C_1s\left(R_2 + \frac{1}{C_2s} + \frac{1}{C_1s}\right) - \frac{1}{C_1s}\right]I_2$$

$\Rightarrow$ $$Y(s) = \frac{1}{C_2s}I_2 = \frac{1}{C_2s}\frac{X(s)}{(R_1C_1s + 1)\left(\frac{R_2C_1C_2s + C_1 + C_2}{C_1C_2s}\right) - \frac{1}{C_1s}}$$

$$H(s) = \frac{1}{(R_1C_1s + 1)\left(\frac{R_2C_1C_2s + C_1 + C_2}{C_1}\right) - \frac{C_2}{C_1}}$$

$$= \frac{C_1}{(R_1C_1s + 1)(R_2C_1C_2s + C_1 + C_2) - (C_2)}$$

If $C_1 = C_2 = 100\mu f$ and $R_1 = R_2 = 2000\Omega$

$$H(s) = \frac{0.0001}{(.2s + 1)(.00002s + .0002) - .0001}$$

$$H(s) = \frac{25}{s^2 + 15s + 50}$$

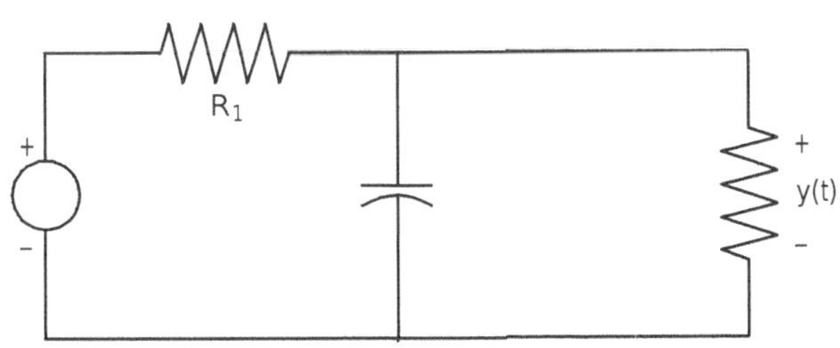

Example 3. *Find the transfer function of the following circuit where $R_1 = R_2 = 1000\ \Omega$ and $C = 100\mu f$.*

(*a*)

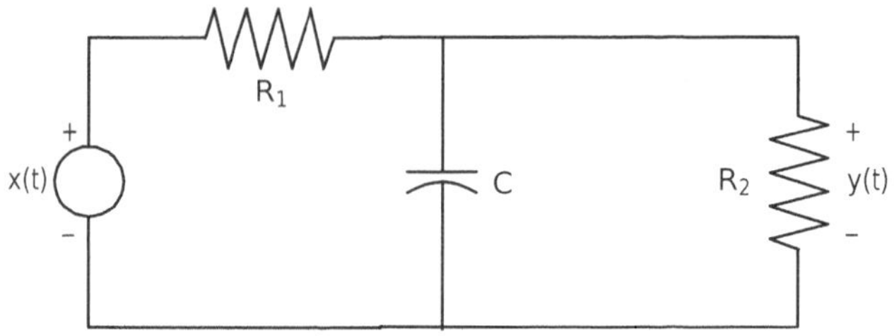

Solution.

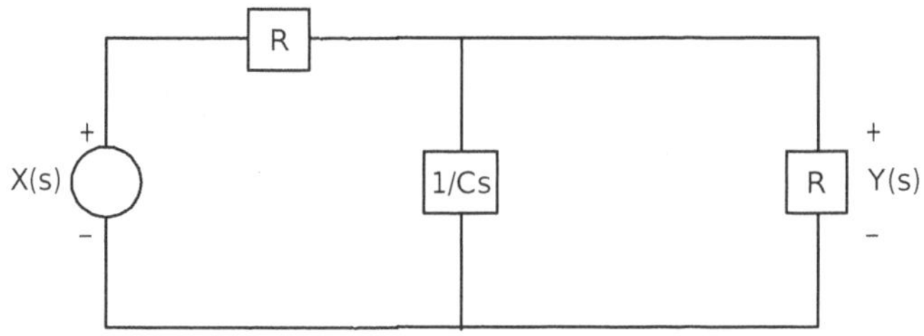

$$R \| \frac{1}{Cs} = \frac{1}{1/R + Cs} = \frac{R}{1 + RCs}$$

$$1/(s) = X(s)\frac{\dfrac{R}{1+RCs}}{R + \dfrac{R}{1+RCs}} \text{ by voltage divider law}$$

$$H(s) \approx \frac{R}{R + R^2Cs + R}$$

$$= \frac{1000}{2000 + 100s} = \frac{10}{s + 20}$$

9.3 POLES AND ZEROS

The function I(s), say of (9.9) determines the nature of the response $i(t)$. The nature of the response depended upon the terms $(s^2 + 2)$, $(3s + 2)$ in the denominator of (9.9); the roots of the denominator are called the poles. The name pole is got from complex variable theory. At a pole, the function becomes infinite, because when the denominator becomes zero (when $s = a$ root of the denominator polynomial), the function is infinite. In general, a function I(s) can be of the form:

$$I(s) = \frac{K(s - s_1)(s - s_2)(s - s_3)...}{(s - s_4)(s - s_5)(s - s_6)} \tag{9.11}$$

The poles, s_4, s_5, s_6 determine the nature of the response–*i.e.*, they give rise to terms

$$\exp(s_4t),\ \exp(s_5t),\ \exp\ (s_6t) \text{ etc.,}$$

in the time function $i(t)$. The values of s_1, $s_2...s_6$ determine the magnitudes of these exponential terms. The roots of the numerator *i.e.*, s_1, s_2, s_3 are called zeros.

s_1, s_2, s_3 ... are called Zeros of I(s)

s_4, s_5, s_6 ... are called Poles of I(s)

The complete response $i(t)$ is known if all the poles and zeros are known. These poles and zeros can be plotted in a complex-plane plot. The poles are plotted with a cross × and zeros by a thick dot. For example, (9.9) has the poles at $s = \pm j2$ and $s = -3/2$ and zeros at $s = \pm j/\sqrt{3}$.

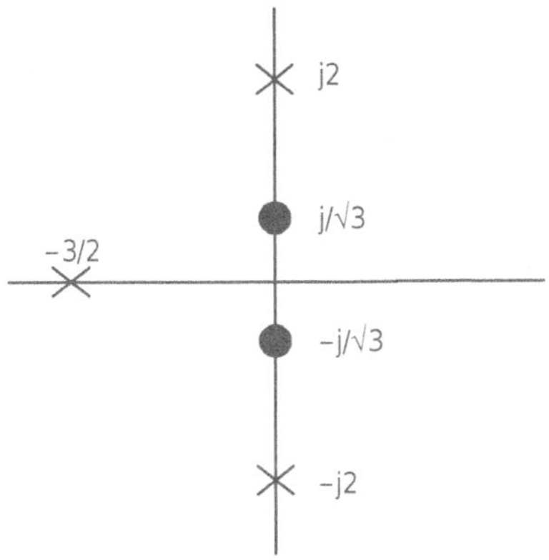

Fig. 9.3 Poles zeros plot.

Since the poles represent the behaviour of the response let us consider the same in detail.

Poles on the imaginary axis

These poles require a factor $(s^2 + \omega^2)$ in the denominator and hence they give rise to a partial fraction expansion term of the form

$$\frac{As + B}{s^2 + \omega^2}$$

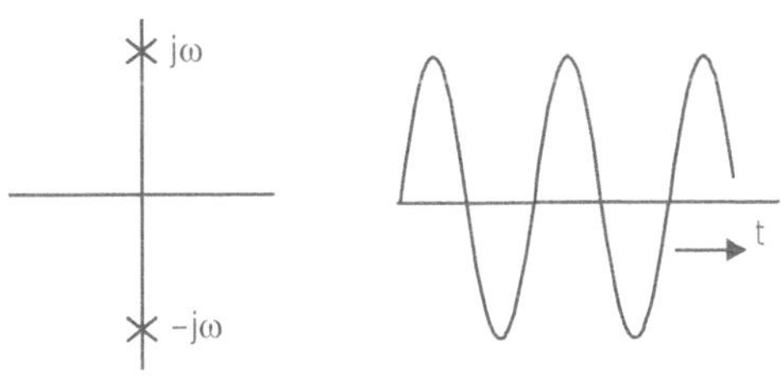

Fig. 9.4

whose inverse transform gives a time function

$$i(t) = A \cos \omega t + (B/\omega) \sin \omega t \tag{9.12}$$

which is an undamped sinusoidal oscillation.

Poles of the negative real axis

A pole at $s = -a$ requires a factor $(s + a)$ in the denominator of I(s) and the corresponding partial fraction expansion is

$$A/(s + a)$$

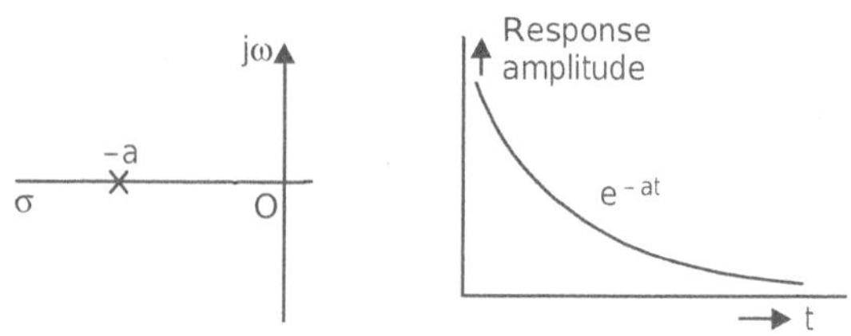

Fig. 9.5

and the inverse transform gives $i(t) = \mathrm{A}\ e^{-at}$. (9.13)

Poles on negative half plane, complex conjugate pair

These poles $-a \pm jb$ form a complex conjugate pair. The factor giving rise to these poles is

$$(s + a)^2 + b^2$$

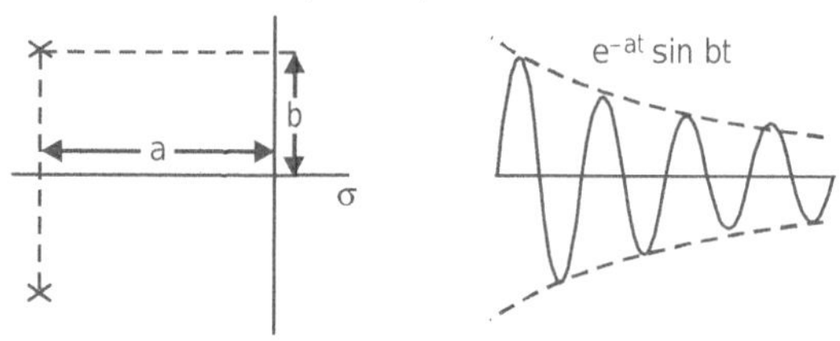

Fig. 9.6

and the corresponding partial fraction expansion

$$\frac{Cs + \mathrm{D}}{(s + a)^2 + b^2}$$

gives the inverse transform

$$i(t) = e^{-at}\ [\mathrm{C} \cos bt + \mathrm{D}/b \sin bt]$$

$$= \mathrm{A}\ e^{-at} \sin (bt + \varphi). \quad (9.14)$$

Poles on the positive real axis

For a pole at $(a + j0)$, the response is e^{at} and goes on increasing with time and so the response is unstable.

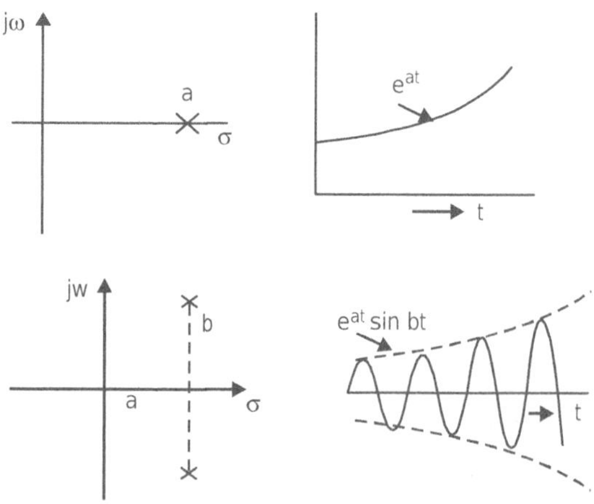

Fig. 9.7

Poles on the right half of the j axis

The response for conjugate poles ($a \pm jb$) is e^{at} cos bt or e^{at} sin bt.

The Oscillations go on increasing and hence it is unstable

Multiple poles. With a second order pole, such as $(s + 1)^2$, we will have partial fraction expansion terms of $A_1/(s + 1)^2$ and $A_2/(s + 1)$ and thus the response has e^{-t} and $t\,e^{-t}$.

For practical networks, poles lie only to the left of the imaginary axis. Complex poles always occur only as conjugate pairs.

Zeros on Right Half of j-axis

So far we considered poles only. If a zero is on the right half plane, then the network is called a non minimum phase network.

9.4 HARMONIC RESPONSE FROM POLES AND ZEROS

Suppose the driving point impedance (input impedance) of a circuit has one pole and one zero. Then, the transform Z(s) can be written as

$$K\left(\frac{s-z}{s-p}\right) \tag{9.15}$$

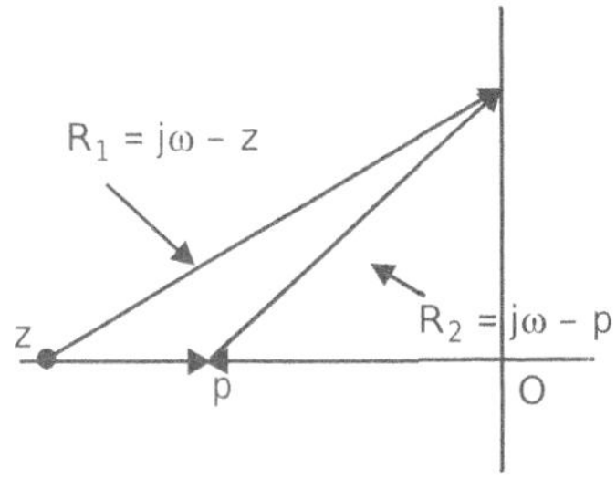

Fig. 9.8

where z is the zero and p is the pole. Suppose we want to find the response of the circuit for a frequency ω, *i.e.*, for an input voltage V sin ωt. Then we have to calculate Z(jω).

$$Z(j\omega) = K\frac{j\omega - z}{j\omega - p}$$

Then ($j\omega - z$) and ($j\omega - p$) can be calculated, given the values ω, z, p in numbers. The phasors ($j\omega - z$) and ($j\omega - p$) are shown in Fig. 9.8.

Hence
$$Z(j\omega) = K\,R_1\angle\alpha / R_2\angle\beta$$
$$= (KR_1/R_2)\angle(\alpha - \beta) \tag{9.16}$$

Knowing this impedance in magnitude and phase angle, the current can be found.

Example 1. *Show that the driving-point impedance at the terminals of A and B has 3 poles and 3 zeros. Find their values.*

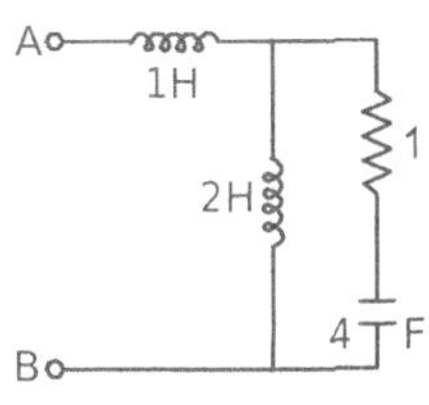

Fig. 9.9

Solution. The parallel branch has the impedance function

$$\frac{Ls(R+1/Cs)}{Ls+R+(1/Cs)} = \frac{Ls(1+CRs)}{LCs^2+RCs+1}$$

For L = 2, C = 4, R = 1,

$$= \frac{2s(1+4s)}{8s^2+4s+1}$$

Adding the series inductance Ls or $1s$,

$$Z(s) = s(8s^2 + 4s + 1) + 2s(4s + 1) \div (8s^2 + 4s + 1)$$

$$= \frac{s(8s^2+12s+3)}{8s^2+4s+1}$$

$Z(s)$ is zero for $s = 0$, and $s = (-12 \pm \sqrt{144-96}) \div 16$

i.e., $s = 0, -0.75 \pm 0.433j$

$Z(s)$ is infinite for $s = (-4 \pm \sqrt{16-32})/16$

and also when $s \to \infty$

Hence poles are at $s = -.25 \pm 0.25j$ and at ∞.

Example 2. *Plot the poles and zeros of the network 9.10. Find the input impedance of the network for frequencies $s = j10/\sqrt{3}$ and $j10\sqrt{3}$ and then find the current in the steady-state for,*

$$e(t) = 10 \sin (10/\sqrt{3})t + 10 \cos(10\sqrt{3}t)$$

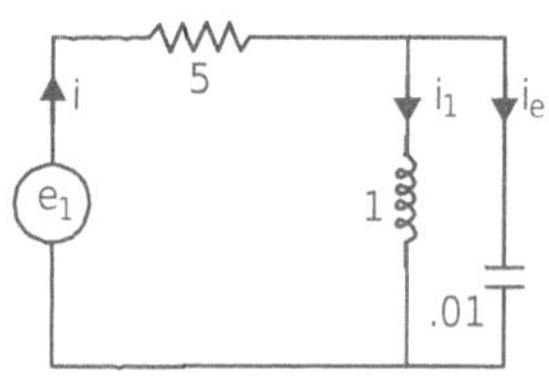

Fig. 9.10

Solution. What factors will the transient current $i(t)$ contain in addition?

$$Z(s) = R + \frac{Ls.(1/Cs)}{Ls+(1/Cs)} = \frac{RLCs^2+Ls+R}{LCs^2+1}$$

$$= \frac{0.05s^2+s+5}{0.01s^2+1} = \frac{5(s+10)^2}{(s+10j)(s-10j)}$$

Zeros of $Z(s)$ occur at $s = -10, -10$ (two zeros).

Poles of $Z(s)$ are at $s = \pm 10j$

Since we want the current, we have to find E(s) / Z(s) or E(s) / Y(s) = I(s). Hence let us consider the admittance function Y(s). Since Y = 1/Z, the poles of Y(s) are the zeros of Z(s) and vice versa.

The input $e(t)$ has the two frequencies $j\,10\sqrt{3}$ and $j\,10/\sqrt{3}$ and we want to find the admittance at these frequencies. We can substitute $s = j\,10\sqrt{3}$ and get the admittance for $\omega = 10\sqrt{3}$. It can also be found graphically. Mark the point $s = j\,10\sqrt{3}$. The polar distances from the poles and zeros must be found.

$$Y(s) = \frac{0.2(s+10j)(s-10j)}{(s+10)^2}$$

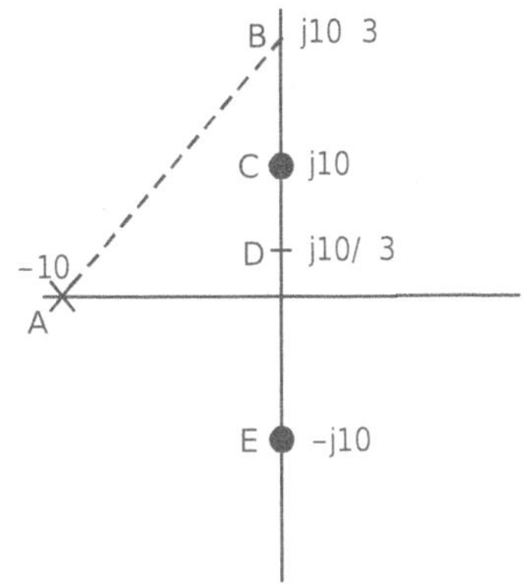

Fig. 9.11

The factor $(s + 10j)$ is CB which is $j(10\sqrt{3} - 10) = 7.32\angle 90°$ and $(s - 10j)$ is CE $= j(10 + 10\sqrt{3}) = 27.32\angle 90°$.

The factor $(s + 10)$ is the phasor AB which is 20 in length and phase angle 60°.

Hence $$Y(j10\sqrt{3}) = \frac{0.2\times 7.32\times 27.2\angle 180°}{20\times 20\angle 120} = 0.1\angle 60°$$

Similarly, for the other frequency $j10/\sqrt{3}$, we get

$$Y(j10/\sqrt{3}) = \frac{0.2\times j10(\sqrt{3}-1/\sqrt{3})\times 10(\sqrt{3}+1/\sqrt{3})j}{(20/\sqrt{3})\angle 30\times 20/\sqrt{3}\angle 30°}$$
$$= 0.4\ \angle(180 - 60) = 0.4\ \angle 120°$$

For $$e(t) = 10 \sin (10/\sqrt{3})t + 10 \cos(10\sqrt{3}t),$$
$$i(t) = 10 \times 0.4 \sin (\omega_1 t + 120°) + 10 \times 0.1 \cos (\omega_2 t + 60°)$$
$$= 4 \sin (\omega_1 t + 120°) + \cos (\omega_2 t + 60°)$$

The transient part of the current will have terms exp $(s_p t)$ where s_p are poles of the function $Y(s)$. Thus there are 2 poles at -10 and so the transient current will have $(At + B)e^{-10t}$, A and B being constants.

Example 3. *The input impedance of the following network has the poles and zeros plotted as shown. If the input resistance for direct current is 1 ohm, find the values R, G, L and C.*

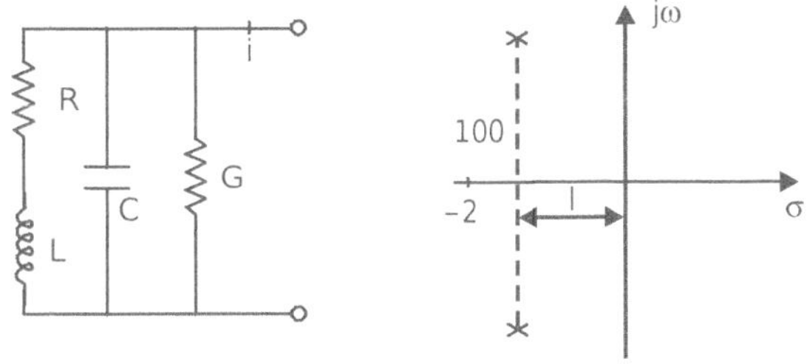

Fig. 9.12

Solution. The input admittance is written easily as

$$Y(s) = G + Cs + [1/(R + Ls)]$$

$$Z(s) = \frac{R+Ls}{(R+Ls)(G+Cs)+1} = \frac{R+Ls}{LCs^2+(GL+RC)s+(1+RG)}$$

For D.C., $s = 0$ (zero frequency) and $Z(0) = 1$ so that

$$Z(0) = 1 = R/(1 + RG) \quad (a)$$

The zero of Z is at $s = -R/L = -2$(from fig.) (b)

From the denominator quadratic in $Z(s)$, we get

Product of roots = Product of poles = (1 + RG)/LC = R/LC

$$\therefore \quad R/LC = +2/C = (-1 + 100j)(-1 - 100j) \simeq 100^2.$$

$$\therefore \quad C = 2 \times 10^{-4} \text{ or } 200\mu\text{ F}.$$

The sum of roots = sum of poles = $-1 + 100j + (-1-100j) = -2$

Hence, $(GL + RC)/LC = 2$

$$G/C + R/L = 2; \; G/C = 2 - 2 = 0$$

giving G = 0 (open circuit of G in fig.)

Hence $R = 1$ from (a) and $L = 0.5$ H from (b).

Additional Problems on Transfer Functions (Zero on RHP)

Example 1. *Find the transfer function of the "Phase Shift network"*

The network shifts the phase as frequency varies without amplitude change.

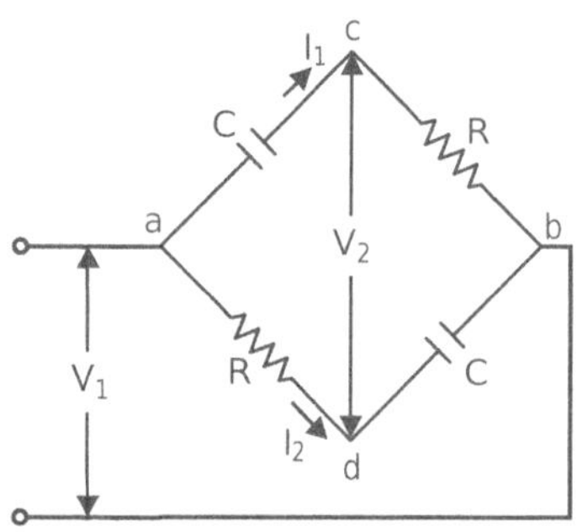

Solution. The Laplace transforms of I_1, I_2 are:

$$I_1 = I_2 = \frac{V_1}{R + 1/sC} = \frac{V_1 sC}{1 + sCR}$$

The voltages at point c and d are:

$$v_c = v_a - I_1(1/sC) = v_a - \frac{V_1}{RCs+1}$$

$$v_d = v_a - I_2R = v_a - \frac{V_1 RCs}{RCs+1}$$

The transform of the output voltage is:

$$V_2 = v_c - v_d = \frac{V_1}{RCs + 1}(RCs - 1)$$

Then
$$k(s) = \frac{V_2(s)}{V_1(s)} = \frac{RCs - 1}{RCs + 1} = \frac{s - 1/RC}{s + 1/RC}$$

$$= \frac{s - a}{s + a}$$

The function $k(s)$ has a zero at $s = a = 1/RC$, that is, in the right half plane. Therefore the phase shifter is called a non-minimum phase network. Furthermore,

$$k(j\omega) = (j\omega - a)/(j\omega + a)$$

$$k(\omega) = |(j\omega - a)| / |(j\omega + a)| = 1 \qquad \text{(amplitude ratio is 1)}$$

$$\theta(\omega) = \tan^{-1}\left(-\frac{\omega}{a}\right) - \tan^{-1}\left(\frac{\omega}{a}\right) = \tan^{-1}\left(\frac{2a\omega}{\omega^2 - a^2}\right)$$

Thus the network is a phase shifter is amplitude response unchanging with frequency, while its phase depends on frequency.

Example 2. *Find the transfer function of the bridge network shown.*

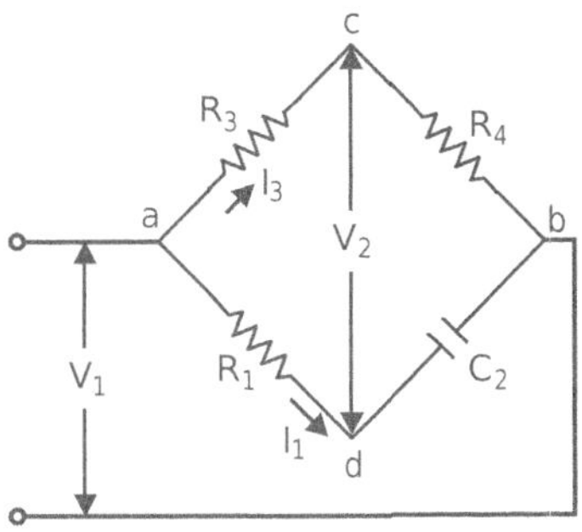

Solution. The currents I_1 and I_3 have transforms

$$I_1 = \frac{V_1}{R_1 + \frac{1}{sC_2}} \qquad I_3 = \frac{V_1}{R_3 + R_4}$$

The voltages at c and d:

$$v_c = v_a - I_3R_3 \qquad v_d = v_a - I_1R_1$$

The output voltage V_2 is

$$V_2 = v_c - v_d = I_1R_1 - I_3I_3 = V_1\left[\frac{R_1}{R_1 + 1/sC_2} - \frac{R_3}{R_3 + R_4}\right]$$

Then the transfer function $V_2(s)/V_1(s) = k(s)$ is

$$k(s) = V_2/V_1 = \frac{sR_1R_4 - R_3/C_2}{(R_3 + R_4)(sR_1 + 1/C_2)}$$

$$= R_4\left\{\frac{s - R_3/R_1R_4C_2}{(R_3 + R_4)(s + 1/R_1C_2)}\right\}$$

The zero lies at $s = R_3/R_1R_4C_2$, *i.e.*, in the RHP. Therefore the circuit is a non-minimum phase network.

9.5 TRANSIENT RESPONSE FROM POLE ZERO PLOTS — RESIDUE EVALUATION

Let the transform I(s) for current have the zeros z_1, z_2 etc., and poles p_1, p_2, p_3, etc. Let us assume that all the poles are simple poles, *i.e.*, they are not multiple poles. Then

$$I(s) = \frac{K(s-z_1)(s-z_3)}{(s-p_1)(s-p_2)(s-p_3)} \tag{9.17}$$

To find the transient current $i(t)$, we have to find the inverse Laplace transform of the above. For this, we must expand I(s) in partial fractions as

$$\frac{A}{(s-p_1)} + \frac{B}{s-p_2} + \frac{C}{s-p_3} \tag{9.18}$$

and the transient $i(t)$ is got as A exp (p_1t) + B exp (p_2t) + ...

The values of A, B, C ... are called the *Residues* of I(s) at the poles p_1, p_2, p_3 – the name is from complex variable theory. To find their values, we have a simple method which can be adopted for graphical calculation also.

Multiply (9.18) and (9.19) throughout by $(s - p_1)$ and put $s = p_1$ in the identity so obtained.

$$\frac{K(p_1-z_1)(p_1-z_2)}{(p_1-p_2)(p_1-p_3)} = A + \frac{B(p_1-p_1)}{(p_1-p_2)} + \frac{C(p_1-p_1)}{(p_1-p_3)} \tag{9.19}$$

which gives A on the right hand side and its value on the L.H.S.

Thus $\qquad A = [I(s)(s - p_1)]$ at $(s - p_1)$

Similarly $\qquad B = [I(s)\ (s - p_2)]$ at $(s - p_2)$ etc. $\qquad$ (9.20)

The graphical evaluation is indicated in Fig. 9.13.

Thus $$A = \frac{OA.DA}{AB.AC}$$

where the distances are treated as phasors. Likewise

$$B = \frac{OB.DB}{BA.BC}$$

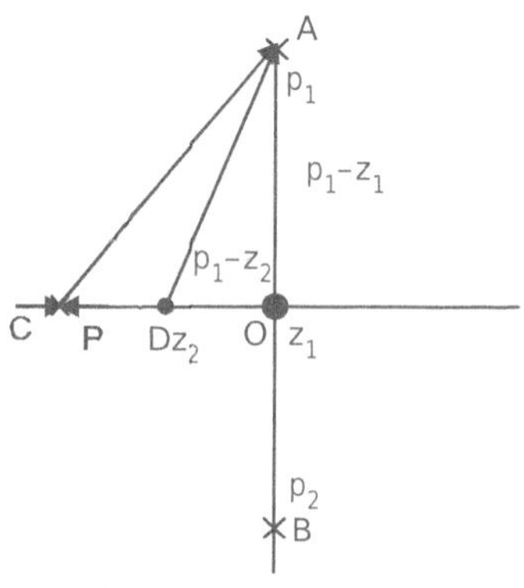

Fig. 9.13

Example 1. *An R–L–C series circuit has poles and zeros as shown (9.14). If L = 1 henry, find the transient response to an input e(t) = 10 cos t.*

The impedance function $Z(s) = R + Ls + (1/Cs)$

Solution. The voltage 10 cos t is Re $[10\ e^{jt}]$. Its transform is $10/(s-j)$

$$\therefore \quad \left(R + Ls + \frac{1}{Cs}\right) I(s) = \frac{10}{s-j}$$

$$I(s) = \frac{(10/L)s}{(s-j)(s^2 + R/Ls + 1/LC)}$$

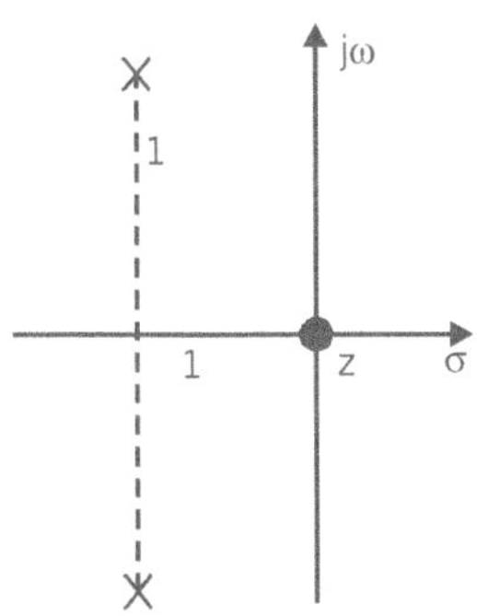

Fig. 9.14

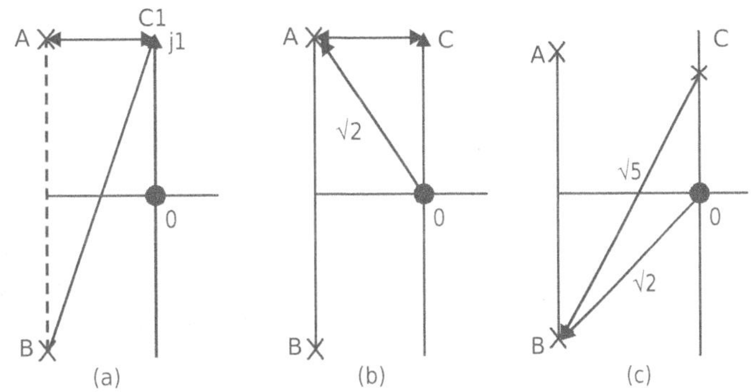

Fig. 9.15

Given L = 1, the above equals

$$\frac{10s}{(s-j)(s-p_1)(s-p_2)}$$

where, p_1 and p_2 are given poles in Fig. 9.14. The values of the residues at these poles $s = j$, $s = p_1$, $s = p_2$ are found using (9.20). The graphical constructions are shown in Fig. 9.15 *a* to *c*.

The residue at pole $j1$ (co-efficient of e^{jt}) (Fig. a)

$$A_j = K.OC/CA.AB = 10 \times 1 \angle 90°/(1 \times \sqrt{5} \angle \tan^{-1} 2)$$

$$= 10/\sqrt{5} \angle \tan^{-1}\frac{1}{2} = 2\sqrt{5} \angle \tan^{-1}\frac{1}{2}.$$

The residue at pole A (co-efficient of $e^{p_1 t}$) (Fig. *b*)

$$A_1 = K.\,OA/CA.BA = 10\sqrt{2}\ \angle 135°/(-1 \times 2\ \angle 90°) = 5\sqrt{2} \angle 45°$$

The residue at pole B(co-efficient of $e^{p_2 t}$) (Fig. *c*)

$$A_2 = K.\ OB/CB.\ AB$$

$$= 10\sqrt{2}\ \angle -135°/(\sqrt{5} \times 2\ \angle -180 - \tan^{-1}\frac{1}{2})$$

$$= \sqrt{5}\sqrt{2} \angle -45° + \tan^{-1}\frac{1}{2}$$

Let $\tan^{-1}\frac{1}{2} = \varphi$. Then

$$i(t) = \text{Real}\,[2\sqrt{5}\, e^{j(t+\phi)} - 5\sqrt{2}e^{(-1+j)t+\pi/4} + \sqrt{10}e^{(-1-1j)t+j(-\pi/4+\phi)}]$$
$$= 2\sqrt{5}\cos(t+\varphi) - 5\sqrt{2}e^{-t}\cos(t+\pi/4) + \sqrt{10}\cos(t - \tfrac{1}{4}\pi + \varphi)e^{-t}.$$

Example 2. *Find the poles of the transfer function V_0/V_t of the following network. (Fig. 9.16(a))*

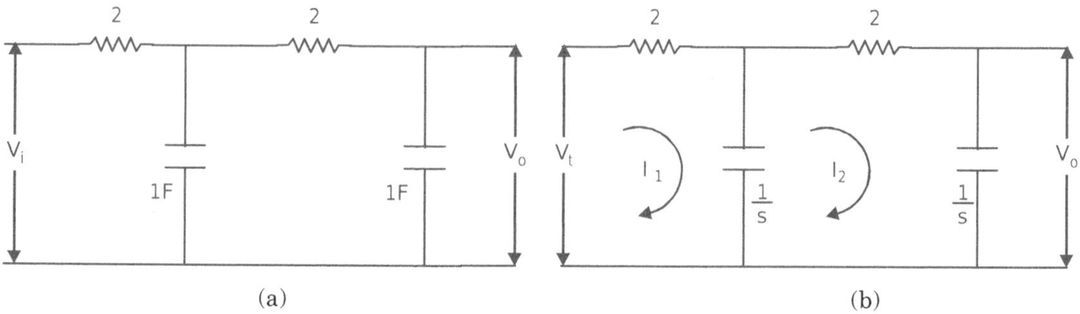

Fig. 9.16

Solution. Assume currents I_1 and I_2 as shown in Fig. 9.16(*b*). Writing the Kirchoff's Loop equations in terms of complex impedences,

$$V_{in} = \left(2+\frac{1}{s}\right)I_1 - \frac{1}{s}I_2$$

$$0 = -\frac{1}{s}I_1 + \left(2+\frac{2}{s}\right)I_2$$

$$\frac{1}{s}\left[V_{in} + \frac{I_2}{s}\right] \div \left(2+\frac{1}{s}\right) = \left(2+\frac{2}{s}\right)I_2$$

Since $V_0 = (1/s)I_2$

$$\left(2s+2-\frac{1}{s+2}\right)V_0 = \frac{V_{in}}{s+2}$$

$$\frac{V_0}{V_{in}} = \frac{1}{(2s+2)(s+2)-1}$$

The poles are got from $2s^2 + 6s + 3 = 0$

as $s = -3/2 \pm \sqrt{3/2}$

and hence real and lying on the negative real axis.

System Response : Additional Problems

Example 1. *Sketch the response of each of the systems below to a step input.*

(*a*) $H(s) = \dfrac{10}{s+2}$

(*b*) $H(s) = \dfrac{0.2}{s+0.2}$

Solution. (*a*) $H(s) \;=\; \dfrac{10}{s+2}$

Steady-state value for step input 1/s is $\lim\limits_{s \to 0} s.\dfrac{1}{s}.\dfrac{10}{s+.2}$;

$$y_{ss} \;=\; \lim_{s \to 0} H(s) = 5, \; \tau \;=\; ½ \; \text{sec}$$

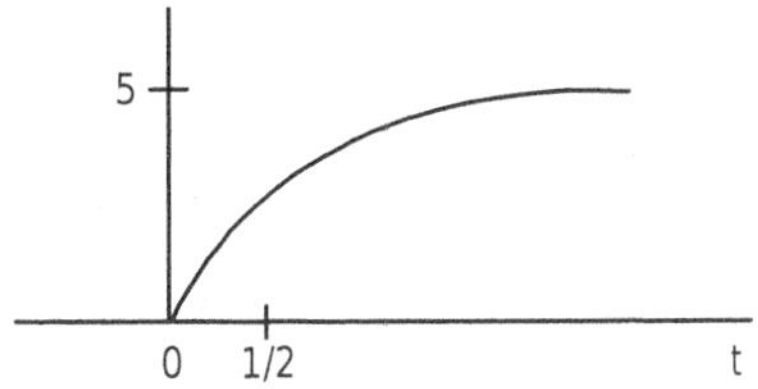

(*b*) $H(s) \;=\; \dfrac{0.2}{s+0.2}$

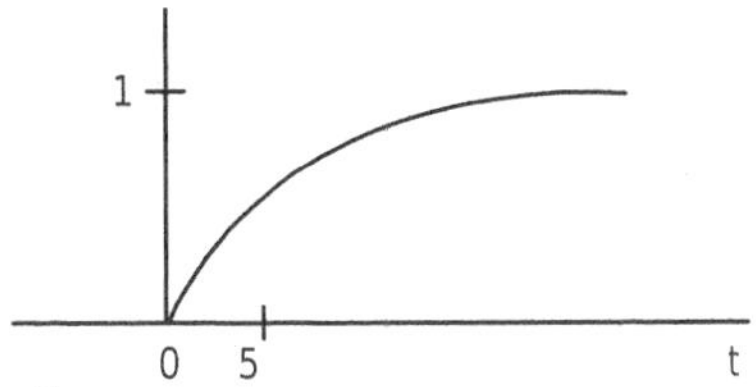

$$y_{ss} \;=\; \lim_{s \to 0} \; s.\frac{1}{s}.\frac{0.2}{s+.2}; \; y_{ss} \;=\; 1, \; \tau \;=\; 1/0.2 \;=\; 5 \text{ sec}$$

Example 2. *Given, the two step responses shown below, the first one is a first order system and the second one is a second order system. Determine the transfer functions for both systems.*

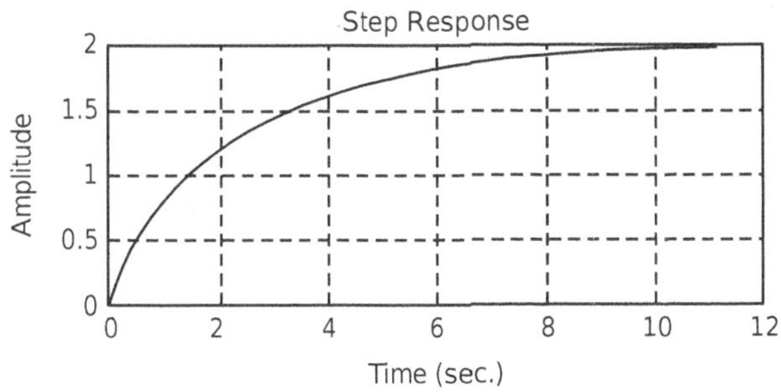

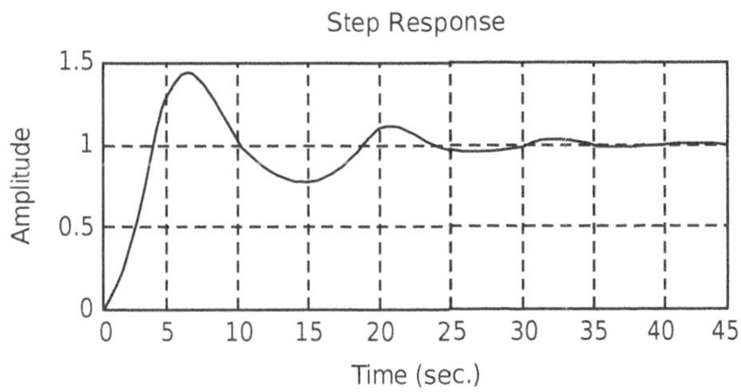

Solution. First order system (top plot) has general form $H(s) \;=\; \dfrac{k}{s+a}$. The time-constant is the time that the response is equal to $2(1 - e^{-1}) = 63\%$ of $2 = 1.26$, so $\tau \approx 2$ sec. $a = 1/\tau = 0.5$. The steady-state value (due to a unit step input) is $H(0) = k/a = k/0.5$. From the plot $2 = k/0.5$ so $k = 1$.

Final answer: $H(s) \;=\; \dfrac{1}{s+0.5}$

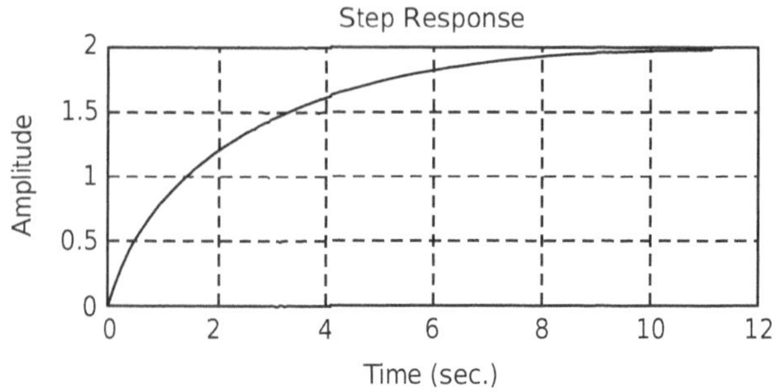

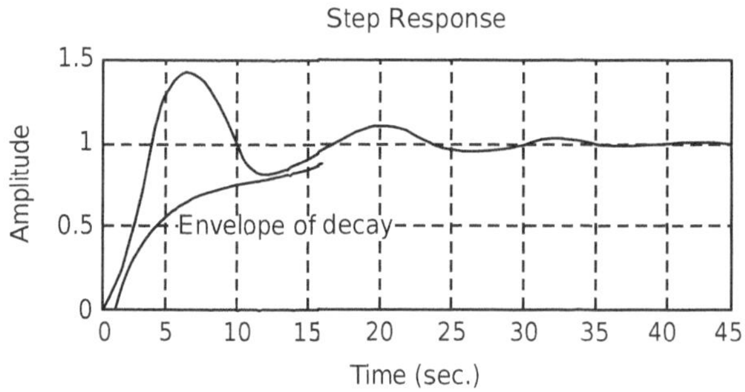

Second order system (bottom plot) has general form $H(s) = \dfrac{k}{s^2 + 2\zeta\omega_n s + \omega_n^2}$ or $H(s) = \dfrac{k}{(s+\zeta\omega_n)^2 + \omega_d^2}$ where the real part of the pole is at $-\zeta\omega_n$ is the real part of the pole (it governs the envelope of decay) and ω_d is the imaginary part of the pole (it governs the frequency of the oscillations, $\omega_d = 2\pi/T$) ζ is called damping factor.

From the plot, $T \approx 12$ sec, so $\omega_d = 2\pi/13$. The time constant of the envelope of decay is about $\tau \approx 7$ sec, so $\zeta\omega_n = 1/7$. k is found from the steady-state value $1 = H(0) = \dfrac{k}{(\zeta\omega_n)^2 + \omega_d^2}$. Solving for k yields $k \approx 0.254$.

Final answer :

$$H(s) = \frac{0.254}{s^2 + 0.286s + 0.254}.$$

$$H(s) = \frac{k}{s^2 + 2\zeta\omega_n s + \omega_n^2}$$

As another example, consider figure below for a 2nd order system.

$$\omega_d = \frac{2\pi}{T} = 9.7$$

$$\tau \approx .35 \text{ sec} \Rightarrow \frac{1}{\zeta\omega_n} = \tau \Rightarrow \zeta\omega_n = 2.9$$

$$H(s) = \frac{k}{(s+2.9)^2 + 9.7^2}, \ H(0) = 0.2 \text{ (steady final value)}$$

$$\Rightarrow \qquad k = 20.5$$

$$H(s) = \frac{20.5}{(s+2.9)^2 + 9.7^2} \text{ from plot}$$

Actual is $$\text{H}(s) = \frac{22}{(s+3)^2 + 10^2}$$

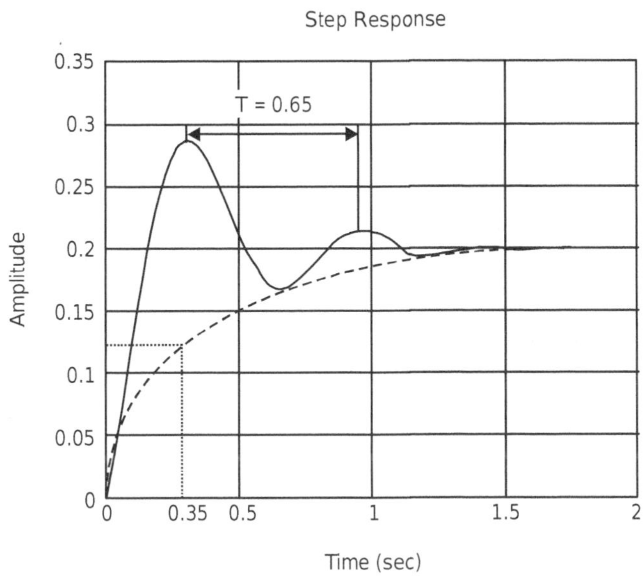

Example 3. *Plot the pole positions for each of the following systems, determine the values for ζ and ω_n for the stable second order systems with complex poles.*

(a) $H(s) = \dfrac{1}{s+4}$

(b) $H(s) = \dfrac{1}{s+10}$

(c) $H(s) = \dfrac{1}{s-2}$

(d) $H(s) = \dfrac{1}{s^2+4s+16}$

(e) $H(s) = \dfrac{1}{s^2+4s+3}$

(f) $H(s) = \dfrac{1}{s^2+4s+2}$

(g) $H(s) = \dfrac{1}{s^2-4s+16}$

Solution. (*a*) $\text{H}(s) = \dfrac{1}{s+4}$

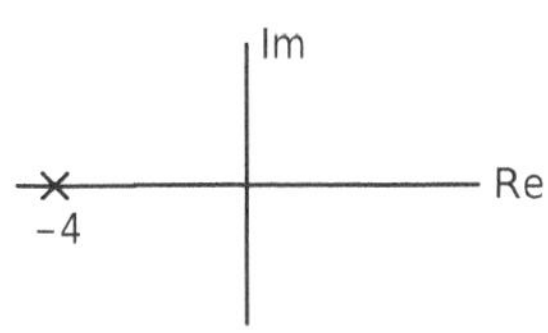

(*b*) $$\text{H}(s) = \frac{1}{s+10}$$

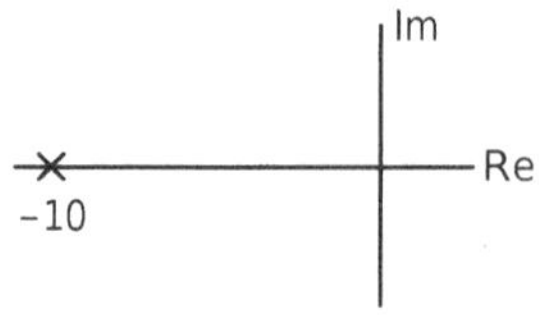

(*c*) $$\text{H}(s) = \frac{1}{s-2}$$

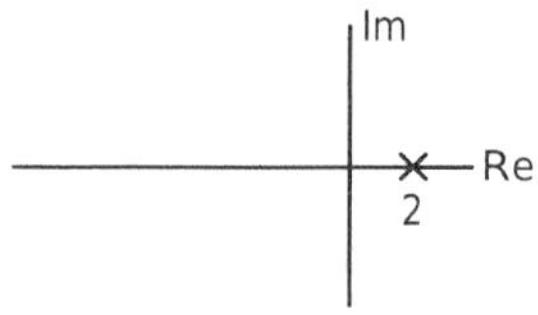

(*d*) $$\text{H}(s) = \frac{1}{s^2+4s+16} = \frac{1}{(s+2)^2+12}$$

$$\omega_n^2 = 16 \Rightarrow \omega_n = 4$$

$$2\zeta\omega_n = 4 \Rightarrow \zeta = 0.5$$

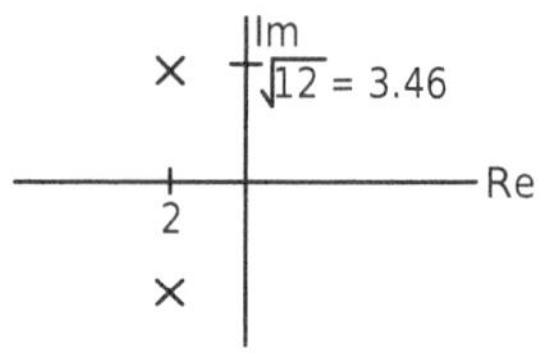

(*e*) $$\text{H}(s) = \frac{1}{s^2+4s+3} = \frac{1}{(s+1)(s+3)}$$

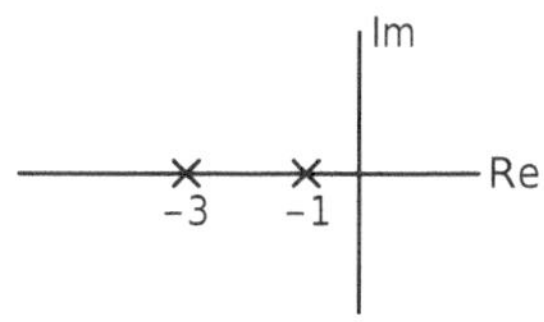

(*f*) $$\text{H}(s) = \frac{1}{s^2+4s+2} = \frac{1}{(s+3.41)(s+0.586)}$$

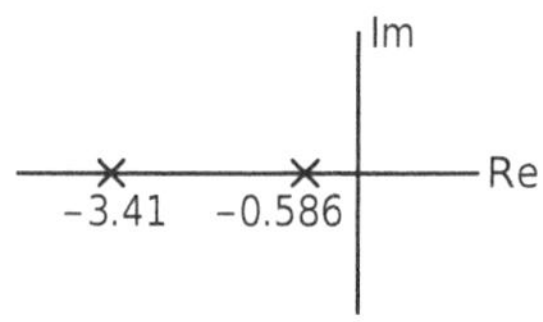

(*g*) $$\text{H}(s) = \frac{1}{s^2-4s+16}, \qquad \omega_n = 4$$

$$= \frac{1}{(s-2)^2+12} \qquad 2\zeta\omega_n = -4$$

$$\text{So } \zeta = -0.5$$

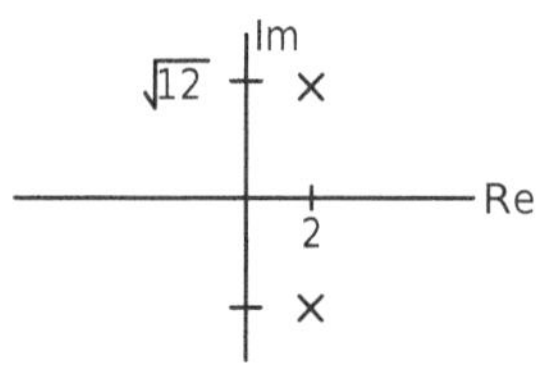

Example 4. *Give the general form of the response of the systems in Problem 3 to a step input.*

Solution. (*a*) $$y(t) = k_1 + k_2 e^{-4t}, \ t \geq 0$$

$$\left(\text{Note, } k_1 \text{ is found from steady-state} = \text{H}(0) = \frac{1}{4}\right) \quad k_2 = -1/4$$

(*b*) $$y(t) = k_1 + k_2 e^{-10t}, t \geq 0, \quad \text{k1} = \frac{1}{10}, \ \text{X}_2 = -1/10$$

(*c*) $$y(t) = k_1 + k_2 e^{2t}, t \geq 0$$

(*d*) $$y(t) = k_1 + k_2 e^{-2t}\cos\left(\sqrt{12t} + \theta\right),\ t \ge 0$$

(*e*) $$y(t) = k_1 + k_2 e^{-t} - k_3 e^{-3t},\ t \ge 0$$

(*f*) $$y(t) = k_1 + k_2 e^{-3\ 4t} + k_3 e^{-0.586t},\ t \ge 0$$

(*g*) $$y(t) = k_1 + k_2 e^{2t}\cos\left(\sqrt{12t} + \theta\right),\ t \ge 0$$

in all cases, $k_1 = \text{H}(0)$

Example 5. *Determine the steady-state response of the systems in Problem 3(a), (d), and (f) to an input of x(t) = 2 cos (4t – 20°)u(t).*

Solution. $\text{A}\cos(\omega t + \theta)u(t) \rightarrow y_t(t) + y_{ss}(t)$

$$\text{A}|\text{H}(\omega)|\cos(\omega t + \theta + \angle\text{H}(\omega))$$

$$\text{let } x(t) = 2\cos(4t - 20°)\,u(t)$$

(*a*) $$\text{H}(s) = \frac{1}{s+4} \text{ so } \text{H}(\omega) = \frac{1}{j\omega + 4}$$

$$\text{H}(4) = \frac{1}{4j+4} = 0.176e^{-45°}$$

$$y_{ss}(t) = 0.352\cos(4t - 20° - 45°)$$
$$= 0.352\cos(4t - 65°)$$

(*b*) $$\text{H}(s) = \frac{1}{s^2 + 4s + 16},\ \text{H}(\omega) = \frac{1}{(16 - \omega^2) + 4j\omega}$$

$$\text{H}(4) = \frac{1}{16j} = 0.0625e^{-j90°}.$$

$$y_{ss}(t) = 0.125\cos(4t - 110°)$$

(*c*) $$\text{H}(s) = \frac{1}{s^2 + 4s + 2},\ \text{H}(\omega) = \frac{1}{(2 - \omega^2) + 4j\omega}$$

$$\text{H}(4) = \frac{1}{-14 + 16j} = 0.047e^{-j131°}.$$

$$y_{ss}(t) = 0.094\cos(4t - 151°)$$

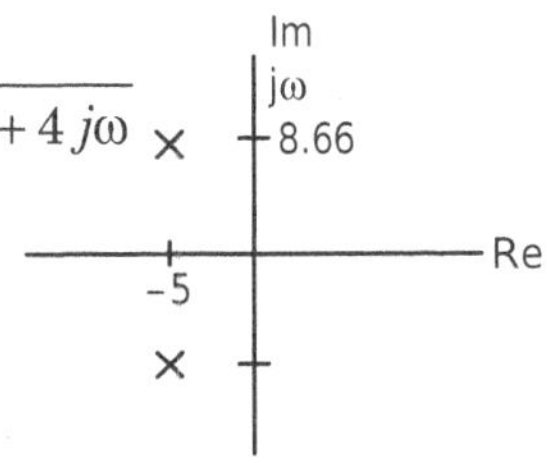

Example 6. *Given the following system:*

$$H(s) = \frac{10}{s^2 + 10s + 100}$$

(a) Plot the poles. Identify the values of ω_n and ζ.

(b) Sketch the step response.

(c) What is the steady-state response of the system to the following input?

$$x(t) = \cos(10t)u(t)$$

Solution. (*a*)

$$\omega_n = 10,\ 2\zeta\omega_n = 10 \rightarrow \zeta = 0.5$$

$$P_1,\ P_2 = -5 \pm 8.66j \quad \omega_d = 8.66$$

(*b*) $\tau = \dfrac{1}{\zeta\omega_n} = \dfrac{1}{5},\ T = \dfrac{2\pi}{\omega_d} = 0.726$ sketch shows $\tau = 0.2,\ \tau = 0.726$

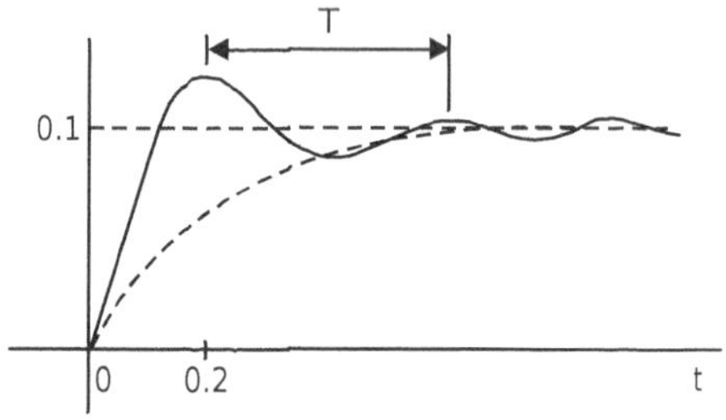

$$y_{ss} = H(0) = 0.1$$

(*c*)

$$H(10) = \frac{10}{(j10)^2 + 10j\omega + 100} = \frac{10}{100j} = 0.1e^{-j\pi/2}$$

$$y_{ss}(t) = 0.1 \cos(10t - \pi/2)$$

EXERCISES

1. In the impedence below (Fig. 1) what must be the relationship between the circuit elements if Z(*s*) is independent of *s* ?

[$R_1 = R_2 = \sqrt{L/C}$].

2. Find the natural frequencies of the following circuits. (Fig. 2 and 3).

[Fig. 2: Pole at $s = \pm j/\sqrt{2}$, zero at $s = \pm j\sqrt{\dfrac{-3 \pm \sqrt{5}}{4}}$

Fig. 3: $s = -2.86,\ s = 1.235,\ s = -0.19 \pm j2.768$].

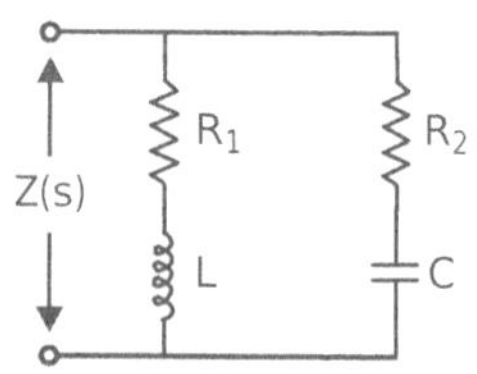

Fig. 1

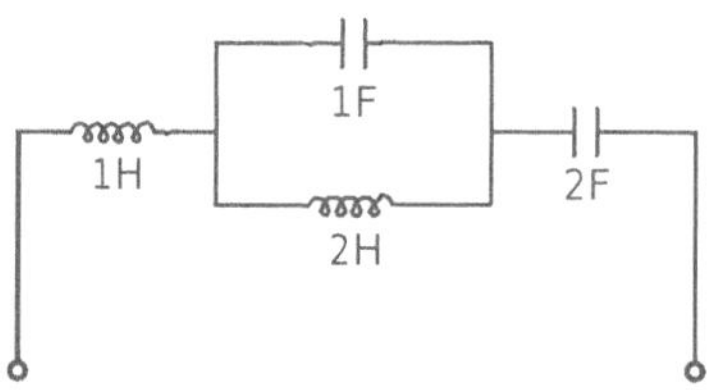

Fig. 2

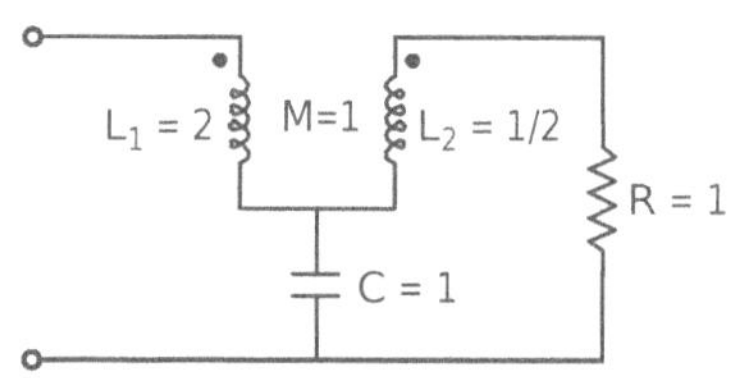

Fig. 3

3. What are the poles and zeros of the driving point impedance at port *ab* of Fig. 4 ?

$$\left[\text{Poles: } \frac{1}{2}(-1 + j\sqrt{2});\ \text{zeros} = \frac{1}{2}(-1 \pm j\sqrt{7})\right].$$

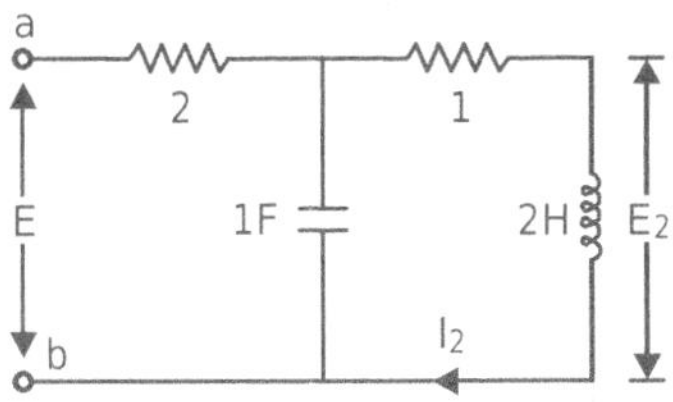

Fig. 4

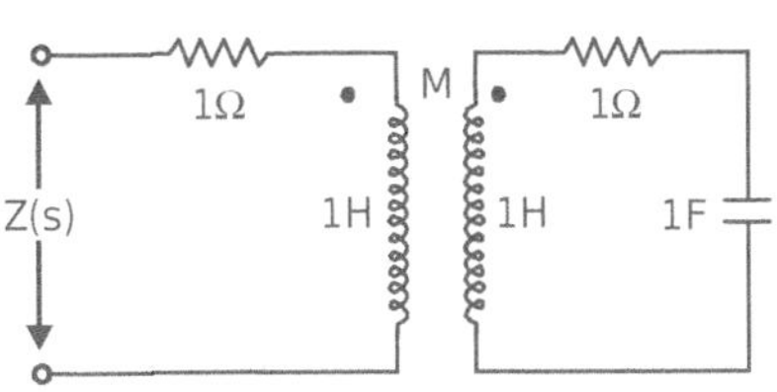

Fig. 5

4. In the coupled circuit (Fig. 5), write the loop equations and hence find the poles for M = 0, 0.1, 0.5 and 1.0. Plot them.

[M = 0 : $s = -1$, $s = -\dfrac{1}{2} \pm j\sqrt{3}/2$

M = 0.1 : $s = -1.12$ and $-0.555 \pm j0.0828$.

M = $\dfrac{1}{2}$: $s = -1.46$, $-0.597 \pm j0.742$.

M = 1 : $s = -1 \pm j$].

5. By calculating the impedance functions for the two network (Fig. 6), show that they are equivalent.

$$\left\{\textbf{Ans.}\ \frac{s^2 + 4s + 3}{s(s + 2)}\right\}$$

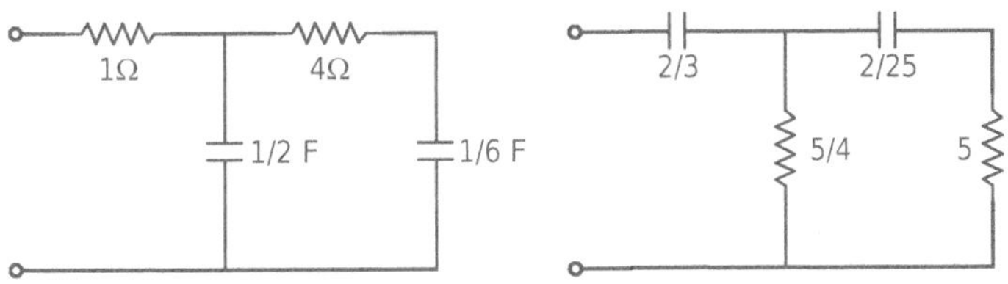

Fig. 6

6. Find the response to a harmonic frequency of ω = 1 if the corresponding network function is

$$\frac{s-2}{s^2 + 4s + 8} \qquad \left\{\textbf{Ans.}\ \frac{1}{\sqrt{13}}\angle 123.7^\circ\right\}$$

by drawing the pole-zero plots. For a 10V d.c. supply, what will be the response in the above?

7. Find the frequency at which the following admittance function is purely reactive. [–5/2]

$$\frac{s^2 + s + 1}{s(s+1)(s+2)} \qquad [\omega = 0.707]$$

8. In Fig. 4 above, find (1) the transfer impedance between I_2 and E. (2). The voltage ratio E_2/E.

$$\left\{\textbf{Ans.}\ \frac{1}{4s^2 + 4s + 3} \text{ and } \frac{2s}{4s^2 + 4s + 3}\right\}$$

9. Find the transient component of the current for an input $2e^{-2t}$ in the network whose admittance is $(s + 1)/(s^2 + 2s + 3)$.

$$\left[\textbf{Ans.}\ -\frac{2}{3}e^{-2t} + \frac{1}{3}e^{-t}\ (2\cos\sqrt{2}\,t + 2\sqrt{2}\sin\sqrt{2}\,t\right].$$

10. A certain network function has a zero at $s = -1$ and a pole at $s = -2$. An input a varying frequency is applied and the response measured. At what frequency will the phase angle be 13° ? What will be the difference if the poles and zeros interchange their values ?

$$\left[\textbf{Ans. } \omega = \frac{1}{2}, \omega = 4, \text{ Lead and lag}\right].$$

SIMPLE EXERCISES

1. Write the Laplace Transform of the differential equation of a R-L-C series circuit. Examine its transient response qualitatively in terms of the roots of the algebraic quadratic equation. Find its frequency of oscillation and the damping. What are the poles and zeros of this circuit ?

$$\left[\omega = \left\{\frac{1}{LC} - \frac{R^2}{4L^2}\right\}^{1/2}, \sigma = -R/2L. \text{ Poles at } \pm j\omega, \text{ zero at } \sigma\right].$$

2. What is the meaning of network function ? Illustrate with an example. [See Sec. 9.2]

3. Given a network function of s, how can one determine the magnitude and phase for a given frequency ω from it ? Given that an impedance is $Z(s) = \dfrac{3s^2 + 12s + 8}{2s + 4}$, find the magnitude and phase of it at $\omega = 2$.

[See example 3, Sec. 9.2, $\sqrt{2} \angle -\pi/4$].

4. Given the differential equation relating the variables x and y as

$y'' + 4y' + 20y = 2x' - x$ (denotes differentiation)

find the transfer function $Y(s)/X(s)$.

[Sec. 9.2.1 Vide Transfer function-Additional Problems-Example 1.(*b*).

5. Explain how you will write loop equation for a circuit in terms of Laplace transform variable s, with an example.

[Vide example 2 on Sec. 9.2.1].

6. What is the principle behind the plotting of poles and zeros of a network function on the real-imaginary s-plane ? Plot the poles and zeros of the $s(s^2 + 4)/(s^2 + 3)$.

[See sec. 9.3, Poles $\pm j\sqrt{3}$, zero $j\sqrt{2}$.]

7. Explain how, from the zeros-poles plot, the transient response of a system can be determined.

[Sec. 9.5].

8. Give an example of a zero on the right half plane in a network. [Sec. 9.5.1].

9. Find the transfer function of the Wien bridge network and find the frequency at null output.

$$\left[f = \frac{1}{2\pi CR}\right].$$

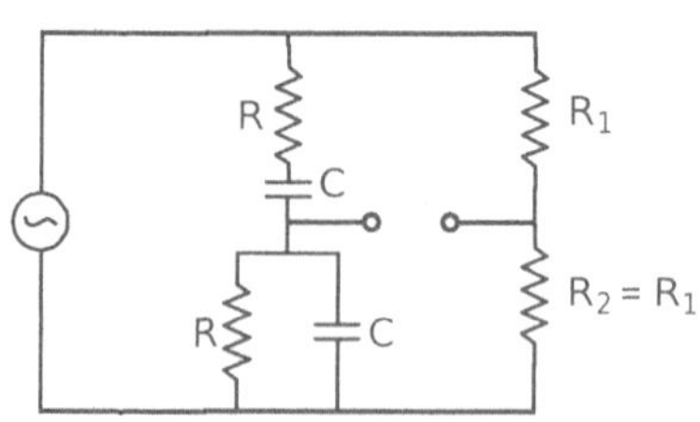

10

NETWORKS INCLUDING TWO-PORTS*

10.1 PROPERTIES OF DRIVING-POINT IMMITTANCES

We have so far been concerned with the *Analysis* problem of Electric Circuits—finding the output for given input and vice versa having the network in hand. In *Synthesis,* a relation between the input and output is given and the network is to be found. For instance, if

$$\frac{s+1}{s} \tag{10.1}$$

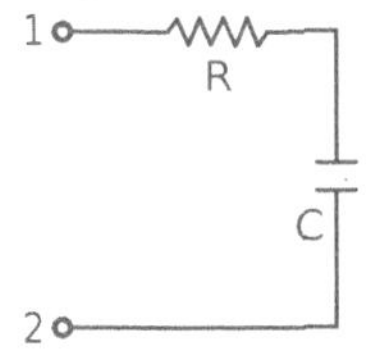

Fig. 10.1

is the impedance of a network, we can say it is

$$1+\frac{1}{s}$$

and hence comprises of a resistor R = 1Ω in series with a C = 1F. The function (10.1) is called the network function of a driving point impedance and it is said to have been "synthesised" at the port (1-2) of Fig. 10.1 above. The driving points are 1-2.

Whereas there is only one unique solution to an analysis problem, synthesis of a network function may be possible in more than one way, *i.e.*, one or more network may have the same network function. Hence, in synthesis, it is first necessary

1. to study about the properties of a network function required to realise the synthesis
2. to determine whether the given network function satisfies these requirements and then find the network.

In our study, we consider only driving-point immittances, *i.e.*, impedances $Z_d(s)$ and admittances $Y_d(s)$.

1. Network Functions are rational functions of s with real Co-efficients

Networks which are constructed from passive circuit elements as R, Ls and 1/Cs will always be rational and as seen from examples in the last chapter, and in general, they are of the form

*This chapter is not included in the syllabus in I year ECE (Anna University).

$$H(s) = \frac{a_m s^m + a_{m-1} s^{n-1} + \dots \dots + a_0}{b_n s^n + b_{n-1} s^{n-1} + \dots \dots + b_0} \tag{10.2}$$

H(s) denotes a general network function, which may be a driving-point impedance, admittance or a transfer function.

The co-efficients of both the numerator and denominator polynomials depend on the network configuration only and hence the *a's* and *b's* are all real. Thus, the co-efficients do not depend on the external excitations.

So, the roots of the numerator or denominator polynomial equation

$$n(s) = a_m s^m + a_{m-1} s^{m-1} + \dots + a_0 = 0$$

$$d(s) = b_n s^n + b_{n-1} s^{n-1} + \dots + b_0 = 0 \tag{10.3}$$

will have either

1. Real roots and/or
2. Complex (imaginary) roots which occur in conjugate pairs, only.

The above property stems from a study of algebraic equations. So, we know that there is no single imaginary root for either $n(s)$ or $d(s) = 0$.

2. A Property concerning the poles/zeros of H(*s*)

Now H(*s*) can be expressed in the form

$$H(s) = \frac{n(s)}{d(s)} = K\frac{(s-z_1)(s-z_2)\dots(s-z_m)}{(s-p_1)(s-p_2)\dots(s-p_n)} \tag{10.4}$$

where $K = a_m/b_n$ and *z's* are the zeros and the *p's* are the poles of H(*s*).

It may be either $m > n$ or $n > m$ *i.e.*, there may be more factors in the numerator than the denominator and vice versa.

$$\text{If } m > n, \text{ then as } s \to \infty, H(s) \to (\infty)^{m-n}$$

$$\text{If } n > m, \text{ then as } s \to \infty, H(s) \to (0)^{n-m}$$

So, in the former we call that H(*s*) has pole of order $(m - n)$ at infinity and in the latter case, it has a zero of order $(n - m)$ at infinity.

But let us consider a real, physically existing network. From (10.4), we know that the gain constant K, the poles and the zeros, completely determine the network. As *s* becomes large, what does the network (which contains, after all, R–L–C elements only) behave like? Capacitors behave like short-circuits and inductors like high reactance, Ls, for example. Thus, as $s \to \infty$, the driving-point impedance $Z_D(s)$ approaches either a constant R (*i.e.*, resistance) or Ls (where L is the inductance). Therefore as $s \to \infty$, $Z_D(s) \to Ls$ (mostly) and so Z_D can have only a simple pole at infinity. Similarly, a driving point admittance, $Y_D(s)$, also may have only a simple pole at $s \to \infty$.

So, either $(m - n) = 1$ (a simple pole)

or, since $Z_D(s) = 1/Y_D(s)$, $(n - m) = 1$

Hence the degree *m* of the numerator polynomial *n*(*s*) can *differ* from the degree *n* of the denominator polynomial *d*(*s*) by at most one.

3. A network function cannot have its poles in the right half plane

This restriction applies to zeros also in the case of driving-point $Z_D(s)$ or $Y_D(s)$. Also, any pole on the $j\omega$ axis must be simple.

The above property stems from the reasoning given in the previous chapter regarding the response *vs* pole locations. Any pole lying on the positive real axis or any conjugate pole pair on the Right half plane, give rise to exponentially increasing responses. Thus with $Z_D(s)$, a step current input to the Z will give a voltage which grows infinitely. Likewise a voltage input to an $Y_D(s)$ will send an infinitely increasing current in it as time $\rightarrow \infty$. For a passive network, physical capacitors and inductors being only dissipative of energy, such conditions are impractical. So the poles must lie on the left half plane. Since Z_D and Y_D are reciprocals, we also find that for driving-point immittances, both *poles* and *zeros* must lie on the L.H.P. (Left half plane).

Poles lying on the $j\omega$ axis must be simple. A complex pole at $j\omega$, say of 2nd order, will have in its response, a term like, say

$$\frac{A}{(s - j\omega)^2}$$

It will give a response te^{jwt} which goes on increasing and hence unstable response.

If a network Z_D or Y_D has all its poles on the $j\omega$ axis only, then it is an L-C network, *i.e.*, contains no resistors. For, any resistive network will be dissipative and so its current or voltage will be decaying for large t in response to a step input. A decaying term of the nature $\exp(-\sigma t)$ must be present, which we know, arises out of a pole having a negative real part, *i.e.*, $-\sigma$; So an L-C network cannot have a pole having a negative real part. Since all the poles perforce lie on the L. H. P. only, the poles of a driving point function lie only on the $j\omega$ axis. What applies to an impedance, applies to an admittance also. As Z and Y are reciprocal, the zeros of one being the poles of the other, we see that zeros also must lie on the $j\omega$ axis.

4. The reflection property

$$H(-j\omega) = H^*(j\omega) \text{ (* denotes conjugate).}$$

$$\begin{aligned} H(j\omega) &= \frac{a_0 + a_1(j\omega) + \ldots + a_m(j\omega)^m}{b_0 + b_1(j\omega) + \ldots\ b_n(j\omega)^n} \\ &= \frac{(a_0 - a_2\omega^2 + a_4\omega^4 - \ldots) + j(a_1\omega - a_3\omega^3 + \ldots)}{(b_0 - b_2\omega^2 + b_4\omega^4 \ldots) + j(b_1\omega - b_3\omega^3 + \ldots)} \\ &= \frac{A + jB}{C + jD} \end{aligned}$$

$H^*(j\omega)$ is thus $(A - jB)/(C - jD)$

$H(-j\omega)$ is got by putting $-\omega$ in the place of every ω and we get $H(-j\omega) = (A - jB)/(C - jD) = H^*(j\omega)$.

The principle of property 4 means that a network is completely specified by its behaviour at positive frequencies only.

From the expression $A + jB/(C + jD)$ got above, we see that the real parts A and C are even functions of ω while the imaginary parts B and D are odd functions of ω.

$$\begin{aligned} H(j\omega) &= \{(AC + BD) + j(BC - AD)\} \div \{C^2 + D^2\} \\ &= \{(\text{even}) + j\ (\text{odd})\}. \end{aligned}$$

The real part is an *even* function, the imaginary part is *odd*.

10.2 SYNTHESIS OF L–C DRIVING POINT FUNCTIONS

We have seen that L–C network function (Z_D or Y_D)

(*i*) Are ratios of polynomials

(*ii*) Degree of $n(s)$ and $d(s)$ can at most differ by 1

(*iii*) Poles and zeros lie on the $j\omega$ axis.

Let us look into simple L–C networks as examples prior to the synthesis problem. A simple L–C series network

$$Z(s) = Ls + \frac{1}{sC} = \frac{L\left\{s^2 + (1/LC)\right\}}{s}$$

has its pole at origin and two zeros at $\pm\ j/\sqrt{LC}$.

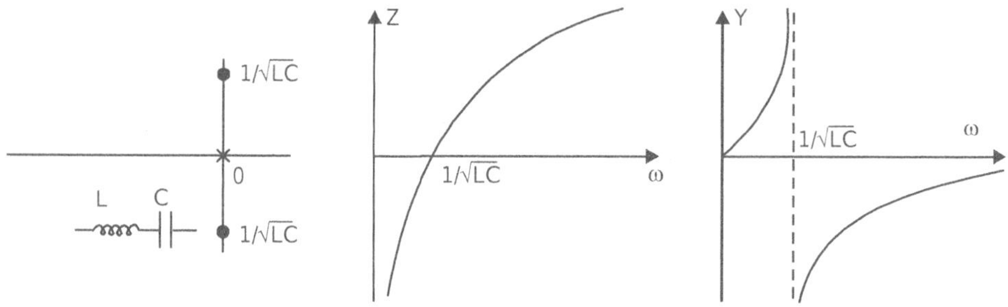

Fig. 10.2 A simple series L–C network

Its impedance function is plotted with ω in the X-axis. The admittance function is also shown. Note that Z becomes zero at $1/\sqrt{LC}$ and Y becomes infinite at $\omega = 1/\sqrt{LC}$.

Next consider a parallel LC network. Its behaviour is similarly represented in Fig. 10.3

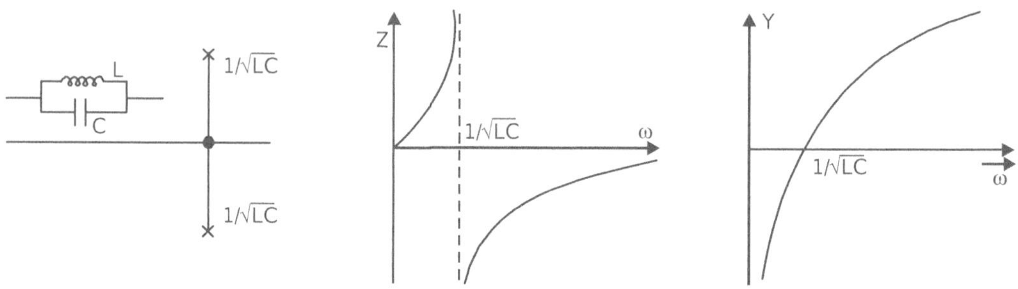

Fig. 10.3 A parallel L–C network

When we consider a most general LC network, we can get one such network either by

(*i*) Connecting a number of LC series branches in *parallel*

or

(*ii*) Connecting a number of LC parallel branches in series.

Note that connecting a number of series branches in series or a number of parallel branches in parallel is trivial–*i.e.*, amount to one single series branch or parallel branch only, by combination. The above two possibilities are shown in Fig. 10.4.

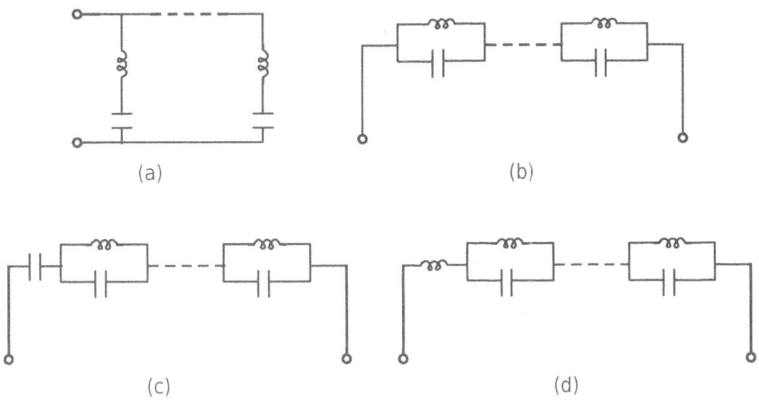

Fig. 10.4

The first network behaves as an open circuit for zero frequency and as an inductor Ls at infinity frequency; *i.e.*, it has a Z_D function having a pole at $s = 0$ and tends to infinity as $s \to \infty$. Its impedance versus frequency plot is shown in Fig. 10.5(*a*) below.

The second network of Fig. 10.4 behaves as a short circuit for zero frequency and as $s(=j\omega) \to \infty$, its capacitors short circuit the input. So it has a zero at $s = 0$ and tends to zero as $s \to \infty$. The plot is shown in Fig. 10.5(*b*) below.

Take the second network of Fig. 10.4. We can get still an LC network by adding either a series capacitor or a series inductor (or both) (Fig. 10.4(*a*) and Fig. 10.4(*d*)).

Figure 10.4(*c*) behaves, like on open circuit for $s = 0$ and a short circuit as $s \to \infty$. Its plot is as in Fig. 10.5(*c*). Figure 10.4(*d*) behaves as short circuit for $s = 0$ and behaves like L*s* for $s \to \infty$. (plot 10.5*d*)

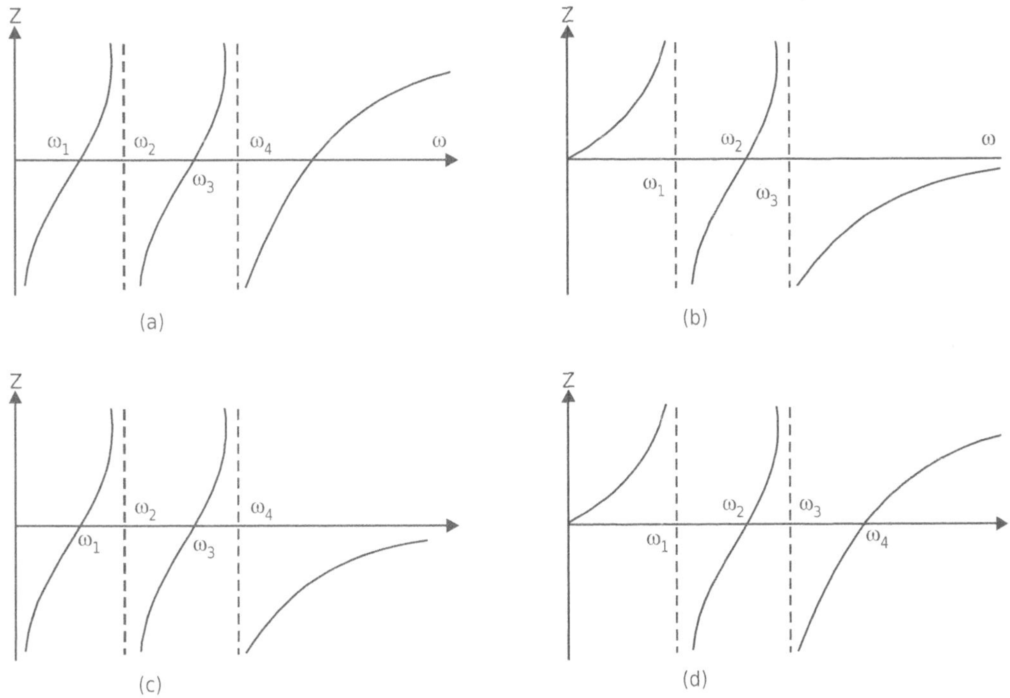

Fig. 10.5 Plots of Z for general LC networks

The second configuration of Fig. 10.4, using series *combinations* of *parallel branches* was first stated by Foster and is known as Foster's *first form*. The Fig. 10.4(*a*) and (*b*) are duals of each other. Fig. 10.4(*a*) is said to be in Foster's second form.

Example 1. *Synthesise the following impedance in F_1 form (Foster's first form).*

$$Z(s) = \frac{s(s^2+2)(s^2+4)}{(s^2+1)(s^2+3)}$$

Solution. This represents an LC network since its poles and zeros are purely imaginary (lie alternatively on the $j\omega$ axis).

Expand Z(s) into partial fractions.

$$Z(s) = \frac{s(s^2+2)(s^2+4)}{(s^2+1)(s^2+3)} = \frac{As}{s^2+1} + \frac{Bs}{s^2+3} + Cs$$

To find A and B, the following procedure is self explanatory.

$$A = \left.\frac{(s^2+2)(s^2+4)}{(s^2+3)}\right\}_{s^2=-1} = \frac{1\times 3}{2} = \frac{3}{2}$$

$$B = \left.\frac{(s^2+2)(s^2+4)}{s^2+1}\right\}_{s^2=-3} \frac{-1\times 1}{-2} = +\frac{1}{2}$$

The constant C is found as follows,

$$C = \lim_{s\to\infty} \frac{1}{s}\frac{s(s^2+2)(s^2+4)}{(s^2+1)(s^2+3)} = 1$$

$$Z(s) = \frac{3s}{2(s^2+1)} + \frac{s}{2(s^2+3)} + s$$

$$= \frac{1}{(2/3)s + 2/(3s)} + \frac{1}{2s+6/s} + s \text{ i.e.,} = \frac{1}{Y_1} + \frac{1}{Y_2} + \frac{1}{Y_3}$$

$Y_1 = (2/3)\,s + 1/(3/2s)$ and represents an admittance combining a capacitor of 2/3 Farad and an inductor of 3/2 henries in parallel.

$Y_2 = 2s + 6/s \rightarrow$ a 2F capacitor in parallel with an 1/6 H inductor.

Y_3 is a capacitor of one farad.

The network is shown in Fig. 10.6 below.

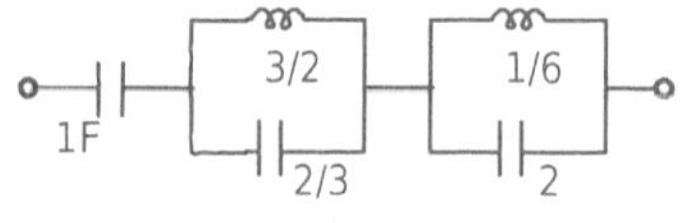

Fig. 10.6

Example 2. *Synthesise the network of example 1 in the Foster's second form (F_2 form).*

Solution. In this case, we use the admittance Y and expand it into partial fractions.

$$Y(s) = \frac{(s^2+1)(s^2+3)}{s(s^2+2)(s^2+4)}$$

$$= \frac{A}{s} + \frac{Bs}{s^2+2} + \frac{Cs}{s^2+4} \quad \text{where}$$

$$A = \left.\frac{(s^2+1)(s^2+3)}{(s^2+2)(s^2+4)}\right\}_{s=0} = \frac{3}{8}$$

$$B = \left.\frac{(s^2+2)}{s} \times \frac{(s^2+1)(s^2+3)}{s(s^2+2)(s^2+4)}\right\}_{s^2=-2} = \frac{(-1)(1)}{(-2)(2)} = \frac{1}{4}$$

$$C = \left.\frac{s^2+4}{s} \times \frac{(s^2+1)(s^2+3)}{s(s^2+2)(s^2+4)}\right\}_{s^2=-4} = \frac{(-3)(-1)}{(-4)(-2)} = \frac{3}{8}$$

Hence $Y(s) = (3/8)\,(1/s) + \frac{1}{4}s/(s^2 + 2) + (3/8)s/(s^2 + 4)$

$= Y_1 + Y_2 + Y_3$

where Y_1 = admittance due to an inductor 8/3

Y_2 = 1/(4s + 8/s), consisting of 4 henry inductor in series with $\frac{1}{8}$ F capacitor

Y_3 = the admittance of series combination having 8/3 henries and 3/32 farads. The realisation of the network is as in Fig. 10.7 shown.

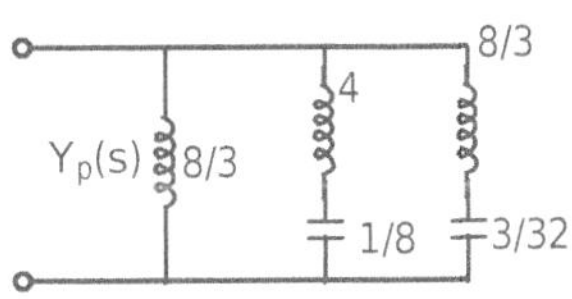

Fig. 10.7

The number of elements in the, F_1 and F_2 forms are equal in this case = 5.

10.3 THE CAUER FORMS OF SYNTHESIS

The Cauer method of synthesis obtains the network as a ladder type, the two Cauer forms of a general driving point synthesis being shown in (*a*) and (*b*) of Fig. 10.8 below.

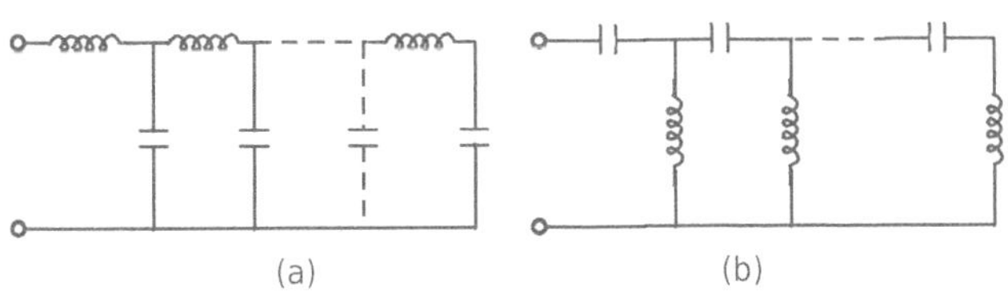

Fig. 10.8

The expansion of Z_D in C_1 form is done by alternately applying the methods of F_1 and F_2 forms in the following manner.

1. Expand $Z(s)$ in F_1 form so that

$$Z(s) = \underbrace{A_1 s}_{\text{Inductor in in } F_1 \text{form}} + \underbrace{Z_1(s)}_{\text{Remainder term}}$$

2. Expand $Z_1(s)$ in F_2 form such that

$$Y_1 = \frac{1}{Z} = \underbrace{B_2 s}_{\text{Capacitor in parallel}} + \underbrace{Y_2}_{\text{Remainder term}}$$

Sometimes step 1 may have to be bypassed, *i.e.*, when the degree of numerator n(s) is equal or less than that of denominator $d(s)$.

The application of the above steps results in the synthesis of figure shown here. Then we find Z_2, the reciprocal of Y_2 and continue the above two steps once again, etc., until the entire network is completed.

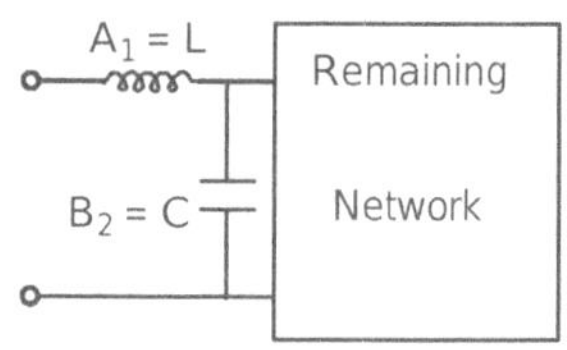

Fig. 10.9

It can be shown that the number of elements required will be equal to the degree of numerator or denominator of $Z_D(s)$, whichever is greater.

Example 3. *Synthesise the following driving-point impedance in the Cauer form.*

$$Z(s) = \frac{s(s^2+2)(s^2+4)}{(s^2+1)(s^2+3)}$$

Solution. *Cauer's* first term. Step 1 is possible since the numerator is one degree more than denominator. So write as

$$Z = A_1 s + Z_1.$$

We expand numerator and denominator and divide:

$$s(s^2 + 2)(s^2 + 4) = s^5 + 6s^3 + 8s$$

$$(s^2 + 1)(s^2 + 3) = s^4 + 4s^2 + 3$$

$$s^4 + 4s^2 + 3)\ s^5 + 6s^3 + 8s\ (1s = A_1 = \text{Inductor 1H}$$

$$\underline{s^5 + 4s^3 + 3s}$$

$$2s^3 + 5s$$

Hence $\quad A_1 = 1$ and $Z_1 = (2s^3 + 5s) \div (s^4 + 4s^2 + 3)$

Now we perform step 2 and

$$Y_1 = s^4 + 4s^2 + 3/(2s^3 + 5s)$$

Hence the first inductor is $A_1 = 1$ henry. Thereafter, the process is continued as shown in the following division table. There are 3 inductors and two capacitors in the network.

$$\frac{1}{2}s \qquad B_1 = \text{Capacitor } \tfrac{1}{2}\text{F}$$

$$2s^3 + 5s)s^4 + 4s^2 + 3$$

$$s^4 + 2\tfrac{1}{2}s^2 \qquad (4/3)s \rightarrow \text{inductor } 4/3\text{H}.$$

$$1\tfrac{1}{2}s^2 + 3)2s^3 + 5s$$

$$2s^3 + 4s \quad 1\tfrac{1}{2}s \rightarrow C = 1\tfrac{1}{2}\text{F}$$

$$s)1.5s^2 + 3$$

$$1.5s^2 \quad \tfrac{1}{3}s \rightarrow L = \tfrac{1}{3}.$$

$$3)s$$

$$s$$

$$0$$

The network is shown. (Fig. 10.10)

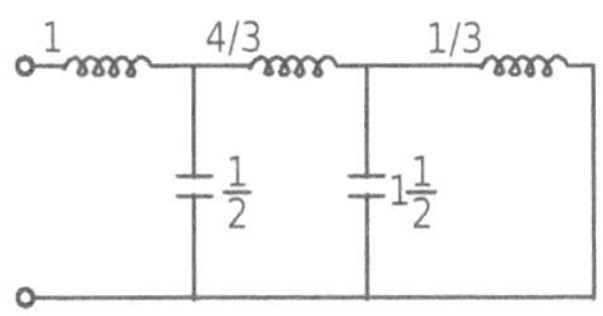

Fig. 10.10

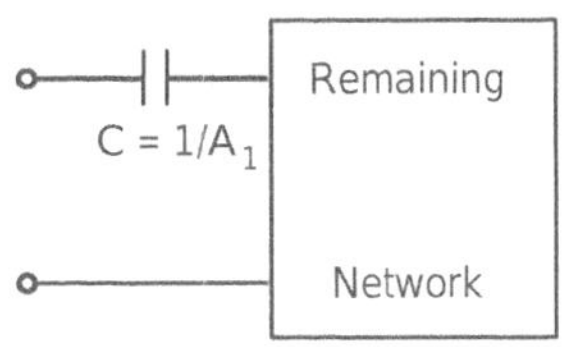

Fig. 10.11

10.4 CAUER'S SECOND FORM

In the second form, a series capacitor is first removed. So, Z is written in the form.

Step 1. $$Z(s) = A_1/s + Z_1(s)$$

and the network is now as in Fig. 10.11 above.

Step 2. Step 1 is possible only if the Z has a pole at the origin $s = 0$. Otherwise, the step 2 is carried out straight away.

$$Y_1 = 1/Z_1 = B_2/s + Y_2$$

B_2/s represents a shunt inductor of value $1/B_2$.

In the practical calculation, to express the above equation, it is necessary to perform division after arranging the co-efficients in ascending powers of s. Step 1 and step 2 repeated alternately, completes the synthesis in the C_2 form. The network is as in Fig. 10.8*b*.

Example 4. *Synthesise in the C_2 form*

$$Z(s) = \frac{s(s^2 + 3)}{s^2 + 2}$$

Solution. The impedance does not have a pole at the origin. So, step 1 is omitted. We start at step 2, after writing $Y(s)$ in ascending powers of s as

$$Y(s) = (2 + s^2)/(3s + s^3)$$

$$
\begin{array}{l}
\quad\quad 2/3s \qquad\qquad\qquad\qquad \rightarrow L = 3/2 \text{ henry} \\
3s + s^3)\ 2 + s^2 \\
\quad\quad 2 + 2/3s^2 \qquad 9/s \qquad\quad \rightarrow C = 1/9F \\
\quad\quad \tfrac{1}{3}s^2)\ 3s + s^3 \\
\quad\quad\quad 3s \qquad\qquad \tfrac{1}{3}/s \qquad \rightarrow L = 3H \\
\quad\quad\quad s^3)\ \tfrac{1}{3}s^2 \\
\quad\quad\quad\quad \tfrac{1}{3}s^2 \\
\quad\quad\quad\quad 0
\end{array}
$$

The network is as shown.

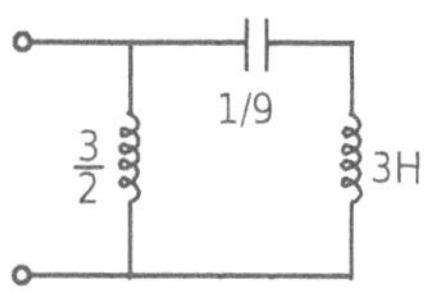

Fig. 10.12

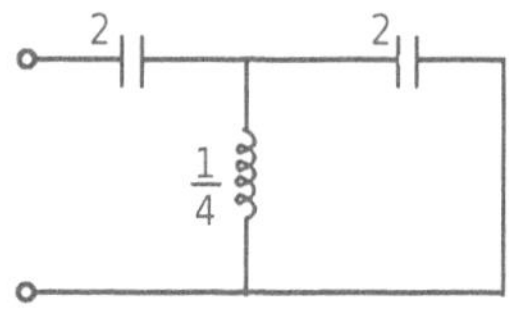

Fig. 10.13

Example 5. Synthesise Z = $(s^2 + 1)/s(s^2 + 2)$ in C_2 form.

As there is a pole at s = 0, step 1 is possible.

Solution.

$$
\begin{array}{l}
\quad\quad 1/2s \qquad\qquad\qquad\qquad \rightarrow \text{Series } C = 2\text{ F} \\
2s + s^3)\ 1 + s^2 \\
\quad\quad 1 + \tfrac{1}{2}s^2 \qquad 4/s \qquad\quad \rightarrow L = \tfrac{1}{2}H \\
\quad\quad \tfrac{1}{2}s^2)\ 2s + s^3 \\
\quad\quad\quad 2s \qquad\qquad \tfrac{1}{2}/s \qquad \rightarrow C = 2F \\
\quad\quad\quad s^3)\ \tfrac{1}{2}s^2 \\
\quad\quad\quad\quad \tfrac{1}{2}s^2 \\
\quad\quad\quad\quad 0
\end{array}
$$

Hence the network appears as shown. (Fig. 10.13)

10.5 FOSTER'S REACTANCE THEOREM AND THE SEPARATION PROPERTY

This theorem states that if the poles and zeros of a driving point LC network (called the *Critical frequencies*) and the value of the gain factor K are known, the network can be realised.

Alternative to the gain factor, it is enough if the reactance at any non-critical frequency (neither a zero nor a pole) is known or the value of $dX/d\omega$, the slope of the curve at any point, is known. The nature of the curves has been shown already in Fig. 10.5.

A look at any of the curves in Fig. 10.5 makes it clear that the impedance (which, of course, represents a reactance, since these are for LC networks) always has a positive slope, and so $dX/d\omega$ is always *positive*. Between any two poles, the impedance increases from $-\infty$

to + ∞, and passes via a zero between the poles. Thus, there is one zero between any two poles. Hence, a driving point reactance function has its poles and zeros alternatively; the poles and zeros *interlace* each other.

If $\omega_1, \omega_3, \omega_5$ ... the poles in order, then we have seen that the partial fraction expansion Z(s) has the form

$$Z(s) = \frac{A_o}{s} + \frac{A_1 s}{s^2 + \omega_1^2} + \frac{A_2 s}{s^2 + \omega_3^2} + \ldots + A_n s$$

Since $$Z(j\omega) = jX(\omega), \qquad \text{(for LC networks)}$$

$$X(\omega) = \frac{1}{j} Z(j\omega) = -\frac{A_o}{\omega} + \frac{A_1 \omega}{\omega_1^2 - \omega^2} + \ldots + A_n \omega$$

Differentiating with respect to ω,

$$\frac{dX}{d\omega} = \frac{A_o}{\omega^2} + \frac{A_1(\omega_1^2 + \omega^2)}{(\omega_1^2 - \omega^2)^2} + \ldots \ldots + A_n$$

Note that for all real values of ω, the above expression is non-negative. Hence, the slope of the curve of X(ω) with respect to ω,

i.e., $$dX(\omega)/d\omega > 0 \tag{10.5}$$

The slope can be positive only if, as we discussed just before, the poles and zeros alternate.

The above property imposes another restriction on the realisation of LC immittances Thus.

$$\frac{s(s^2+2)(s^2+3)}{(s^2+4)(s^2+5)}$$

cannot be realised since its poles are

$$j2, j\sqrt{5}$$

while its zeros are $$s = 0,\ s = j\sqrt{2},\ s = j\sqrt{3}.$$

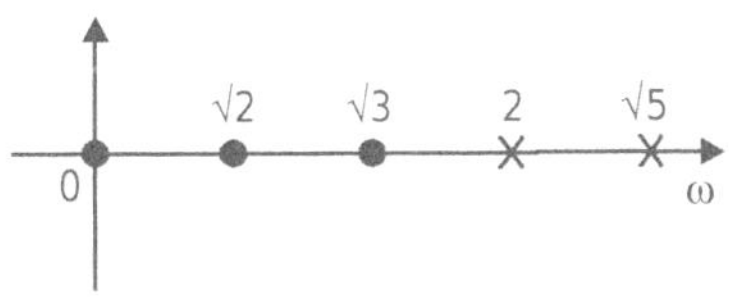

Fig. 10.14

The pattern of these critical frequencies being as shown, clearly reveals that the poles and zeros do not interlace each other.

Example 6. *A reactance function has poles at ω = 0, ω = 2 and ω = 4. At these frequencies, the corresponding susceptance slope has a value equal to unity. Determine the zeros of the reactance and realise it.*

Solution. X(s) is of the from $\dfrac{A(s^2+B)(s^2+C)}{s(s^2+4)(s^2+6)}$

where A, B, C are the unknowns. The susceptance B(s) is

$$B(s) = \frac{s(s^2+4)(s^2+16)}{A(s^2+B)(s^2+C)}$$

$$\frac{dB}{ds} = \frac{[(d/ds)\text{of Numerator}].(\text{Den}) - [(d/ds)\text{Denom}].(\text{Num.})}{(\text{Denominator})^2}$$

But the value of the numerator is zero at $\omega = 0$, 2 or 4.

$\therefore$ dB/ds at ω = 0, 2 or 4 can be calculated from

$$\frac{[d/ds(\text{Num.})] \times \text{Denom.}}{(\text{Denom})^2} = \frac{d/ds(\text{N})}{\text{D}}$$

$$d/ds\ (\text{Numerator}) = [(s^2 + 4)(s^2 + 16) + 2s^2(s^2 + 16) + 2s^2(s^2 + 4)]$$

At $s = 0, \dfrac{dB}{ds} = \dfrac{64}{ABC} = 1$

At $\omega = 2$ or $s^2 = -4, \dfrac{dB}{ds} = \dfrac{-8 \times 12}{A(B-4)(C-4)} = 1$

At $\omega = 4\ s^2 = -16, \dfrac{dB}{ds} = \dfrac{-32(-12)}{A(B-16)(C-16)} = 1'$

from which A = 5, B + C = 12 and B = 10.8, C = 1.2

Hence
$$X(s) = \frac{5(s^2+1.2)(s^2+10.8)}{s(s^2+4)(s^2+16)}$$

The zeros of X(s) are thus $\pm j\sqrt{1.2}$ and $\pm j\sqrt{10.8}$

For realising the reactance, say, in F_1 form, expand as below:

$$\frac{5(s^2+1.2)(s^2+10.8)}{s(s^2+4)(s^2+16)} = \frac{A}{s} + \frac{Bs}{s^2+4} = \frac{Cs}{s^2+16}$$

Cross multiplying,

$$5(s^2 + 1.2)(s^2 + 10.8) = A(s^2 + 4)(s^2 + 16) + Bs^2(s^2 + 4) + Cs(s^2 + 4)$$

Equating co-efficients of like powers of s,

$$A = 1,\ B = 2,\ C = 2$$

$A/s = 1/s$ gives a series capacitor of 1F.

$2s/(s^2 + 4)$ is a || branch of $\frac{1}{2}$ H and $\frac{1}{2}$F.

$2s/(s^2 + 16)$ is a || branch of $\frac{1}{8}$ H and $\frac{1}{2}$ F.

All the three are in series to realise the required X(s).

10.6 R.C. DRIVING-POINT IMMITTANCES

Whereas LC networks are lossless, an RC network has a loss due to the resistance R. The Foster's form of an RC driving point impedance can be drawn by replacing the inductances by resistances. The general F_1 and F_2 forms of an RC immittance looks as in Fig. 10.15.

Considering the figure of the F_1 form, each of the sections having R and C in parallel has a general form

$$\frac{R.(1/Cs)}{R+1/Cs} = \frac{R}{1+sCR} = \frac{A}{(s+\sigma)}$$

and so the total impedance of an RC network is

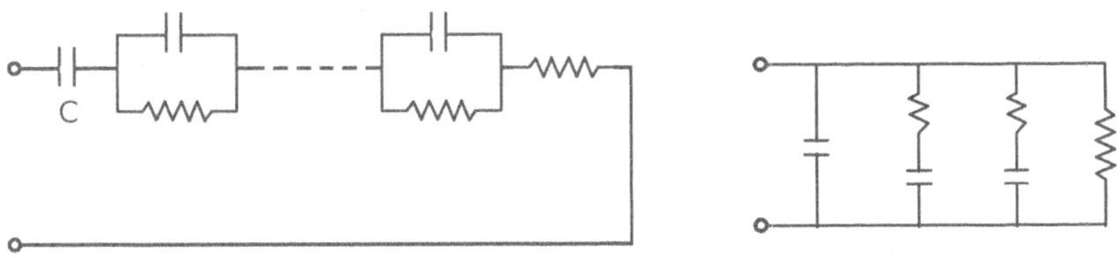

Fig. 10.15 F_1 and F_2 forms of RC networks

$$Z_{RC}(s) = \frac{A_0}{s} + \frac{A_1}{s+\sigma_1} + \frac{A_2}{s+\sigma_2} + \ldots + A_n \quad (10.6)$$

The first and last terms are due to the single capacitance and single resistance shown in Fig. 10.15. The simplification into a common denominator form yields:

$$Z_{RC}(s) = K\frac{(s+\sigma_2)(s+\sigma_4)\ldots}{s(s+\sigma_1)(s+\sigma_3)\ldots} \quad (10.7)$$

A look at the above shows that:

1. The poles are at $s = -\sigma_1, -\sigma_3, \ldots$ and since the σ's are real and positive, ($\sigma = 1/CR$) the poles lie on the negative real axis.
2. As $s \to \infty$, every capacitor behaves as a short circuit, the value of $Z(s)$ for large s cannot become infinite. Hence, there is no pole at $s = \infty$ as in case of an LC impedance. So, the numerator of (10.4) cannot exceed that of the denominator.
3. At low frequencies a capacitor behaves like an open circuit and so the impedance $Z(s)$ will be infinite. This means that the first critical frequency can only be a pole.
4. Plots of $Z_{RC}(s)$ as s varies on the real axis or σ-axis can be made. For the network of Fig. 10.15 shown, the pattern will look as in Fig. 10.16.

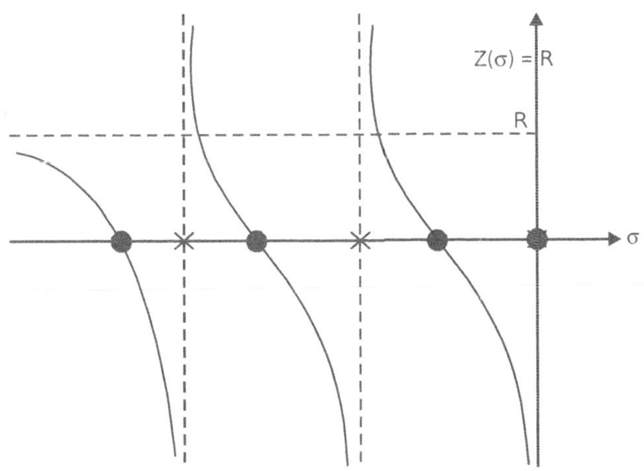

Fig. 10.16

From the Fig. 10.16 we find that the slope of the input resistance of the network is negative when plotted with respect to σ.

$$dZ(\sigma)/d\sigma < 0$$

We also find the poles and zeros interlacing each other.

5. If the series capacitor C is absent in the network of the F_1 form (Fig. 10.12), then at $s = 0$, the network does not have a pole at $s = 0$. In that case, the first pole occurs at $\sigma < 0$, near the origin. So, the two possible pole zero plots look like Fig. 10.17 below.

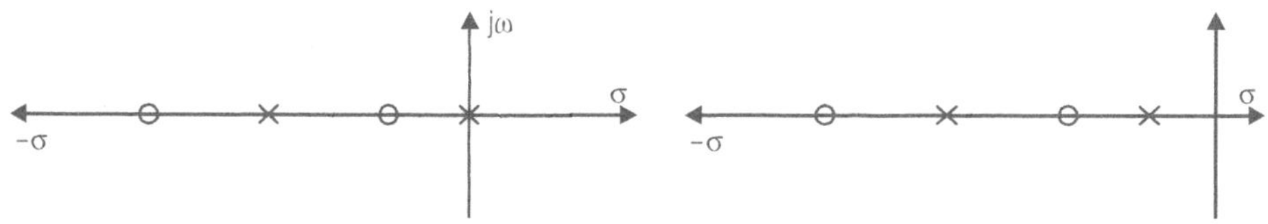

Fig. 10.17 Pole-zero plots of RC networks

10.7 REALISATION IN THE FOUR CANONIC FORMS

The RC driving-point impedance, subject to the realisability conditions, can be realised in the F_1, F_2, C_1 and C_2 forms.

Example 7. *Find the F_1 form of :*

$$Z(s) = \frac{s+3}{(s+1)(s+5)}$$

Solution. The poles are at –1, –5 and zero is at –3. They interlace each other. The first critical frequency (nearest to origin) is a pole. There is a zero at ∞. Thus it is realisable.

$$Z(s) = \frac{A}{s+1} + \frac{B}{s+5}$$

$$A = \left.\frac{(s+1)}{1} \frac{s+3}{(s+1)(s+5)}\right\}_{s=-1} = \frac{1}{2}$$

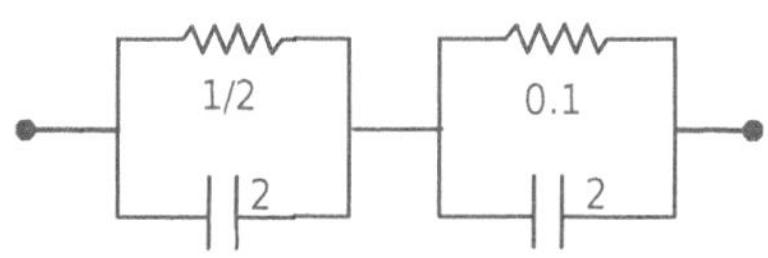

Fig. 10.18

$$B = \left.\frac{(s+5)}{1} \times \frac{s+3}{(s+1)(s+5)}\right\}_{s=-5} = \frac{1}{2}$$

$$Z(s) = \frac{1}{2}\frac{1}{s+1} + \frac{1}{2}\frac{1}{s+5}$$

The general relation is $1/C \div (s + 1/CR)$, so that the first term gives rise to a 2 F capacitor in parallel with a resistor $R = \frac{1}{2}$.

Example 8. *Realise in the F_2 form.*

$$Z(s) = \frac{(s+1)(s+3)}{s(s+2)(s+4)}$$

Solution. Analogous to the LC synthesis, we have to expand Y(s) in partial fractions. Each branch of the F_2 network has a resistor and a capacitor in series. So, each branch has an impedance of the form R + (1/Cs) and hence the admittance is Cs/(1 + sCR) or s/(s + σ) form. But expansion in partial fractions can be made only in terms of A/(s + σ) and not in s/(s + σ) form. So, the *method is to expand Y(s) / (s) in partial fractions*, and not Y(s). This will be clear now.

$$\frac{Y(s)}{s} = \frac{(s+2)(s+4)}{(s+1)(s+3)} = \frac{A}{s+1} + \frac{B}{s+3}$$

where $$A = 3/2 \text{ and } B = \frac{1}{2}.$$

$$Y(s) = \frac{(3/2)s}{s+1} + \frac{\frac{1}{2}s}{s+3}$$

The first term gives R + 1/Cs = (2/3) + (2/3s) so that R = 2/3 and C = 3/2F. Similarly, the second term gives 2 + 6/s which has R = 2 and C = 1/6.

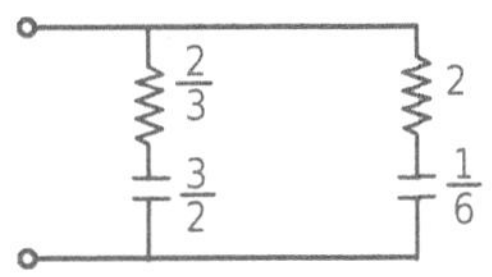

Fig. 10.19

Example 9. *Realise in the Cauer forms:*

$$Z(s) = \frac{(s+1)}{s(s+2)}$$

Solution. The Cauer forms of RC networks are similar to those of LC networks, if inductors are replaced by resistors. The C_1 form has a resistor for its first element while the C_2 form has a capacitor.

C_1 form: Arrange numerator and denominator in ascending powers of s and perform continued division. The quotients are alternatively resistors and capacitors.

$$2s + s^2)\ 1 + s\ (\tfrac{1}{2}/s \qquad \text{gives } C = 2$$

$$1 + \tfrac{1}{2}s$$

$$\tfrac{1}{2}s)\ 2s + s^2\ (4 \qquad \text{gives } R = \tfrac{1}{4}$$

$$2s$$

$$s^2)\ \tfrac{1}{2}s\ (\tfrac{1}{2}/s \qquad C = 2$$

$$\tfrac{1}{2}s$$

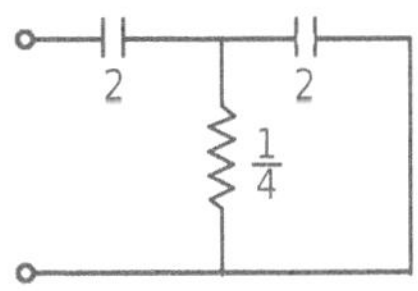

Fig. 10.20

Example 10. *Synthesise in the C_2 form: $Z = (s + 3)/(s + 2)$.*

Writing it as $(3 + s)/(2 + s)$, the first division gives $3/2$, but in the C_2 form the first term must be a capacitor, i.e., must be of the form A/s. So, the first step is omitted and there is no series capacitor.

Solution.

$$3 + s\,)\,2 + s\,(\,2/3 \qquad \text{gives R} = 3/2$$

$$\underline{2 + 2s/3}$$

$$\tfrac{1}{3}s\,)\,3 + s\,(\,9 \qquad \text{gives C} = 1/9$$

$$\underline{3}$$

$$s\,)\,\tfrac{1}{3}s\,(\,\tfrac{1}{3} \qquad R = 3$$

$$\underline{\tfrac{1}{3}s}$$

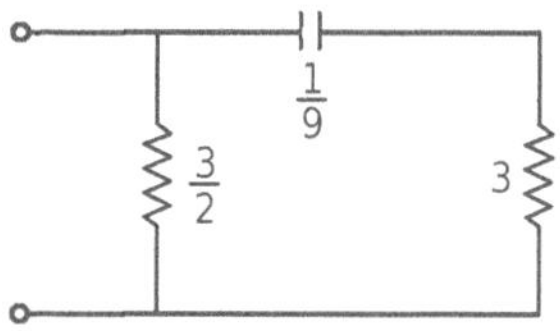

Fig. 10.21

Example 11. *The impedance Z(s) of the network shown has poles at $s = -1$ and -3; its zeros are $s = -2$ and -4. The resistor is $R = 7/3\Omega$. Determine the other elements.*

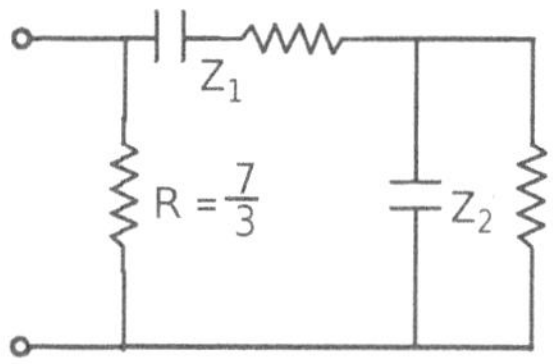

Fig. 10.22

Solution. The impedance $Z_D(s)$ is got from data is

$$Z_D(s) = \frac{A(s+2)(s+4)}{(s+1)(s+3)}$$

where A is the unknown. But the value of the constant A can be determined by looking at the network. Suppose we give a D.C. supply ($\omega = 0$) at the input points. The network behaves, as a resistor R = 7/3, because the capacitor in Z_1 blocks D.C. and so the elements to the right of this capacitor carry no current. Thus for $\omega = 0$ or $s = 0$, the value of $Z_D(s) \rightarrow 7/3$.

$$A \times 2 \times 4/(1 \times 3) = 7/3$$

giving

$$A = 7/8$$

Suppose we remove the resistor R. What will be the network function after removing it? For that find $Y_D(s)$ and subtract the conductance 3/7 from it. We get

$$Y_D(s) - 3/7 = \frac{8(s+1)(s+3)}{7(s+2)(s+4)} - \frac{3}{7}$$

$$= \frac{s(5s+14)}{7(s+2)(s+4)}$$

$$\therefore \qquad Z_1 + Z_2 = \frac{7(s+2)(s+4)}{s(5s+14)} = A + \frac{B}{s} + \frac{C}{5s+14}$$

In the above, $Z_1 + Z_2$ is the impedance of the remaining network, which has been expanded in partial fractions. We get

$$A = \frac{7}{5},\ B = 4,\ C = \frac{12}{5}$$

So $Z_1 = A + (B/s)$ represents a resistor of 1.4 and a capacitor of $\frac{1}{2}$ F.

The impedance Z_2 is

$$\frac{12/5}{5s+14} = \frac{12/(5\times 14)}{1+(5/14)s}$$

giving R = 6/35 and C = 25/12F.

10.8 SYNTHESIS OF R-L ONE-PORTS

We saw that the RC one-port can be synthesised if the poles and zeros lie on the negative real axis alternatively and nearest to the origin we have a pole. Similarly an LC network can be shown to have its poles and zeros on the $-\sigma$ axis, but in this case, it is a zero which is nearest to the origin. The simple reason for this is that an inductor has low impedance at lower frequencies, while a capacitor has high impedance for low frequencies. So, an R-L one port has a zero near origin, but R-C one port has a pole near the origin.

The development of the R-L one port can be made in all the four canonic forms. The synthesis will have inductors in place of capacitors in the figures of 10.14.

Example 12. *Synthesise in the F_1 from:*

$$Z = \frac{2s^2+6s+3}{1+s}$$

Solution. This function has its pole at $s = -1$; its zeros are

$$s = (-3 \pm \sqrt{3})/2,\ i.e.,\ -0.634 \text{ and } -2.366.$$

So, the given function has all its poles and zeros lying alternatively on the $-\sigma$ axis, and it has a zero nearest to the origin. So it can be synthesised as an R-L network.

An F_1 form is that of a series connection of one or more R-L parallel branches. Each R-L branch will have an impedance of the form

$$\frac{RLs}{R+Ls} \text{ or } \frac{As}{s+B}$$

but a partial fraction expansion cannot have a term like $As/(s + B)$ but only as $A/(s + B)$. So, before performing the expansion, we divide by s and then expand. (*cf*: F_2 form synthesis of an R-C network).

$$\frac{Z(s)}{s} = \frac{2s^2+6s+3}{s(1+s)} = P + \frac{Q}{s} + \frac{R}{1+s}$$

We get P = 2, Q = 3, R = 1.

$$Z(s) = 2s \quad + \quad 3 \quad + \quad \frac{s}{(1+s)}$$

↑ An inductor of 2H

↑ A resistor of 3 Ω

↑ A one ohm Resistor in || to one henry inductor.

Example 13. *A certain R-L one port impedance has a pole at s = –1 and has two zeros. It behaves as a resistance of 2/3 ohm for DC. The slope of the admittance at the pole is –3. Find the impedance.*

Solution. Assume that its zeros are at $-a$ and $-b$. The constant term is 2/3, since it is given that Z(*s*) is 2/3 as s → 0.

$$Z(s) = \frac{2(s+a)(s+b)}{3(s+1)}$$

Hence at $s \to 0$, $Z(0) = 2ab/3$, and so $ab = 1$

$$Y(s) = \frac{(3/2)(s+1)}{(s+a)(s+b)}$$

dY/ds can be got from $\dfrac{d/ds(\text{numerator})}{\text{Denominator}}$ at $s = -1$

$$\therefore \qquad -3 = \left.\frac{3/2}{(s+a)(s+b)}\right\}_{s=-1} = \frac{3/2}{(a-1)(b-1)}$$

$$(a - 1)(b - 1) = -\frac{1}{2}$$

$$ab - (a + b) + 1 = -\frac{1}{2}$$

As $ab = 1$, we get $(a + b) = 2\frac{1}{2}$. Thus $a = 2$, $b = \frac{1}{2}$.

The function has zeros at $-\frac{1}{2}$, –2 and a pole at –1.

The network in F_2 form has two branches, one of (2Ω + 1H), another of (1Ω + 2H).

10.9 ABCD PARAMETERS*

For a two port network, (Fig. 10.23), the input voltage and current (V_1, I_1) and output voltage and current (V_2, I_2) are related by the expression.

$$\left.\begin{aligned} V_1 &= AV_2 + BI_2 \\ I_1 &= CV_2 + DI_2 \end{aligned}\right\} \text{ or, } \begin{vmatrix} V_1 \\ I_1 \end{vmatrix} = \begin{vmatrix} A & B \\ C & D \end{vmatrix} \begin{vmatrix} V_2 \\ I_2 \end{vmatrix} \qquad (10.8)$$

The matrix [A B C D] is called the transfer matrix of the two port.

* From this, included in 1 year ECE (Anna University)

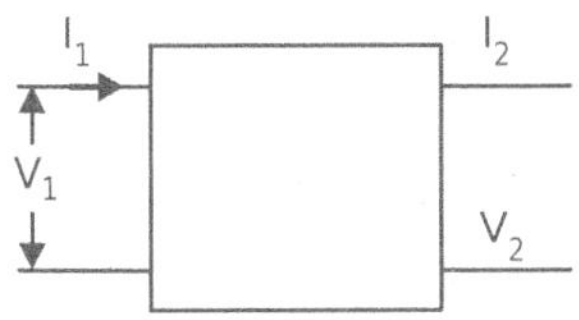

Fig. 10.23

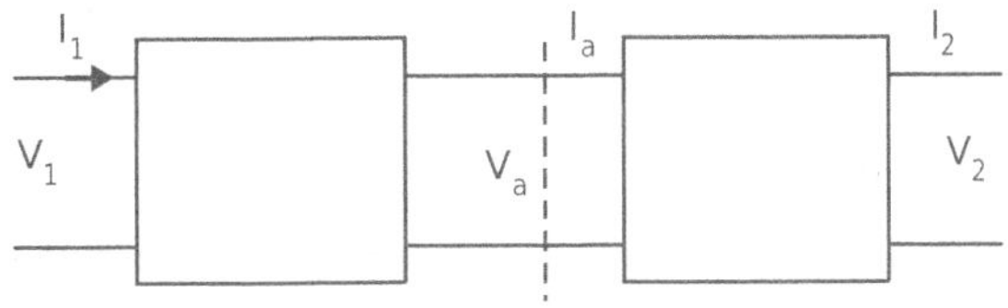

Fig. 10.24

When, two-ports are in cascade as in Fig. 10.24,

$$\begin{vmatrix} V_1 \\ I_1 \end{vmatrix} = \begin{vmatrix} A_1 & B_1 \\ C_1 & D_1 \end{vmatrix} \begin{vmatrix} V_a \\ I_a \end{vmatrix} = \begin{vmatrix} A_1 & B_1 \\ C_1 & D_1 \end{vmatrix} \begin{vmatrix} A_2 & B_2 \\ C_2 & D_2 \end{vmatrix} \begin{vmatrix} V_2 \\ I_2 \end{vmatrix}$$

which means that we get the transfer matrix of the cascaded network by multiplication of the two matrices.

Example 14. *For each of the two ports in Fig. 10.25, find the ABCD parameters.*

Solution.

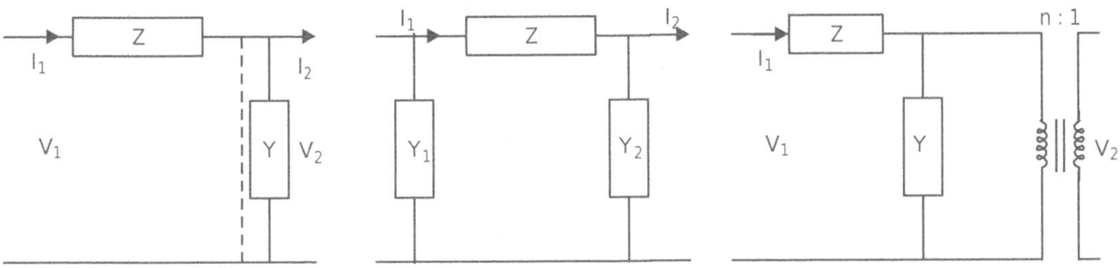

(*a*) (*b*) (*c*)

Fig. 10.25

(*a*) For Fig. (*a*), there are two cascaded parts, one Z and one Y, as shown by a dotted line. For the series Z,

$$\begin{aligned} V_1 &= V_a + ZI_a \\ I_1 &= I_a \end{aligned} \quad \text{or} \quad \begin{vmatrix} V_1 \\ I_1 \end{vmatrix} = \begin{vmatrix} 1 & Z \\ 0 & 1 \end{vmatrix} \begin{vmatrix} V_a \\ I_a \end{vmatrix}$$

For the shunt admittance Y, we know that

$$\begin{aligned} V_a &= V_2 \\ I_a &= I_2 + V_2 Y \end{aligned} \qquad \begin{vmatrix} V_a \\ I_a \end{vmatrix} = \begin{vmatrix} 1 & 0 \\ Y & 1 \end{vmatrix} \begin{vmatrix} V_2 \\ I_2 \end{vmatrix}$$

Hence the ABCD parameters of (a) will be

$$\begin{vmatrix} 1 & Z \\ 0 & 1 \end{vmatrix} \begin{vmatrix} 1 & 0 \\ Y & 1 \end{vmatrix} = \begin{vmatrix} (1+YZ) & Z \\ Y & 1 \end{vmatrix}$$

(*b*) For (*b*), we consider it as a series connection of a shunt Y and a two port of Fig. (a). So,

$$\begin{vmatrix} 1 & 0 \\ Y & 1 \end{vmatrix} \begin{vmatrix} (1+YZ) & Z \\ Y & 1 \end{vmatrix} = \begin{vmatrix} 1+YZ & Z \\ 2Y+Y^2Z & 1+YZ \end{vmatrix}$$

since the product of the two matrices gives the overall transfer matrix for cascaded 2 ports.

(*c*) In (c), we have a transformer in cascade with a two-port of Fig.(a) For a transformer,

$$\begin{aligned} V_1 &= nV_2 \\ I_1 &= I_2/n \end{aligned} \quad \text{or,} \quad \begin{vmatrix} V_1 \\ I_1 \end{vmatrix} = \begin{vmatrix} n & 0 \\ 0 & 1/n \end{vmatrix} \begin{vmatrix} V_2 \\ I_2 \end{vmatrix}$$

Cascading this with the matrix of Fig. (*a*), we get

$$\begin{vmatrix} V_2 \\ I_2 \end{vmatrix} = \begin{vmatrix} 1+ & YZ & Z \\ Y & & 1 \end{vmatrix} \begin{vmatrix} n & 0 \\ 0 & 1/n \end{vmatrix} = \begin{vmatrix} n(1+YZ) & Z/n \\ nY & 1/n \end{vmatrix} \begin{vmatrix} V_2 \\ I_2 \end{vmatrix}$$

10.10 RELATION BETWEEN THE PARAMETERS

If the network is passive, we may show that

$$AD - BC = 1 \tag{10.9}$$

Consider the two port being short circuited at port 2.

Since $V_2 = 0$ we get, under this condition,

$$V_1 = BI_2 \tag{10.10}$$

$$I_1 = DI_2$$

Now, remove the short circuit at port 2 and short circuit port 1. Now $V_1 = 0$.

$$0 = AV - BI_2'$$

$$-I_1' = CV - DI_2' \tag{10.11}$$

We put $-I_1'$ in the above eqn. since, I_1 now flows right to left. Applying reciprocity theorem, we know that $I_1' = I_2$. Hence from (10.10) and (10.11),

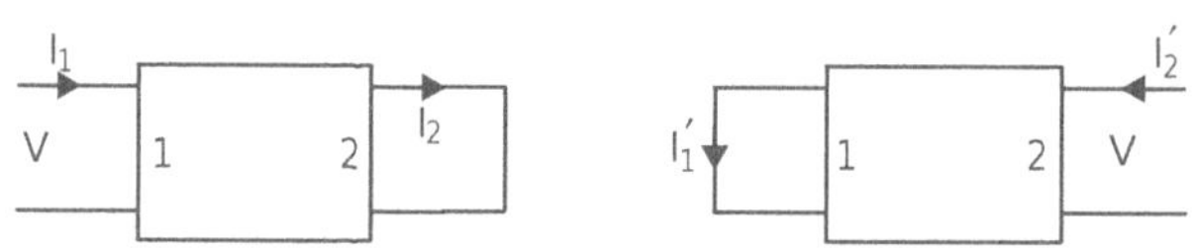

(*a*) Port 2 shorted (*b*) Port 1 shorted

Fig. 10.26

$$-I_2 = CV - DAV/B$$

$$-V/B = CV - DAV/B, \text{ which yields (10.9).}$$

If the 2-port is symmetrical, we can show that

$$A = D \tag{10.12}$$

Consider the input impedance at port 1 in Fig. 10.26(a).

$$Z_{1in} = V/I_1 = B/D.$$

The impedance at port 2 in Fig. 10.26(b) will be

$$Z_{2in} = V/I_2' = B/A,$$

but both Z_1 and Z_2 will be equal for a symmetric two-port. Hence A = D.

10.11 IMAGE IMPEDANCE AND ITERATIVE IMPEDANCE

When power is transferred from a source to a load via a 2-port network (used as a filter or equaliser), we require that the power be maximum; so, we have to match the resistance seen at the load terminals to the source resistance. (Maximum power transfer theorem).

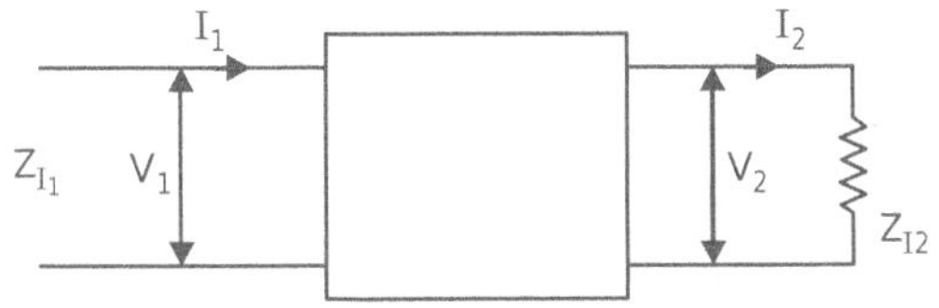

Fig. 10.27

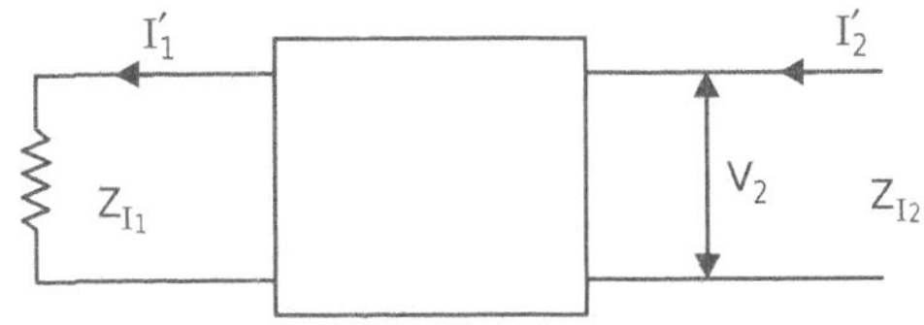

Fig. 10.28

Image impedances are defined for Figs. 10.27 and 10.28. The 2-port is loaded by Z_{I_2} at port-2 and the impedance at port 1 is Z_{I_1}. When the port 1 is loaded by Z_{I_1}, the impedance at port 2 is Z_{I_2}. (Image-like property). For Fig. 10.27.

$$V_1 = AV_2 + BV_2/Z_{I_2}$$

$$I_1 = CV_2 + DV_2/Z_{I_2}$$

So
$$R_{I1} = \frac{V_1}{I_1} = \frac{A + B/Z_{I_2}}{C + D/Z_{I_2}} = \frac{AZ_{I_2} + B}{CZ_{I_2} + D} \qquad (a)$$

From Fig. 10.28,

$$V_1 = AV_2 - BI_2' = (AV_2/I_2' - B)\, I_2' = (AZ_{I_2} - B) I_2'$$

$$-I_1 = CV_2 - DI_2' = (CV_2/I_2' - D)\, I_2' = (CZ_{I_2} - D) I_2'$$

Dividing, we get

$$-Z_1 = (AZ_2 - B)/(CZ_2 - D) \qquad (b)$$

From (a) and (b), we get

$$(AZ_2 + B)/(CZ_2 + D) = -(AZ_2 - B)/(CZ_2 - D)$$

from which
$$Z_{I_2} = \sqrt{BD/AC}$$

and using (a),
$$Z_{I_1} = \sqrt{AB/CD} \qquad (10.13)$$

If the 2-port is symmetrical, since A = D, we get the two image impedances as equal.

Example 15. *A filter shown is used to match a source of 100V to an impedance. Find the generator resistance and the load resistance for proper matching.*

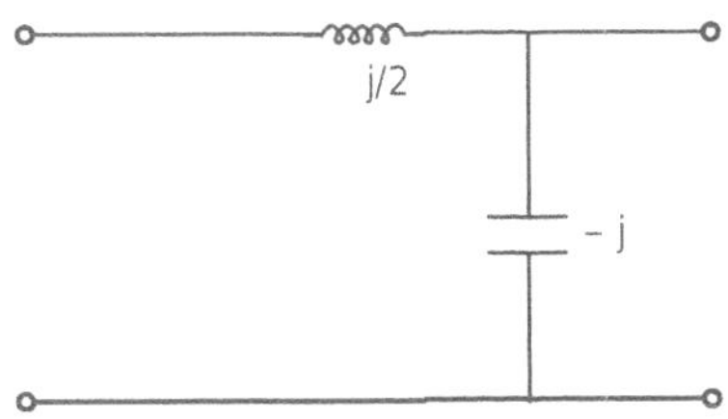

Fig. 10.29

Solution. The matrix for the filter is, (from Fig. 10.25*a*),

$$\begin{vmatrix} 1+YZ & Z \\ Y & 1 \end{vmatrix} = \begin{vmatrix} 1+j.\frac{1}{2}j & \frac{1}{2}j \\ j & 1 \end{vmatrix}$$

Hence $$Z_{I_1} = \sqrt{AB/CD} = \left(\tfrac{1}{2}\times\tfrac{1}{2}j\right)/(j\times 1) = \tfrac{1}{2}\Omega$$

$$Z_{I_2} = \sqrt{BD/AC} = \left(\tfrac{1}{2}j\times 1\right)/\left(\tfrac{1}{2}\times j\right) = 1\Omega$$

For matching, Z_{I_1} will be equal to the source internal resistance and the load will have a resistance equal to Z_{I_2}.

Characteristic Impedance: For a symmetrical 2-port, the image impedance as seen from both ports are equal. Then it is called *Characteristic impedance*, Z_0.

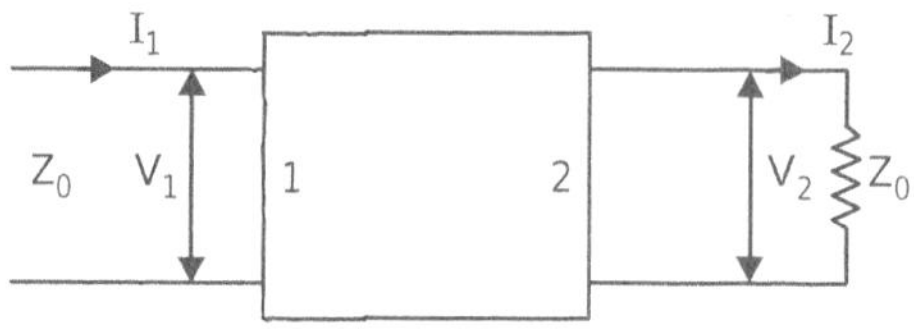

Fig. 10.30

When terminated in its characteristic impedance, a two port has the power input : output ratio of

$$P_1/P_2 = V_1I_1/V_2I_2 = V_1^2/V_2^2 \tag{10.14}$$

since the impedance is the same (Z_0) at both the ports. The power ratio is generally expressed in *decibels*.

$$\text{Decibels} = 10 \log_{10}(P_1/P_2) = 20 \log V_1/V_2$$

The logarithmic ratio of power is more convenient than the numerical ratio. Multiplication by 10 is to convert from the bigger unit *bel* to decibels.

10.12 TYPES OF NETWORKS

Port of a network: A pair of terminals of a network between which a source of voltage of current is applied is called a port, derived from the word 'portals'. Networks are characterised as one-port networks, two-port networks and *n*-port networks.

A simple two-port network is shown in Fig. 10.31. Sometimes, a two port network is drawn showing only its external terminals Fig. (*b*). An *n*-port network may be similarly

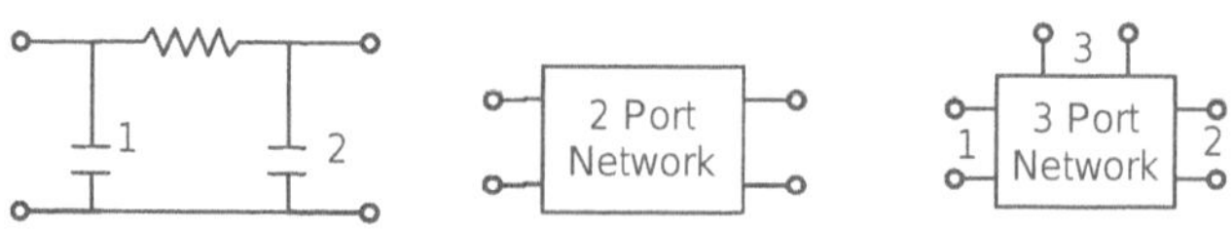

(a) & (b) Two-port network (c) A3-port network.

Fig. 10.31

represented as in Fig. (*c*), which shows a 3-port network. Two common two-port networks are the ladder and the lattice networks. Many such networks can be connected in series and these are common as filter circuits in communication work.

Two-port parameters: In chapter 1, we defined the Input resistance between the two terminals of a passive network, *i.e.*, for a one-port network. For a one-port network, the

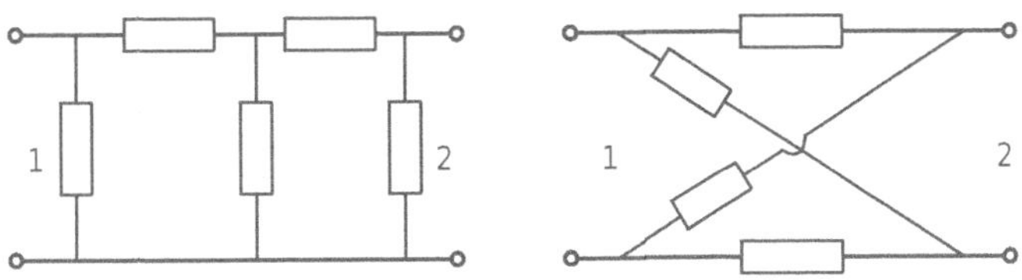

Fig. 10.32 Ladder and lattice networks.

input impedance is the ratio of V to I shown in Fig. 10.32. The *input admittance* is its reciprocal.

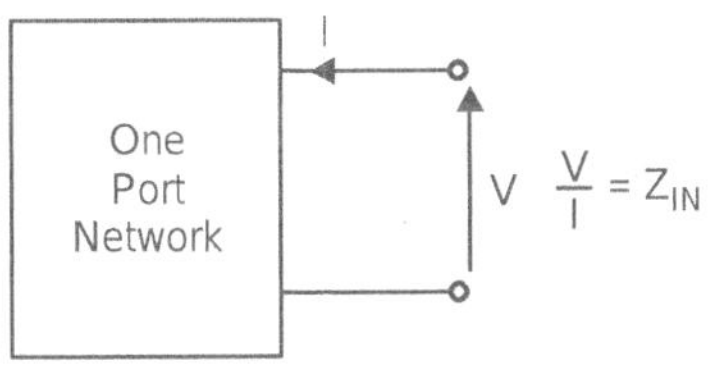

Fig. 10.33

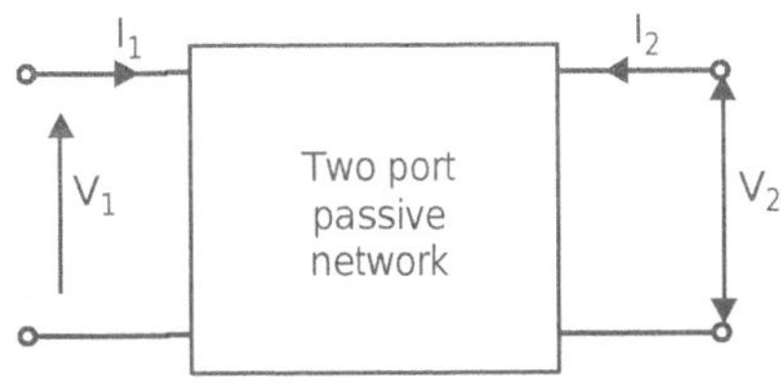

Fig. 10.34 parameters.

Similarly, for a two-port network we may define some parameters (Fig. 10.34)

$$V_1 = h_{11}I_1 + h_{12}V_2$$
$$I_2 = h_{21}I_1 + h_{22}V_2$$

In matrix form,
$$\begin{vmatrix} V_1 \\ I_1 \end{vmatrix} = \begin{vmatrix} h_{11} & h_{12} \\ h_{21} & h_{22} \end{vmatrix} \begin{vmatrix} I_1 \\ V_2 \end{vmatrix} \quad \text{or} \quad \begin{vmatrix} V_1 \\ I_2 \end{vmatrix} = \begin{vmatrix} h_i & h_r \\ h_f & h_o \end{vmatrix} \begin{vmatrix} I_1 \\ V_2 \end{vmatrix}$$

(*i*) With second port shorted ($V_2 = 0$) the ratio $\frac{V_1}{I_1}$ (Input impedance, h_{11})

(*ii*) With second port short circuited the ratio $\frac{I_2}{I_1}$ (Current ratio h_{21})

(*iii*) With first port open ($I_1 = 0$), the ratio $\frac{V_1}{V_2}$ (Voltage ratio h_{12})

(*iv*) With first port open, the ratio $\frac{I_2}{V_2}$ (Admittance h_{22})

These ratios are given the names h_{11}, h_{21}, h_{12} and h_{22}. A knowledge of these four quantities enable us to determine the currents I_1, I_2 for any other applied voltages V_1, V_2 to the ports (provided the circuit is passive, *i.e.*, contains no sources inside).

These co-efficients are also represented as:

$$h_i = h_{11} = \textit{input}\ \text{impedance}$$
$$h_f = h_{21} = \text{short circuit}\ \textit{forward}\ \text{current gain}$$
$$h_r = h_{12} = \textit{reverse}\ \text{voltage ratio}$$
$$h_o = h_{22} = \textit{output}\ \text{admittance.}$$

and are called the *hybrid parameters* of a two-port, very much used in transistor circuit theory. Other parameters called open-circuit and short circuit parameters are also defined, but they will not be dealt with here. The name hybrid is because of the mixing of V_1 and I_2 on the left; quantities of port 1 and port 2 are mixed up.

Example 16. *Find the hybrid parameters of two-port shown.*

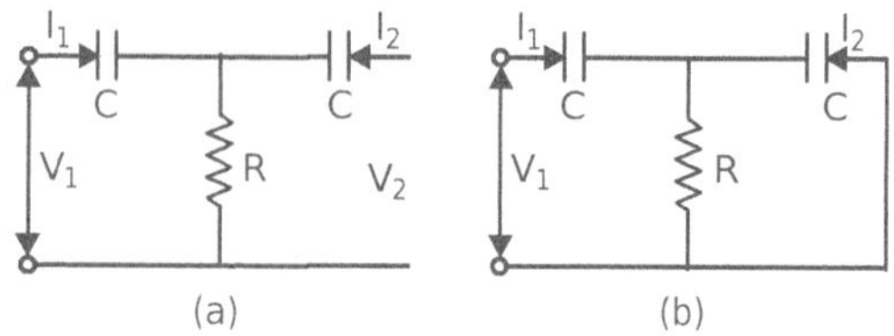

Fig. 10.35 A symmetrical two-port network.

Solution. $h_i = h_{11}$ = Input impedance with port 2 shorted. (Ref. Fig. 10.35(*b*)) Assume V_1 to be of frequency ω. Then h_i consists of the resistor R and C in parallel together with a C in series.

$$h_i = -j/C\omega + \frac{R(-j/C\omega)}{R - j/\omega C} = \frac{-j/C\omega(2 - j/\omega CR)}{1 - j/\omega CR}$$

$$h_f = h_{21} = I_2/I_1 \text{ with port 2 short circuited.}$$

$$I_2 = \frac{-RI_1}{R - j/\omega C}$$

$$\therefore \quad h_f = \frac{-R}{R - j\omega C}$$

To find $h_r = h_{12}$, determine V_1/V_2 from Fig. 10.35(*a*). Since port 1 is open, if V_2 is applied to port 2, we get

$$V_1 = \frac{RV_2}{R - j/\omega C} \text{ so that } h_r = \frac{1}{1 - j/\omega CR} = -h_f$$

We now find $h_o = h_{22}$ with port 1 open. Referring to Fig. 10.35(*a*) since port 1 is open, $I_1 = 0$ and so I_2 R = V_1

$$\therefore \quad h_0 = I_2/V_1 = R$$

It is seen that $h_r = -h_f$. For a symmetrical network, it can be shown that $h_{12} = -h_{21}$. A *symmetrical network* is one which appears identical when looked at from either of its two ports.

Impedance Matrix for 2 Ports

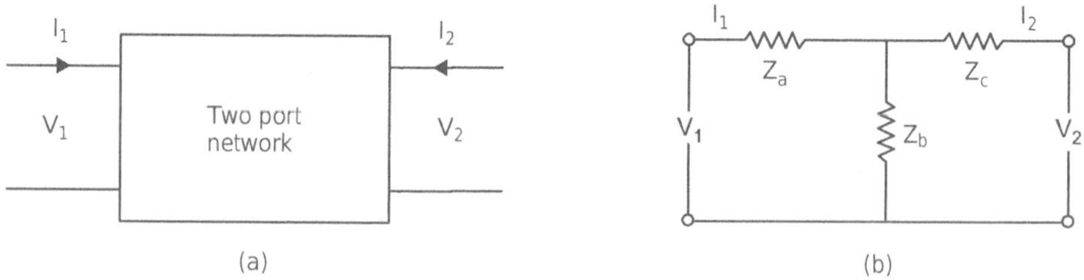

Fig. 10.36

If we consider the entire 2-port Network as a composite impedance,

$$V = Z\,I,$$

$$\begin{vmatrix} V_1 \\ V_2 \end{vmatrix} = \begin{vmatrix} Z_{11} & Z_{12} \\ Z_{21} & Z_{22} \end{vmatrix} \begin{vmatrix} I_1 \\ I_2 \end{vmatrix}$$

gives the impedance Matrix Z.

Example 17. *Find the impedance parameters Z_{11}, Z_{12}, Z_{22} for the circuit shown in Fig. (10.36b).* *[Anna University, May 2007]*

Solution.

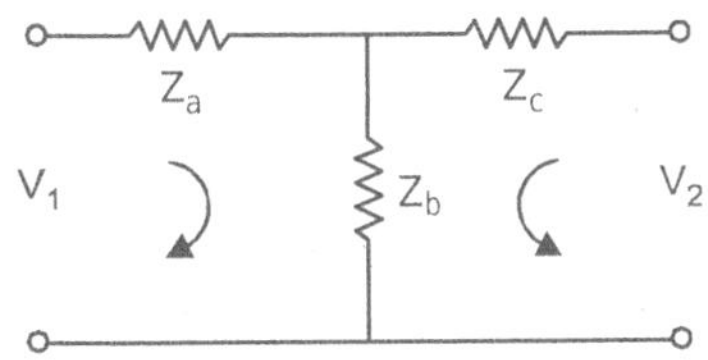

Writing loop equation,

$$V_1 = I_1 Z_a + (I_1 - I_2)Z_b$$
$$V_2 = Z_b\,(I_2 - I_1) + Z_c\,I_2$$

In matrix form,

$$\begin{vmatrix} V_1 \\ V_2 \end{vmatrix} = \begin{vmatrix} Z_a + Z_b & -Z_b \\ -Z_b & Z_b + Z_c \end{vmatrix} \begin{vmatrix} I_1 \\ I_2 \end{vmatrix}$$

So $\quad Z_{11} = Z_a + Z_b\ ,\ Z_{12} = -Z_b,\ Z_{22} = Z_b + Z_c.$

EXERCISES (ELEMENTS OF SYNTHESIS)

1. State which of the following are L-C driving-point impedances:

(*i*) $\dfrac{s(s^2 + 1)}{s^2 + 6s^2 + 8}$ (*ii*) $\dfrac{s^4 + 3s^2 + 2}{s(s^2 + 3)}$

(*iii*) $\dfrac{s^4 + 4s^2 + 3}{s(s^2 + 2)}$ (*iv*) $\dfrac{s}{4s^2 + s^2 + 1}$

(*v*) $\dfrac{s^2 + 2}{(s^2 + 1)\,(s^2 + 3)}$ (*vi*) $\dfrac{s + s^2}{(s^3 + s^2 + s + 1)}$

[**Ans.** Except (*iii*) all are not LC *d.p.* functions. For (*i*) and (*ii*), poles and zeros don't alternate. (*iv*) has pole on $j\omega$ axis; (*v*) has degree difference $n(s)$ and $d(s) \neq 1$; (*vi*) common factor occurs.]

2. Realise (*iii*) in Example 1 above in all the four canonic forms.

[**Ans.** F_1 *form* : 1H + 3/2F + (| | combination of 2F and 1/4 H)

F_2 *form* : (2H and 1/2F) in parllel with (2H, 1½ F).

C_1 *form* : First series inductor 1H, second 4H. Shunt capacitors 1/2 and 1/6.

C_2 *form* : First series cap 2/3, second 0.04 F. Shunt inductors are 1.25 and 5H.]

3. A reactance function has zeros at $\omega_1 = 10^6$ and infinity. It has poles at $\omega_1 = 0$ and $\omega_2 = 1.1 \times 10^6$. It is required that the slope of the reactance curve at $\omega = 10^6$ should be 0.1Ω/rad/sec. Find the network meeting this requirement in F_1 form.

[**Ans.** 115 pF in series with a || branch of 1510 μH and 546 pF].

4. A reactance function vanishes at the angular frequencies of 10^6 rad/sec and 4×10^6 r/sec. It has 3 poles at $\omega = 0$, 2×10^6 and 8×10^6. At very high frequencies, the value of the reactance function is approximately equal to the reactance of a 1μF capacitor. Determine the reactance network as a ladder with shunt capacitors and series inductors.

[**Ans.** 3 shunt capacitors from input end 1, 0.242 and 11.5 μF 2 series inductors 0.0196 and 0.067 μH; 1 μF capacitor is at input end.]

5. A reactance function has poles at the origin and at $\omega = \sqrt{2}$. The slopes of the susceptance curve at the above ω values are 2/3 and 4 respectively. Find the F_1 form of the reactance.

[**Ans.** 1½ F in series with a combination of 2F and 1/4H.]

6. State which of the following can be RC driving-point immittances.

(*i*) $\dfrac{4s + 10}{3s + 6}$ (*ii*) $\dfrac{(s + 2)(s + 4)}{(s + 1)(s + 3)}$

(*iii*) $\dfrac{s^2 + 3s + 2}{s^2 + 7s + 12}$ (*iv*) $\dfrac{s^2 + s + 1}{(s + 1)(s + 2)}$

(*v*) $\dfrac{(s + 1)(s + 3)}{(s + 2)(s + 4)}$ (*vi*) $\dfrac{(s + 1)(s + 3)}{s(s + 2)(s + 4)}$

[**Ans.** Only (*i*), (*ii*) and (*vi*) are realisable. In (*iii*), factors cancel. (*iv*) has poles imaginary. (*v*) pole not nearest to origin].

7. Realise the immittance functions in Ex. 6 as : (*i*) in F_1 and F_2 forms, (*ii*) in C_1 and C_2 forms and (*vi*) in F_2 form.

[**Ans.** (*i*) F_1 : $1^1/_3$ Ω in series with C = 3/8 and R = $1^1/_3$ in ||.

F_2 : 5/3 Ω in || with R (= 20/3) + C (= 0.6).

(*ii*) First series R = 1, second is 4/3, third 1/3. Shunt C_1 = 1/2, C_2 = 3/2, third = 0.

(*vi*) 3 parallel branches : 1F; 3/2 F + 2/3 Ω; and 1/6 F + 2Ω].

8. The impedance at 1-2 has poles at $-1 \pm \frac{1}{2}\sqrt{2}$ and zeros at –1/5 and –1. Find the unknown elements.

[**Ans.** $R_1 = R_2 = 2$, $C_1 = 1$, $C_2 = 2$.]

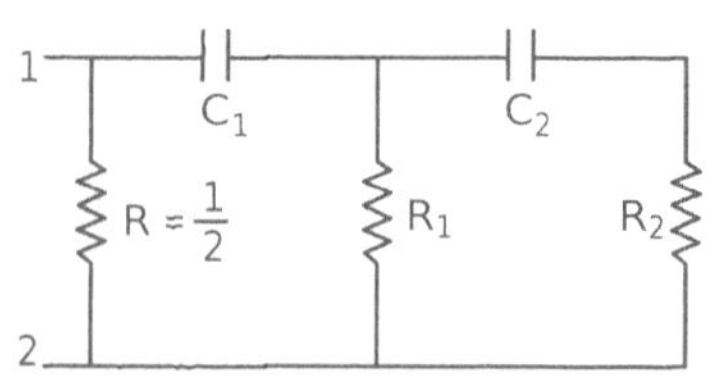

Fig. 1

EXERCISE (TWO PORT NETWORKS)

1. A transmission circuit has a T network having series impedances Z = 30 ∠60°, and shunt admittance 0.5×10^{-3} mho. Determine the general circuit constant A, B, C, D.

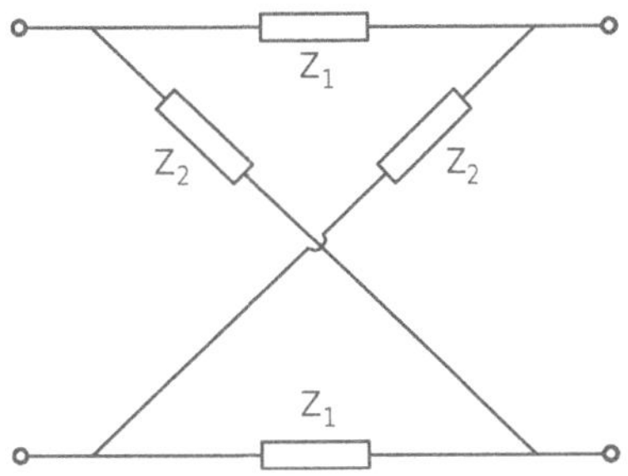

Fig. 1 A lattice network

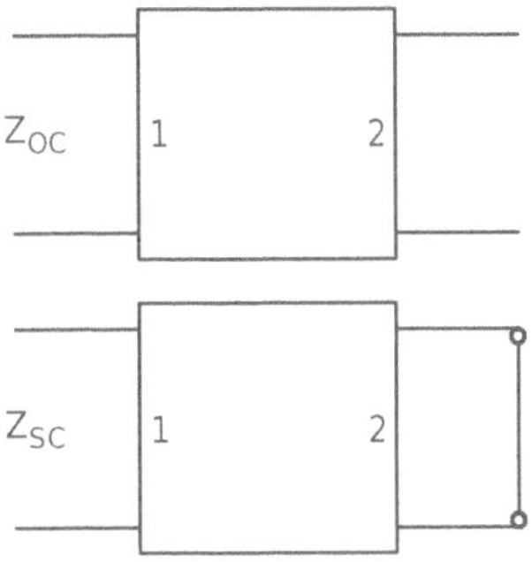

Fig. 2

[**Ans.** 0.885 ∠4.3°, 56.4 ∠62°, 0.5×10^{-3} ∠90°]

2. Show that the image impedance is determinable by finding open and short circuit impedances Z_{0c1} and Z_{sc1} as $Z_{I_1} = \sqrt{Z_{0c}Z_{sc}}$.

3. In the lattice of Fig. 1, if $Z_1 = 2j$, $Z_2 = -2j$, find Z_0. [**Ans.** 4Ω]

SIMPLE EXERCISES

1. Explain the terms Immittance, Driving-point, one-port, two-port, transfer function. [Sec. 10.1 and General]

2. Show that any network function will have only rational terms of *s* in the numerator and denominator. [Sec. 10.2]

3. For an immittance function, show that there can have one simple pole only. [Sec. 10.3]

4. Show that no network can have its poles on the right half plane in the *s*-plane. [Sec. 10.3]

5. State the reflection property of a network function and its significance. [Sec. 10.4]

6. What is Foster's first form of one-port synthesis ? Draw the frequency response plot of a network in this form. [Fig. 10.5, Sec. 10.5]

7. What is Foster's second form of one-port realisation ? [Example 2, Sec. 10.5]

8. What is the difference between Cauer's form and Foster's form of one-port synthesis ? [The latter is a ladder, Sec. 10.6]

9. Why is the continued division method found useful for Cauer's synthesis ? [See example 3, Sec. 10.6]

10. What is the difference between Cauer's first and second forms of one port RC synthesis ? [C_1 form has an R and C_2 a C in front]

11. What are ABCD parameters for a 2-port ? Find the relation between them.

12. Define characteristic impedance for a 2-port network.

13. Define hybrid parameters of a 2-port network and state the advantage in its use.

INDEX

D

E

F

G

H

I

K

L

M

N

O

T

U

V

W

Z